Sarah Said

Image registration between MRI and spot mammograms for X-ray guided stereotactic breast biopsy

Image registration between MRI and spot mammograms for X-ray guided stereotactic breast biopsy

by
Sarah Said

KIT Scientific Publishing

Karlsruher Institut für Technologie
Institut für Prozessdatenverarbeitung und Elektronik

Image registration between MRI and spot mammograms
for X-ray guided stereotactic breast biopsy

Zur Erlangung des akademischen Grades einer Doktorin der Ingenieurwissenschaften (Dr.-Ing) von der KIT-Fakultät für Elektrotechnik und Informationstechnik des Karlsruher Instituts für Technologie (KIT) genehmigte Dissertation

von Sarah Said, M.Sc.

Tag der mündlichen Prüfung: 8. April 2024
Hauptreferentin: PD. Dr. Nicole Ruiter
Korreferentin: Prof. Dr.-Ing. Maria Francesca Spadea
Korreferent: Assoc. Prof. Priv.-Doz. Dr. Pascal Baltzer

Impressum

KIT Scientific Publishing

Karlsruher Institut für Technologie (KIT)
KIT Scientific Publishing
Straße am Forum 2
D-76131 Karlsruhe

KIT Scientific Publishing is a registered trademark
of Karlsruhe Institute of Technology.
Reprint using the book cover is not allowed.

www.ksp.kit.edu

Print on Demand 2024 – Gedruckt auf FSC-zertifiziertem Papier

ISBN 978-3-7315-1366-7
DOI 10.5445/KSP/1000170686

Abstract

Among women, breast cancer is the most prevalent cancer type. 24.5% out of the total number of cancer cases were diagnosed as breast cancer worldwide in 2020. While digital mammography plays a crucial role in the early diagnosis of breast cancer, many tumors are not detectable in mammography, for instance, in women with dense breast tissue. Lesions that are not evident on mammography can frequently be found with contrast-enhanced magnetic resonance imaging (CE-MRI). MRI-guided biopsy must be used to further evaluate suspicious lesions. However, MRI-guided biopsy is costly, time-consuming, and not universally accessible. This thesis proposes a novel method for a matching tool between MRI and spot mammograms. The goal is to enable X-ray guided biopsy when the lesion is only visible in MRI by transferring its location onto mammographic spot projections. Two registration methods are used: a biomechanical model based registration between MRI and full X-ray mammograms, followed by an image based registration between full and spot mammograms. A virtual deformation of the breast relative to the deformation of X-ray mammography is estimated using a patient-specific breast geometry of 3D MRI. To generalize robustness and accelerate the workflow, two novel methods have been developed and evaluated: segmentation of breast MRI tissue using unsupervised learning and simulation of the mechanical response of breast tissue under mammographic compression using machine learning. The image based registration method entails three steps: preprocessing, identifying a region of interest (ROI) from the full X-ray mammogram, and then applying image similarity metrics between the ROI and the spot mammogram while sliding the spot mammogram incrementally over the ROI. To robustify the proposed method, multiple image similarity metrics are combined and clustered using K-means. The proposed methods have been tested using 51 patients from the Medical University of Vienna. The first and second registration methods achieved a median target registration error (TRE) of $31.8\,mm$ and $20.2\,mm$, respectively. The number of overlapping annotations of lesions after registration is 22 and 26 cases, respectively. For the accumulative registration of the two methods, the median TRE achieved is $35.6\,mm$ and 18 cases have an overlap between the two annotations. Characteristics that correspond to the variations in datasets have been used to analyze the three methods. It is found that the methods work better with cranio caudal (CC) views and with the location of lesions in the center of the breast. Clinical applicability has been tested by simulating stereotactic biopsy. 11 and 14 datasets could be clinically used when extracting 12 and 24 specimens with a standard needle biopsy, respectively. For the analyzed dataset, the proposed methods showed not only promising results but also the feasibility of clinical use. In the future, the method could enable performing the widely accessible and less expensive X-ray guided biopsy even if a suspicious lesion can only be seen in MRI.

Zusammenfassung

Bei Frauen ist Brustkrebs die am weitesten verbreitete Krebsart. 24.5% aller Krebsfälle weltweit wurden im Jahr 2020 als Brustkrebs diagnostiziert. Die digitale Mammographie spielt zwar eine entscheidende Rolle bei der Früherkennung von Brustkrebs, aber viele Tumore sind in der Mammographie nicht zu erkennen, zum Beispiel bei Frauen mit dichtem Brustgewebe. Läsionen, die in der Mammographie nicht erkennbar sind, können häufig mit der kontrastverstärkten Magnetresonanztomographie (CE-MRT) der Brust gefunden werden. Zur weiteren Abklärung verdächtiger Läsionen muss eine MRT-gesteuerte Biopsie durchgeführt werden. Die MRT-gestützte Biopsie ist jedoch kostspielig, zeitaufwändig und nicht überall zugänglich. In dieser Arbeit wird eine neuartige Methode zur Registrierung von MRT und Spot-Mammogrammen vorgeschlagen. Ziel ist es, eine röntgengeführte Biopsie zu ermöglichen, wenn die Läsion nur im MRT sichtbar ist, indem ihre Lage auf mammographische Spot-Projektionen übertragen wird. Es werden zwei Registrierungsmethoden verwendet: eine auf einem biomechanischen Modell basierende Registrierung zwischen MRT- und Vollbild-Röntgenmammogrammen, gefolgt von einer bildbasierten Registrierung zwischen Vollbild- und Spot-Mammogrammen. Eine virtuelle Verformung der Brust relativ zur Verformung der Röntgenmammographie wird anhand einer patientenspezifischen Brustgeometrie aus dem MRT-Volumen geschätzt. Zur Verallgemeinerung und Beschleunigung, wurden zwei Neuerungen entwickelt und evaluiert: die Segmentierung von MRT-Brustgewebe durch unüberwachtes Lernen und die Simulation der mechanischen Verformung von Brustgewebe unter mammographischer Kompression durch maschinelles Lernen. Die bildbasierte Registrierungsmethode umfasst drei Schritte: Vorverarbeitung, Identifizierung einer Region von Interesse (ROI) aus dem vollständigen Röntgenmammogramm und anschließende Anwendung von Bildähnlichkeitsmetriken zwischen der ROI und dem Spot-Mammogramm, während das Spot-Mammogramm schrittweise über die ROI geschoben wird. Um die vorgeschlagene Methode robust zu gestalten, werden mehrere Bildähnlichkeitsmetriken kombiniert und mit Hilfe von K-means geclustert. Die vorgeschlagenen Methoden wurden an 51 Patienten der Medizinischen Universität Wien getestet. Die erste und zweite Registrierungsmethode erreichten einen medianen Zielregistrierungsfehler (TRE) von 31.8 *mm* bzw. 20.2 *mm*. Die Anzahl der sich überschneidenden Läsionen nach der Registrierung beträgt 22 bzw. 26 Fälle. Bei der kumulativen Registrierung der beiden Methoden beträgt der mittlere TRE 35.6 *mm*, und in 18 Fällen überschneiden sich die beiden Annotationen. Zur weiteren Analyse der Methoden wurde eine Korrelation des TRE mit Merkmalen durchgeführt, die verschiedene Variationen der Datensätze abbilden. Es hat sich gezeigt, dass die Methoden bei kranio-kaudalen (CC) Ansichten und bei der Lokalisierung von Läsionen in der Mitte der Brust besser funktionieren. Die klinische Anwendbarkeit wurde durch Simulation einer stereotaktischen Biopsie getestet. 11 bzw. 14 Datensätze konnten klinisch verwendet werden, wenn 12 bzw. 24 Proben mit einer Standard-Nadelbiopsie entnommen wurden. Für den analysierten Datensatz zeigten die vorgeschlagenen Methoden nicht nur vielversprechende Ergebnisse, sondern auch die Machbarkeit der klinischen Anwendung. In Zukunft könnte die Methode die Durchführung der weithin zugänglichen und kostengünstigeren röntgengeführten Biopsie ermöglichen, selbst wenn eine verdächtige Läsion nur im MRT zu sehen ist.

List of Abbreviation

Att-BLSTM	bidirectional long short-term memory with an attention layer.
BCFCM	Bias Field Correction based on Fuzzy C-means.
BI-RADS	Breast Imaging-Reporting and Data System.
BLSTM	Bidirectional long short-term memory.
BM	Biomechanical modeling.
CB	Core Biopsy.
CC	Cranio Caudal.
CCW	counterclockwise.
CE-MRI	Contrast-enhanced magnetic resonance imaging.
CGAL	Computational Geometry Algorithms Library.
CLAHE	contrast limited adaptive histogram equalization.
CNN	Convolutional Neural Network.
COG	center of gravity.
Conv	Convolution.
Corr	Correlation Coefficient.
CPU	Central Processing Unit.
CT	Computed tomography.
DBT	Digital breast tomosynthesis.
DICOM	Digital Imaging and Communications in Medicine.
DIV	Difference in Variance.
DNNs	deep neural networks.
DSC	DICE Similarity Coefficient.
DT	decision tree.
ERGAS	Erreur Relative Globale Adimensionnelle de Synthèse.
ERT	extremly randomized tree.
FCM	Fuzzy C-means.
FE	Finite Element.
FEM	Finite Element Method.
FL3D	Fast Low Angle Shot 3D.

FNS	Fine Needle Biopsy.
FOV	field of view.
GANs	Generative Adversarial Networks.
GLCM	gray level co-occurrence matrix.
GPU	Graphics Processing Units.
GQS	glandular quality score.
HVS	Human Visual System.
IE	Internal Energy.
IoU	Intersection over Union.
KE	Kinetic Energy.
LIBRA	Laboratory for Individualized Breast Radiodensity Assessment.
LM	Lateromedial.
LSTM	Long short-term memory.
MAE	Mean Absolute Error.
MI	Mutual Information.
MITK	Medical Imaging Interaction Toolkit.
ML	Mediolateral.
MLO	Mediolateral Oblique.
MQS	muscle quality score.
MR	Magnetic Resonance.
MRI	Magnetic Resonance Imaging.
MS-SSIM	Multiscale Structural Similarity.
MSE	Mean Square Error.
NMI	Normalized Mutual Information.
NN	Neural Network.
PD	Percentage Density.
PDEs	Partial Differential Equations.
PDF	probability distribution function.
PNL	posterior nipple line.
PSNR	Peak Signal to Noise Ratio.
r	Pearson Correlation Coefficient.
ReLU	Rectified Linear Unit.
RF	radiofrequency.
RIU	Ratio Image Uniformity.

RL	Rolledlateral.
RMSE	Root Mean Square Error.
RNN	Recurrent neural network.
ROI	Region of Interest.
SDD	Standard Deviation Difference.
SGD	Stochastic Gradient Descent.
SNR	Signal-to-Noise Ratio.
SOFA	Simulation Open Framework Architecture.
SQL	Structured Query Language.
SSIM	Structural Similarity Index Measure.
SVR	support vector regression.
TE	echo time.
TLED	total Lagrangian explicit dynamics.
TPR	True Positive Rate.
TR	repetition time.
TRE	Target Registration Error.
UIQI	Universal Objective Image Quality Index.
US	Ultrasound.
VAB	Vacuum Assisted Biopsy.
XGBoost	extreme gradient boosting.

List of Symbols

Other Symbols

α	Learning rate
ε	Strain
μ_x	Mean of x
v	Poisson's ratio
σ	Stress
σ_x^2	Variance of x
σ_{xy}^2	Covariance of x and y
σ_x	Standard deviation of x
E	Young's modulus
G	Shear modulus
K	Bulk modulus
R^2	R-squared

Contents

Abstract i

Zusammenfassung iii

List of Abbreviations v

List of Symbols ix

1 Introduction 1

1.1 Motivation 1

1.2 Thesis Topic 2

1.2.1 Aim 3

1.2.2 Challenges 3

1.2.3 Proposed Method 3

1.3 Thesis Layout 5

2 Background 7

2.1 Magnetic Resonance Imaging 7

2.2 X-ray Mammograms 11

2.3 Spot Mammograms 12

2.4 Image Registration Methods 14

2.4.1 Transformations 16

2.5 Machine Learning 17

2.5.1 Tree-Based Ensemble Algorithms 18

2.5.2 Neural Networks 19

3 Biomechanical Model Based Registration 25

3.1 Biomechanical Models 25

3.1.1 Finite Element Methods 26

3.1.2 Biomechanical Model and FEM Simulation 27

3.2 Proposed Method 31

3.2.1 Segmentation 33

3.2.2 Meshing 43

3.2.3 Material Model and Boundary Conditions 44
3.2.4 Experiments for Optimizing Parameters 51
3.2.5 Machine Learning Algorithms for Simulation of Biomechanical Model Deformation 65
3.3 Summary 71

4 Image Based Registration **73**
4.1 Intensity-Based Methods 73
4.2 Proposed Method 79
4.2.1 Preprocessing 81
4.2.2 Multi-Resolution Approach 84
4.2.3 Image Similarity Metrics 86
4.3 Summary 86

5 Clinical Data **89**
5.1 Characteristics of Clinical Data from Metadata 90
5.1.1 Patient Age 90
5.1.2 Combination of Views 90
5.1.3 Resolutions 91
5.1.4 Devices 91
5.2 Characteristics of Clinical Data Calculated or Informed from Radiologists 95
5.2.1 Histopathological Diagnosis 95
5.2.2 Breast Density 96
5.2.3 Sizes of Lesions 97
5.2.4 Location of Lesions 97

6 Evaluation Methods **101**
6.1 Data Processing 101
6.1.1 Generating Ground Truth for Breast MRI Segmentation 101
6.1.2 Generating Ground Truth for Image Based Registration 101
6.2 Evaluation Metrics 102
6.2.1 True Positive Rate 102
6.2.2 DICE Coefficient 102
6.2.3 Jaccard Index 103
6.2.4 Target Registration Error 104
6.3 Evaluation Methods 104
6.3.1 Biomechanical Model Based Registration Method 104
6.3.2 Image Based Registration Method 105

6.3.3 Accumulative Registration of the Two Methods 106

7 Results **107**

7.1 Segmentation 107

7.1.1 Qualitative Analysis 107

7.1.2 Quantitative Analysis 108

7.1.3 Comparison to Literature 114

7.2 Machine Learning Algorithms for Simulation of Biomechanical Model Deformation . . 115

7.2.1 Unloaded State Estimation 116

7.2.2 Compression Step 117

7.2.3 Cross-Validation Analysis 120

7.2.4 Comparison to Literature 122

7.3 Biomechanical Model Based Registration 123

7.3.1 Mesh Quality and Volume Loss by FEM 130

7.3.2 Correlation with Features of the Breast 131

7.4 Image Based Registration 140

7.4.1 Image Quality Assessment for the Contrast of Spot Mammograms 141

7.4.2 Evaluation of Cropping the Muscle Area from MLO Cases 142

7.4.3 Optimization 144

7.4.4 Combination of the Most Robust Image Similarity Metrics 151

7.4.5 Correlation with Features of the Breast 155

7.5 Accumulative Registration of the Two Methods 166

7.5.1 Correlation with Features of the Breast 169

7.6 Summary 176

8 Clinical Use **179**

8.1 Dimensions of the Needle 179

8.1.1 Craniocaudal Approach 181

8.1.2 Lateral Approach 184

8.1.3 Summary 185

9 Discussion, Conclusions, and Outlook **187**

9.1 Discussion and Outlook 187

9.1.1 Biomechanical Model Based Registration 187

9.1.2 Image Based Registration 191

9.1.3 Accumulative Registration of the Two Methods 191

9.1.4 General Limitations 194

9.1.5 Applicability for Clinical Use 194

9.2 Conclusion and Contributions 195

Acknowledgments **197**

A Publications and Supervision **199**

A.1 Journal 199

A.2 Proceedings 199

A.3 Supervision of students 200

B Appendix **201**

B.1 Interpolation 201

B.2 Select ROI Using Window 7.5 x 7.5 202

B.3 Distribution of Patients Regarding Compression Ratio 202

B.4 More Details about Data Characteristics as in Chapter 5 203

B.4.1 Resolution 203

C List of Figures **205**

D List of Tables **211**

E Bibliography **215**

1 Introduction

1.1 Motivation

Cancer is the second most common cause of death in the United States and globally a major public health concern. The United States has seen 1,918,030 new cancer cases and 609,360 cancer-related deaths in 2022 (170). In 2020, 6.9% of all cancer deaths were caused by breast cancer worldwide for women and men (181; 57). For women, 15.5% have died in 2020 (57). Investing in more focused cancer control activities and improving early diagnosis and treatment will enable a decrease in cancer mortality (170). Despite the progress in diagnosis and therapy, breast cancer also remains one of the leading causes of cancer-related deaths. In Europe, 23.5% were diagnosed with breast cancer, and 20.7% died in 2020 (5; 57).

For early diagnosis, several imaging techniques are available for diagnosing breast cancer, such as Magnetic Resonance Imaging (MRI), X-ray mammography, Digital breast tomosynthesis (DBT), Computed tomography (CT), and Ultrasound (US) images. The development of medical imaging modalities leads scientists to implement new medical image-processing algorithms.

The typical clinical workflow for a lesion is screening first using one of the mentioned modalities, and then, based on the degree of the lesion, a follow-up is needed or a biopsy is required. One important requirement is that if screening the breast is done by a specific modality, for example, an X-ray, for further follow-up or a biopsy to be taken, it should also be done in X-ray. The same goes for other modalities (171; 128).

Among the existing clinical imaging modalities, breast MRI has the highest sensitivity for detecting breast cancer and is essential for breast imaging practice (114; 4; 81). It is considered a functional technique (114). Although Contrast-enhanced magnetic resonance imaging (CE-MRI) of the breast has a variable specificity, it is very sensitive for the diagnosis of breast cancer (4). MRI has the potential to detect lesions that are not evident using conventional imaging techniques like X-ray mammography or US (81; 174), especially for dense breasts. Additionally, detected lesions by MRI can be problematic in the clinical management of breast disease. As malignancy is found in approximately every second to a third of such lesions, further workup is necessary (81; 174). These lesions may be either followed up or a biopsy is required based on the degree of suspicion (113). While US can be used to identify and biopsy the suspicious lesion identified by MRI and it may help to make a pre-surgical wire-localization easier (84), a recent meta-analysis made by a team of the Medical University of Vienna showed a limitation of US which has a user-dependent heterogeneity regarding lesion detection rates ranging between 22.6%

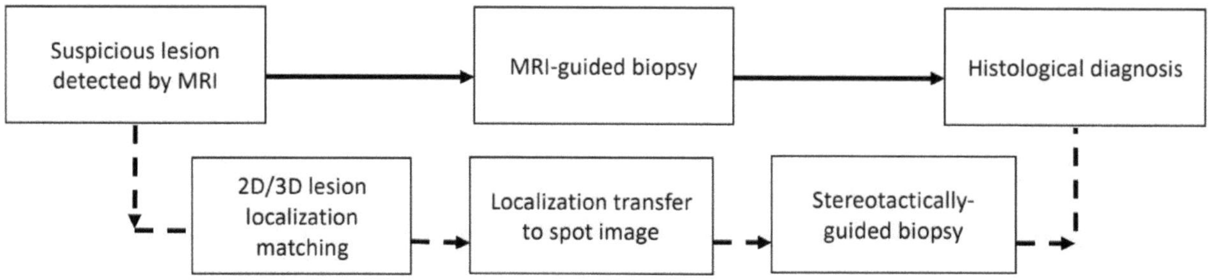

Figure 1.1: Suspicious breast lesions detected by MRI. Top row/solid lines: traditional workflow using costly MRI-guided biopsy. Bottom row/dashed lines: proposed new workflow by image registration (160).

and 82.1% (179). In addition to that, the probability of non-detected lesions or those identified by second look US leading to cancer is up to 54% of these cases (179; 30). In these specific cases, an MRI-guided biopsy is essential (179). MRI-guided biopsies, however, are costly and require a long time to do the procedures, typically in the range of 60 to 90 minutes, and a recent survey showed a significant shortage of MRI-guided breast interventions worldwide (43). A lack of MRI-guided breast interventions is a major limitation for a wider introduction of CE-MRI in clinical practice (96). Without them, MRI findings cannot be translated into a clinical workflow. That is why a lack of MRI-guided breast interventions is considered a serious problem.

1.2 Thesis Topic

The goal of this thesis is to transfer lesions that are detected by MRI to close-up stereotactic biopsy, so-called spot mammograms, which are used for X-ray guided biopsy so that a cheap and widely available lesion workup with X-ray guided biopsy is possible for lesions which are only visible in MRI. To my best knowledge, this is considered a novel approach for a clinical workflow. This transfer will be investigated by developing a matching tool between MRI and spot mammograms as shown in Figure 1.1. Instead of using MRI-guided biopsy, the position of a lesion only visible in MRI will be transferred to X-ray guided biopsy by a model based image registration, allowing an X-ray guided lesion workup. With this method, first, the probability of detecting lesions will be more available. Second, the cost of a biopsy will decrease by 50%. Third, it will reduce a lot of time which is typically in the range from 20 to 30 minutes for lesion workup (82). Fourth, the problem of a lack of MRI-guided breast interventions will be solved. Three opening questions from a clinical point of view have been raised:

(a) Is transferring lesions visible in MRI to spot mammograms applicable?
(b) What are the methods that lead to the current best registration accuracy?
(c) What are the meaningful evaluation methods in order to allow radiologists to estimate whether the proposed workflow can be integrated into clinical practice?

1.2.1 Aim

The clinical challenge raised above is considered an image inter-modality registration problem for patient-specific matching of MRI and spot mammograms. The general aim of this work is to develop, evaluate, and clinically test a registration method between MRI and spot mammogram images taken during X-ray-guided biopsy. There are three sub-aims: test and adapt the methods taking into consideration the huge variance between clinical datasets, explore methods of acceleration keeping the same accuracy in order to integrate into the clinical workflow, and find meaningful evaluation methods for knowing if the methods could be integrated into clinical practice.

1.2.2 Challenges

The challenges are the significant differences in modalities regarding dimensions, view of the breast, patient positioning, compression state, physical basis, and image resolution. Each modality has a different imaging setup. For Magnetic Resonance (MR) images, 3D images of the two breasts in a prone position are acquired with the breast subject to gravity in an uncompressed state as shown in Figure 1.2 (left). Different amounts of tissues may be imaged. MRI usually has pixel sizes between $0.5\,mm$ to $2\,mm$. For full mammograms, a 2D image of one side of the breast in the upright standing position is acquired as shown in Figure 1.2 (middle). The breast is compressed up to 50% between two parallel compression plates in order to enhance image contrast. X-rays pass through the breast to a detector positioned on the opposite side. Full X-ray mammograms provide high resolution with a pixel size of less than $0.1\,mm$. Spot mammograms are taken from an X-ray-guided biopsy device in which the patient lies in a prone position or stands upright (35) as shown in Figure 1.2 (right). A 2D image of a subset from the breast of size 50×50 mm with a detailed view of the compressed breast is taken. Spot mammograms provide resolution with a pixel size of range between $0.04\,mm$ to $0.08\,mm$. Due to the different physical basis of MRI and X-ray, the breast tissue is imaged with different contrast and grey value mapping.

1.2.3 Proposed Method

There may be three ways to transfer the lesion visible in MRI to spot mammograms. The first method is the registration between the full set of MRIs, full X-ray mammograms, and spot mammograms in a two step approach. It consists of two steps of registration methods between MRI and full X-ray mammograms followed by a registration between full X-ray and spot mammograms which will be discussed in detail in this thesis in chapter 3 and 4, respectively. The second method is the registration between MRI, and spot mammograms in a two step approach as well. It consists of the same two steps of registration methods but instead of using a full X-ray mammogram as an intermediate modality, it uses a synthetic X-ray (projected deformed MRI) created from the biomechanical model. The third method is the registration between MRI and spot mammograms in one step by creating synthetic spot mammograms. The advantages and disadvantages of each method will be discussed in detail in chapter 9.

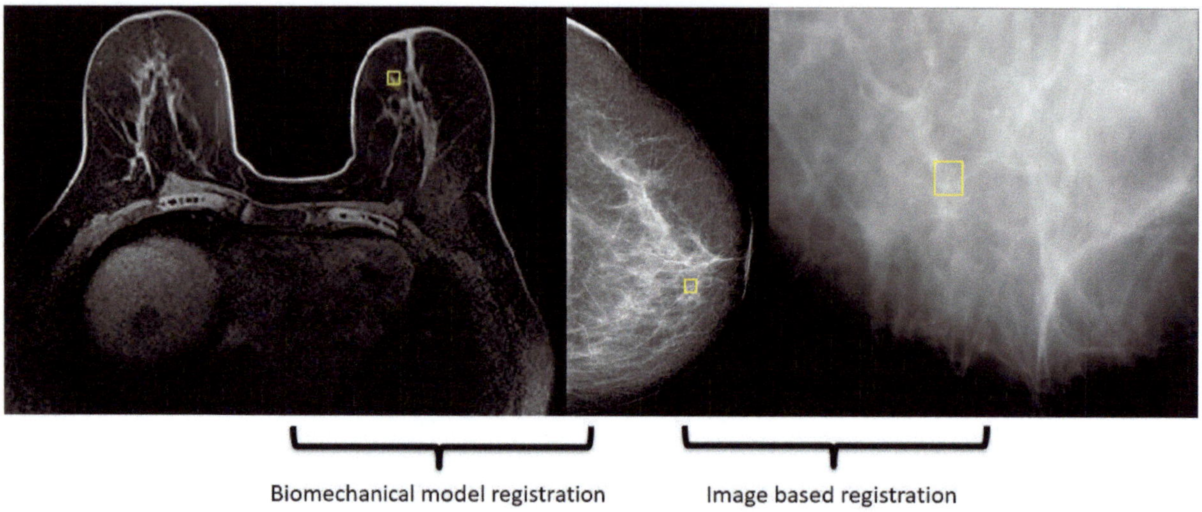

Figure 1.2: Two step image registration between MRI and spot mammograms for X-ray guided stereotactic breast biopsy. To register MRI (left) with full view mammograms (middle), a biomechanical model based registration is used. To register spot mammograms (right) with full view mammograms (middle), an image based registration is applied. Yellow rectangles indicate the marked lesions in all modalities by an expert radiologist (160).

As a proof of principle, the proposed method is a two step registration using three modalities MRI, full X-ray, and spot mammograms as shown in Figure 1.2.

First, a biomechanical model based registration is used for aligning the lesion detected by MRI with the full X-ray mammogram.

It estimates a breast configuration relative to its form in X-ray mammography using patient-specific breast geometry as obtained with MRI in 3D. The deformation applied virtually to the MRI is simulated by a biomechanical model to account for non-linear deformation and incompressible tissue behavior during the deformation process for the breast (160). The method originates from an earlier work presented by Ruiter et al. and Hopp et al. (156; 158; 76). It will be discussed in detail in Chapter 3. Some questions are raised in this method in order to achieve an accurate, robust, and automatic simulation of the biomechanical behaviour of the female breast:

- Is it applicable to have accurate, robust, and automatic biomechanical modeling to cover the clinical variability?
    - With robust segmentation: fatty, glandular, and muscular tissue
    - With a good quality mesh for convergence
    - Leading to a stable biomechanical model
    - Optimization of parameters to adapt patient-specific characteristics and uncertainties in meta-data such as compression thickness which could have an error of $5\,mm$ range (201), Young's modulus for fatty and glandular tissues, positioning of compression plates, and rotation around three axis
    - Reasonable computation time for clinical applicability

Second, the localization of the lesion in the full X-ray mammogram is transferred to the spot mammogram, which is taken during an X-ray guided biopsy, using an image based registration which will be discussed in detail in chapter 4.
Some questions are raised in this method in order to achieve an accurate, robust, and automatic image-based algorithm regardless of the difference in imaging situation and the deformation of the female breast:

- What is the correct image transformation to be applied?
- How to adapt the difference in contrast and grey value mappings?
- Which methods of image similarity metrics are to be used?
- How to select the most robust image similarity?
- How to combine the most robust image?
- Optimization: regarding computation time and accuracy

1.3 Thesis Layout

This thesis is composed of nine chapters. Chapter 2 gives an overview of three modalities and a general view of the image registration methods. Then, a lot of terms for machine learning and deep learning algorithms are explained from which some blocks are used in the proposed registration method proposed. Chapter 3 gives a detailed literature review of the biomechanical model based registration followed by applying the methods of biomechanical model based registration to register between MRI and X-ray mammogram. Chapter 4 gives a detailed explanation of intensity-based methods followed by the proposed methods of image based registration between full X-ray mammograms and spot mammograms. Chapter 5 discusses the variance and the characteristics of the clinical datasets received from the Medical University of Vienna. Chapter 6 presents the evaluation methods used for testing the proposed methods. Chapter 7 presents and discusses the results of each method separately and then discusses the accumulative registration of the two methods. Additionally, it investigates if the accuracy of each method separately or the accumulative registration of the two methods correlates with the characteristics of the clinical datasets. Chapter 8 discusses how it could be evaluated for clinical use. Then afterward, it presents how many datasets could be clinically used based on this evaluation. Chapter 9 concludes the contribution of this thesis and the results achieved. Several suggestions for future work are also presented.

2 Background

This chapter introduces the challenges and the difference in imaging situations between the three main modalities: MRI, full X-ray mammograms, and spot mammograms, followed by a broader view of the basics of image registration methods and concepts needed to be known for applying machine and deep learning algorithms for optimization or accuracy purposes.

2.1 Magnetic Resonance Imaging

Breast MRI has been first introduced in 1980s by Heywang-Köbrunner et al. (72) and Kaiser and Zeitler (86). It is considered an essential modality in addition to X-ray mammography and US images for early breast diagnosis. MRI is considered to be a non-invasive technology (132).

Strong magnetic fields are used in MRI scanners to stimulate the hydrogen nuclei (protons) found in human body tissue, evaluate their responses, and provide images of the body's organs (132; 50). Spin is an inherent characteristic of protons (107). The patient is first temporarily exposed to energy from an oscillating magnetic field at a resonance frequency adequate for the body's protons. These protons tend to align in the same phase of the magnetic field. Then radiofrequency (RF) pulses perpendicular to the magnetic fields are generated which enforce the proton to move against its nature by either 90° or 180°. After the RF pulses are turned off, protons tend to emit energy to return to their equilibrium and to realign again with the magnetic field. A receiving coil measures the RF signal that these protons release and transforms them into images (50). The emitted signal is affected by the relaxation period of protons, which is the amount of time required to allow the proton to return to equilibrium. It can therefore be utilized to identify the contrast between various molecules, and consequently, the tissue types (132).

For getting images of diagnostic quality, a specific breast coil must be used (114). The recommended minimum number of channels for a breast coil is four. More channels on a coil typically result in a higher Signal-to-Noise Ratio (SNR). Additionally, the employment of more parallel imaging factors is made possible by having more channels, which helps speed up the acquisition of images (114).

Numerous parameters, such as the kind of sequence and numerous sequence parameters per acquisition, can be used to control MR images. A particular image's contrast or weighting selection is crucial since it affects the image's potential medical applications and, thus, its diagnostic value. Proton density (PD) weighting, T1 weighting, and T2 weighting are the three main differences that are being identified. Weighting is utilized in the sense that the acquisition parameters were selected so that image contrast

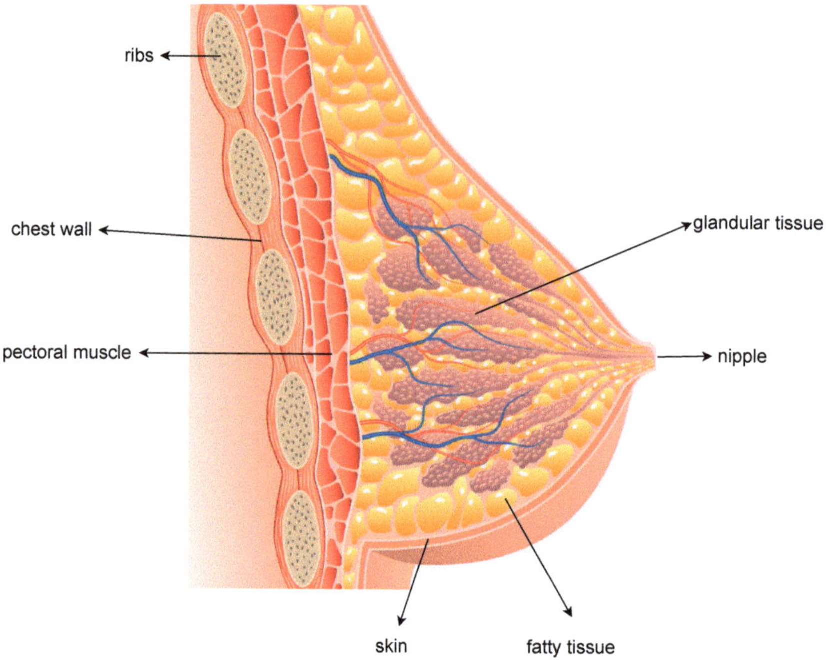

Figure 2.1: This image illustrates the anatomy of the breast and the available tissues (1) and was modified by the author.

primarily reflects fluctuations resulting from one of these tissue-inherent features, such as spin-spin relaxation (T2) (107).

The echo time (TE) and the repetition time (TR) are the variables that determine how a spin echo sequence is weighted. TE is the period of time between an RF pulse being generated and the RF signal being detected. While TR stands for the elapsed time between subsequent RF pulses (107).

A short TE and a short TR at which the longitudinal magnetization has not yet retrieved define T1 weighting. For defining T2 weighting, a long TR can be used to lessen the T1 impact while a long TE can be used to let the variations in T2 decay to appear. For T1-weighted MRI series, the fatty tissue is bright, while the fluid, glandular, and muscular tissues are darker compared to the fatty tissue. For T2-weighted MRI series, the fatty tissue is presented as intermediate-bright. The fluid, glandular, and muscular tissues are dark. The air is ideally black in both series.

The recommended method for acquiring images with a high spatial resolution is to utilize a field strength of at least 1.5 T (114).

Nowadays, MRI is a multiparametric technique in which for the same patient, it has two or more sequences with its subtraction series for different times (114). In this thesis, it is common to have one T1-weighted series and one T2-weighted series and if T2-weighted series do not exist, then two T1-weighted series are available.

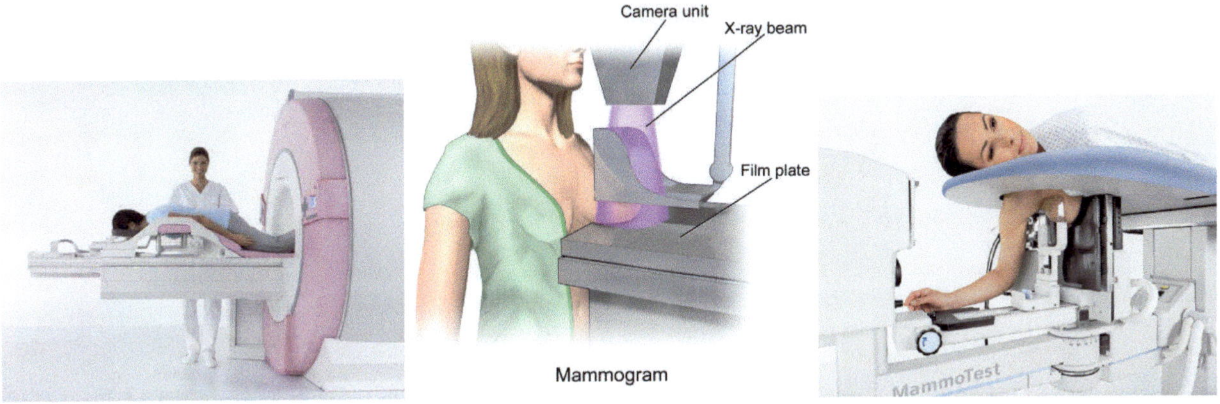

Figure 2.2: Imaging situation of each modality: MR image in a prone position (left) (118), X-ray mammography in an upright standing position (middle) (37), and spot mammogram in a prone position (right) (83).

In MRI, the patient lies in a prone position in which the breasts hang freely without compression as shown in Figure 2.2 (left). With this design, it permits the breast tissue to spread, which would make finding abnormalities easier and help in decreasing motion artifacts caused by breathing (114).
The gradient coils, which enable us to apply a linear change of the homogeneous magnetic field B_0, are a crucial part of an MRI system. A variance of this kind in any spatial direction is enabled by a weighted combination of the three gradient coils that are positioned in three orthogonal directions: head-feet, left-right, and anterior-posterior (107). Slice selection and spatial encoding are the two concepts that must be understood to enable the spatial localization of emitted RF waves based on the gradient coil system. For slice selection, the slice's thickness can be set to give a suitable compromise between spatial resolution and SNR by producing a wave with a variety of frequencies. For spatial encoding, the spins' transversal plane phase information is employed. In order to obtain an approximate representation of the underlying hydrogen density distribution, multiple gradients must be applied, i.e., different patterns must be created in the spins' phase orientation. Then, the net magnetization must be measured. This intermediate representation can subsequently be used to recreate the real distribution within a slice if done correctly. Only spins within the slice of interest are subject to phase encoding, as this step comes after slice selection (107). By those two concepts, 3D MR images can be created. There are two ways: slice selective and volume selective. For slice selective, 2D spatial encoding is used for localization inside each slice, and 2D slices of the desired 3D volume are acquired, reconstructed, and stacked by using slice selection. For volume selective, the entire volume without slice selection is excited, 3D spatial encoding is used for localization, and then rebuild the volume in 3D. Each way has its benefits and drawbacks. Slice selective acquisitions are frequently non-isotropic and have high inplane resolutions of $1\,mm$ or less, but they also have thick slices that are several millimeters thick in order to reduce image recording time (140). While volume selective acquisitions have a good SNR and they have an isotropic resolution. In this thesis, slice selective is used. The available MR images are non-isotropic, i.e., with a slice thickness larger than the inplane resolution (107).

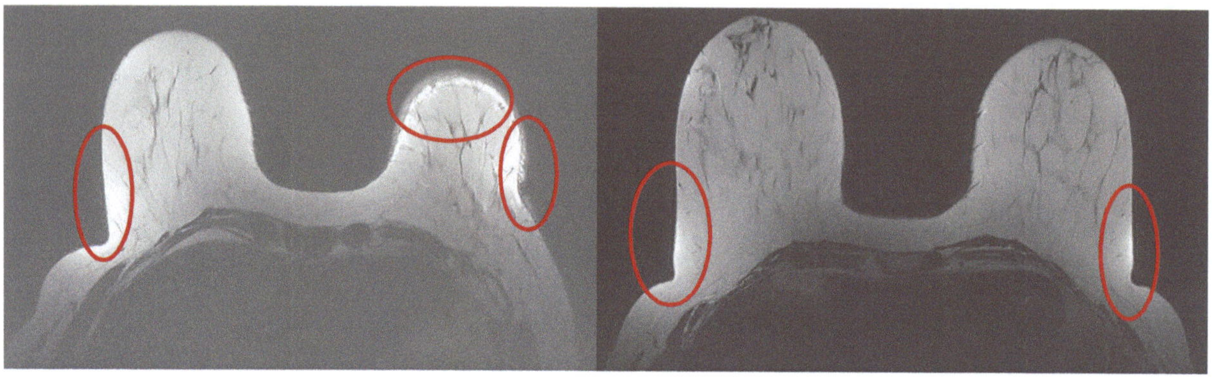

Figure 2.3: Examples of the two categories of MRI artifacts that are available in datasets from the Medical University of Vienna. The first image is from signal processing artifacts, which have a partial volume (left). The second image is from hardware-related artifacts, which is inhomogeneity (right). Red circles emphasize those two problems.

There are a variety of artifacts that might appear in an MRI that have an impact on the diagnostic quality. The distinction between artifacts and noise, however, is that artifacts are precisely replicated by repeating a scan with the same item and scan parameters, whereas noise effects vary according to a probabilistic system (20). The artifacts can be split into three different categories: patient-related, signal processing-dependent, and hardware-related (53).

In the patient-related category, there are motion, flow, and metal artifacts. In the signal processing category, there is chemical shift, partial volume, wrap around, and ringing artifacts. In the last category hardware-related, there are several artifacts such as RF quadrature, external magnetic field B_0, gradient field B_1 inhomogeneity, RF inhomogeneity, asymmetrical brightness, RF noise, RF tip angle inhomogeneity, bounce point, surface coil artifact (attenutation of signal), and slice to slice interference (53).

There are two recognized artifacts in the existing MRI datasets as shown in Figure 2.3: one from the signal processing category and one from the hardware-related category. The partial volume artifact is the first artifact when objects smaller than voxels lose their identity. This results in a loss of detail and is brought on by the resolution anisotropy that was previously discussed. Second, the datasets have a significant amount of intensity distortion because of the inhomogeneity of the external and gradient magnetic field as well as the nonuniform sensitivity of the receiving coil. Furthermore, surface coil artifacts cause further intensity distortions and a reduction in image brightness. A bias field is a term used to describe this corrupted image. It results in the issue of varied gray-level distributions within one tissue in the same image. Additionally, it lessens the image's high-frequency material, such as the image's edges. Since fatty tissue makes up the majority of the non-black tissue, its intensity should be consistent. The bias field causes the breasts to have both strongly darkened and overlit patches, and the connection between the breasts is almost invisible.

2.2 X-ray Mammograms

X-rays were first discovered by Wilhelm Conrad Röntgen in 1895. X-ray is considered to be in the group of electromagnetic rays in the range of wavelengths between 0.01 and 10 nm and range of 100 keV down to 100 eV in terms of energy. It uses photons and waves to carry radiant energy through space. Although X-rays can pierce through matter, the amount of penetrating X-ray photons depends on the material. They can be utilized to obtain information about internal organs due to their capacity to penetrate human tissue. High or low energy X-ray spectra result from various tube voltages between the cathode and the anode. X-rays can be dangerous because high energy radiation given to the body during an X-ray can cause ionization which raises the possibility of cancer developing (20). That is why it commonly uses low energy to enhance soft tissue contrast. It is considered the main modality that is used worldwide not only in diagnosis but also in interventions (20). High spatial resolution and a straightforward digital readout of the X-ray image are the two primary advantages of X-rays (20).
X-ray photons either completely absorb, scatter elastically, or inelastically. As a result, the radiation intensity decreases, which is a decrease in the amount of photons arriving at the detector, i.e. attenuation is caused by a variety of physical phenomena, such as changes in the number, direction, or energy of photons.
The primary principle used for X-ray imaging is absorption, which results in a reduction of energy. X-rays passing through a material are attenuated based on the properties of the material. The initial X-ray source intensity, the material's absorption characteristics, and the thickness of the material that the X-ray travels in all affect the received X-ray intensity at a detector (117). The Beer-Lambert Law is used to express this relationship. The definition of the Beer-Lambert Law is differently represented for linear homogeneous and inhomogeneous materials as shown in equation 2.1 and 2.2, respectively (20). The main difference between the two equations is that μ in the first equation is the material attenuation coefficient, while the second equation is the linear attenuation coefficient at every point along the ray path. X-rays can be described as a map of the density of tissues in the body; white areas on X-ray images indicate high density structures. For breast images, high intensities represent, e.g., glandular tissues. The amount of energy that is absorbed is directly correlated with the acquisition dose given to the patient (20).

$$I = I_o.e^{-\mu x} \tag{2.1}$$

where I_o is the X-ray intensity at the X-ray source, μ is the absorption coefficient (material attenuation coefficient), and I is the intensity that was detected in relation to the X-ray and the object's thickness, or the intersection length x (20).

$$I = \int_0^{E_{max}} I_o(E)e^{-\int \mu(x,E)\,dE} \tag{2.2}$$

For X-ray mammography, a full 2D projected image of the patient in the upright standing position is obtained as shown in Figure 2.2 (middle), while the breast is compressed up to 50% between two parallel compression plates. X-ray mammography uses low energy and low doses to increase the contrast of images. The breast is compressed to reduce the distance the X-rays need to penetrate the tissue. This leads to a lot of deformation for the tissues that make structures appear different than in MRI. The resolution of full X-ray mammograms is typically higher than 0.1 mm/pixel. Two views are taken in the clinical routine: Cranio Caudal (CC) and Mediolateral Oblique (MLO). For CC, the rotation angle of the X-ray tube and detector is $0°$ while MLO is of rotation angle $45°$ or $-45°$ based on which side of the patient is imaged.

There are several artifacts in X-rays, like vignetting and distortion on image detectors (20). A decrease in brightness that appears at the edges of the screen is referred to as vignetting. It results from light scattering, which shifts light photons in the output phosphor from its exterior to its interior. Image distortion is a typical artifact. It is well known that magnetic or electric fields can affect the electron optics of an image. The output image is considered distorted, even by the earth's magnetic field (20). The distortion field must be measured regularly through the measurement of predetermined calibration objects in order to correct distortion artifacts. If the images have been digitized, the distortion can be fixed by either modifying the electron optics appropriately or by doing additional image processing (20).

2.3 Spot Mammograms

A spot mammogram is a special kind of mammogram which is taken during an X-ray guided biopsy. A biopsy is taken to determine the degree of suspicion for malignancy. For the cases with Breast Imaging-Reporting and Data System (BI-RADS) category 4 (suspicious) or category 5 (highly suggestive malignancy), having a biopsy of the lesion is mandatory in clinical workflow, while also category 3 (probably benign) might need a biopsy as well (82).

Over time, a needle biopsy has shown to be more reliable and acceptable than a surgical biopsy. It is readily available, less invasive, and carries very low risks (82). There are three types of needle biopsy nowadays, as shown in Table 2.1: Fine Needle Biopsy (FNS), Core Biopsy (CB), and Vacuum Assisted Biopsy (VAB). FNS allows studying cells (cytological examination), while CB and VAB allow studying tissues (histopathological examination) which is the case in this thesis. FNS and CB, however, should be used with US guidance, while VAB should be used with mammography or tomosynthesis or MRI guidance (23).

Table 2.1: Types of needle

Difference	Fine Needle Sampling	Core Needle Biopsy	**Vaccum Assisted Biopsy**
Examination	cytological	histopathological	**histopathological**
Guidance	US	US	**Mammography** or tomosynthesis or MRI

It has been proven that VAB performs better or is considered more accurate than CB. The fact that the amount of tissue taken during the procedure in VAB is more, leads to a more reliable histological classification. In addition, a reduced rate of targeting errors along with lessened rebiopsy and underestimation rates is provided in VAB. Moreover, it uses a suction and rotating cutter which can move in an angular range of 360° without the need to remove the needle from the lesion (145; 82). The procedure of a biopsy takes around 20 to 30 minutes.

A spot mammogram in this thesis is a type of image that is taken during a stereotactical biopsy using a VAB which is called a scout image if it is taken at 0°. It uses a subset 2D image of size 50×50 mm with a detailed view of the compressed X-ray mammography to identify the location of suspicious areas within the breast. There are two types of imaging situations: either in a prone position as shown in Figure 2.2 (right) or in an upright position (35).

In the prone position, a separate device like Fischer Mammotest (Siemens) is used. There are a lot of benefits to this type of position. It causes less anxiety and less motion. In addition to that, physicians have more room to handle everything since the breast is located freely hanging under the table (82). One of the limitations of the prone position is that the table could bear a certain limit of weight (136 -156) kilograms which lead to the use of the table in the upright position (82). The upright position biopsy unit is an attachable stereotactic biopsy attached to an existing mammography device (82). There are several disadvantages to this position. First, it causes more anxiety for the patient because she can see the process of taking a biopsy. Second, there is a lot of motion caused by the patient which can be reduced by adding pillows or cushions. However, the resolution and quality of images are usually superior (82).

The biopsy procedure consists mainly of four parts: first positioning the paddles where the suspicious lesion occurs based on the patient's characteristics such as breast thickness.After making sure from the correct position, a couple of images, which are called stereo images, are taken at -15° and +15° in order to get the depth of the lesion using parallax shift (82; 35). Parallax shift takes into account the condition that the lesion is in the center of 0° images and the three X-rays are intersecting at a point which is the lesion. The lesion position is different than the center of rotation. The lesion's x (horizontal), y (vertical), and z (depth from the skin surface) are determined through the measurement of the parallax shift of the target area from defined angles of views (+15° and -15°) from the midline and along the x-axis (82). Then pre-fire images are taken to make sure that the needle will be in the correct place. Last, the marker clip positioning in which samples are removed from the breast is carried out. This step consumes most of the time around 15 minutes to 20 minutes (8; 35). Once the specimen retrieval is completed, a localizing post-biopsy marker clip is positioned, it is a $3mm$ titanium clip, which is placed at the biopsy site. The purpose of this clip is to guide future surgical excision and to permit future mammography surveillance of the area.

The needle could be inserted using three different views for taking a biopsy: CC, Mediolateral (ML), and Lateromedial (LM) where the rotation angle of the X-ray tube and detector is 0°, 90°, and -90°, respectively. There are also three possible available images for making a biopsy (0° and -15°) or (0°

and +15°) or (-15°, 0°, and +15°). The last option (-15°, 0°, and +15°) is the most frequently used in this thesis. For the successful cases, lesions are clearly identified in the three images: scout and stereo images. By that, it is possible to get the x, y, and z coordinates of a lesion center by equating X-ray pass through lesions in the stereo images (82; 35).

Some problems might appear while taking spot mammograms. First lesions might fail to appear on both stereo images as the lesion may be obscured by other tissues or because of the geometric configuration of the imaging system. Second, some parts could appear in the +15° or in the -15° only and not in the scout image. In addition, some parts might have blind spots due to the location of the center of rotation and the size of the biopsy window. It has been realized that the deeper the lesion, the more it will disappear. Third, other geometry-related errors might occur. X-ray source is not exactly at +15° or -15°, which leads to a wrong assumption of angles resulting in errors in the depth of estimation (35).

To sum up, some challenges might occur in the X-ray-guided biopsy because of (82)

- small breasts
- thin breasts
- superficial lesion close to the skin
- very small size of the lesion
- deep lesions
- lesions adjacent to the chest wall
- lesions located in the very superior portions (inner or outer quadrant)

Proper patient positioning is the key to technical success. In the case of the prone table, in order to fully reach the upper or posterior breast, the patient's arm may be inserted through the table's aperture in addition to the target breast. The arm may be supported by a sling or lying on a stand. To restrict the breast from moving, the affected hand may additionally grip a step stool's handrail. The receptor plate can be angled to improve access to posterior breast tissue and lesions so that the compression paddle can be pressed against the sternum or rib cage as closely as feasible (82).

To sum up, there are challenges regarding the differences between the three modalities. First, compression leads to different deformations in the breast structures. Second, patient positioning leads to different applied forces. The other differences are in resolutions, acquisition, physical basis, image contrast, dimensions, and breast shapes, as shown in Table 2.2. This leads to applying registration methods to transfer the position of the lesion from one modality to another, as will be explained in the next section.

2.4 Image Registration Methods

Transferring the lesion from one modality to another is mandatory for early diagnosis. Hence, registering between images is required. In image processing, registration is a fundamental operation used to align two or more images that were captured, for instance, at various times, with various sensors, or from

Table 2.2: Difference between modalities

Characteristics	MRI	Full X-ray mammograms	Spot mammograms
Imaging Situation Position	Prone	Upright	Prone or Upright
Dimensional	3D	2D	2D
Resolution (mm)	(0.5 - 0.93) / (0.8-5)	0.05 - 0.1	0.048 - 0.085
Deformation - Compression state	Uncompressed	around 50% Compression	around 50% Compressed
Range of Image Intensities	0 - 4082	0 - 16383	0 - 4095
Breast shape	two breasts	one side of breast	subset from the breast

various perspectives (29). To register between the same modality is called intra-modality and between different modalities, is called inter-modality.

Image registration as a general concept is the process of mapping the spatial and intensity differences between two images (29; 73). Two concepts of registration need to be known.

The first definition is finding a transformation model that can connect a feature's position in one image or coordinate space to the position of the matching feature in another image or coordinate space. In other words, the feature space extracts the desired information from the images to match the images. The class of transformation known as the search space can align the images (216; 29; 101).

The second definition of registration allows to compare the intensity at those related positions, for example, subtraction image intensity values as well as relate the position of corresponding features. This second definition of registration includes resampling and interpolation (73). In other words, in order to find the optimal transformation, a search strategy must be used to select the next transformation from this space. Then, the relative merit of each test is established by the similarity metric. Until a transformation is found whose similarity measure is sufficient, the search is carried out following the search strategy (29).

The purpose of image registration is to determine how points in a moving image I and those in either a fixed reference image R or the intervention's physical space correlate (75).

Thus, it could be illustrated as five elements commonly present in registration algorithms (75). They are all influenced by knowledge of the features of each form of variation (29):

$$\psi = D[I(T_\theta(x)), R(x)] + \lambda(T_{\theta(x)}) \tag{2.3}$$

- R are the fixed points in the reference image
- I are the points which need to be determined in the moving image
- A transformation model $T_\theta(x)$ is utilized to limit the distortion that occurs between images or between images and physical space within the given spatial domain, x, which is determined by a set of parameters θ
- Cost function ψ to calculate the similarity between the images and the result can optionally be regularized λ
- An optimization method to reduce the cost function arg $min_\theta(\psi)$

There are three major types of variations in which registration methods are needed. The first kind of variation is the misalignment of the images as a result of the differences in acquisition which is called corrected distortions. The second category of variations includes those that also result from different acquisition methods but are difficult to be modeled, which is called uncorrected distortions, such as lighting, shadows, and environmental factors. However, they may also be spatial, such as perspective distortions, and they typically impact intensity values. The third sort of variation is when the images that are of interest differ, whether it is due to object movement, scene changes, or growth or deformations. The second and third sorts of variations are not usually eliminated by registration and they make registration more challenging because an exact match can not be achieved based on the complexity of the transformation. Depending on the types of variations between images, the expected transformation for registration is chosen.

2.4.1 Transformations

The most common general transformations are rigid, affine, projective, and curved (197). In this section, the rigid transformation will be illustrated, which will be further used in the proposed registration methods.

The transformation could be global or local. When a change in any one of the matching parameters affects how the image is transformed overall, the matching transformation is a global one. A modification that only affects a small portion of the image is called local matching transformation (197). The size and shape of an object are preserved throughout rigid transformations. Translation and rotational transformations are included. In contrast, a non-rigid transformation modifies' size or shape. Scale, affine, projective, and shear transformations are all included (29; 111; 197).

Rigid Transformation The rigid transformation can be expressed by rotation and translational matrices. The rotation matrix is a rotation about one of the axes of a coordinate system by an angle θ, as shown in equation 2.4. In this equation, each matrix represents the rotation around x,y, and z, respectively, in the cartesian coordinate system. It is a clockwise representation. A translation matrix is shifting objects along one or more of the three axes by a value d, as shown in equation 2.5 (12).

$$R(\theta) = \begin{bmatrix} 1 & 0 & 0 \\ 0 & \cos(\theta_x) & -\sin(\theta_x) \\ 0 & \sin(\theta_x) & \cos(\theta_x) \end{bmatrix} \begin{bmatrix} \cos(\theta_y) & 0 & \sin(\theta_y) \\ 0 & 1 & 0 \\ -\sin(\theta_y) & 0 & \cos(\theta_y) \end{bmatrix} \begin{bmatrix} \cos(\theta_z) & -\sin(\theta_z) & 0 \\ \sin(\theta_z) & \cos(\theta_z) & 0 \\ 0 & 0 & 1 \end{bmatrix} \tag{2.4}$$

$$Translation = \begin{bmatrix} 1 & 0 & 0 & dx \\ 0 & 1 & 0 & dy \\ 0 & 0 & 1 & dz \\ 0 & 0 & 0 & 1 \end{bmatrix} \tag{2.5}$$

Affine Transformation An affine transformation is the extension of rigid transformation. It is a non-rigid transformation that includes more degrees of freedom. It consists of four matrices: rotation, translation as explained, scaling, and shear. A scaling transformation enlarges or reduces an object's size by varying all of its voxels or vertices along its three axes by three different scalar values of S_x or S_y or S_z. A shear transformation tilts objects. This is accomplished in the upper 3 by 3 submatrix by non-zero off-diagonal components (115).

Elastic Transformation Challenges occurred not only by global deformations but also, by local non-rigid deformations. This leads to extending registration to more degrees of deformation such as elastic transformation to overcome non-linear deformation (11). This type of transformation depends on elasticity theory and local deformations. It models the geometric difference between images as the result of elastic deformation, treating images as continuous bodies (151). Models from fluid mechanics could be utilized as an alternative. These models are more adaptable and are used to explain 3D projections onto 2D images (3D–2D registration), like the biomechanical registration model in this thesis. However, constraining the mapping in such registration approaches is a common issue (151).
In general, there are two types of registration approaches: landmark-based and intensity-based. Landmark-based algorithms generate a transformation based on features extracted from images, such as points, curves, and surfaces. Intensity-based techniques compute the transformation by directly utilizing the image intensities (151). In this thesis, landmark-based is used which is the case for the biomechanical registration, and intensity-based is used which is the case for image based registration.

2.5 Machine Learning

Recently machine learning algorithms have been used for improving registration and segmentation methods. Problems involving classification and regression may benefit from its utilization. When the result is a category, for example, a "dog" or "cat," classification is taking place. Regression only occurs when the output is an actual value. The data is trained until it performs acceptably with unseen data. Machine learning methods include supervised, unsupervised, and semi-supervised algorithms.
For supervised machine learning algorithms, an input and an output are both present. In unsupervised machine learning, there is no labeled data, simply the input data is provided for training. In order to better understand the data, the goal is to model the underlying structure of the data. It could be applied to clustering and association. The goal of clustering is to identify the most significant categories in the data. However, the association is utilized to identify patterns that describe significant portions of data. In semi-supervised machine learning, only a small portion of the large used dataset gets labeled. Both supervised and unsupervised methods could be employed in this area; supervised methods would offer the best predictions for untrained data. Unsupervised methods attempt to discover the input variable's structure.

One of the machine learning algorithms that is used in this thesis is the tree-based ensemble algorithm which will be discussed in the next section.

2.5.1 Tree-Based Ensemble Algorithms

The tree-based ensemble algorithm is a subset of the tree-based algorithms but is generally considered a more accurate model than the traditional tree-based algorithms since it can have better performance (64). It is based on building multiple trees together in an ensemble learner, where the features are randomly split in each tree and the predictions of all the trees are combined. Three classes of ensemble learning techniques have been developed for the tree models: bagging, stacking, and boosting. In this thesis, extremly randomized tree (ERT) is the one used from the family of the bagging tree models. It has been proposed recently by Martínez-Martínez et al. (116; 159). Another model from the boosting family is used, which is extreme gradient boosting (XGBoost) (64).

Extremely Randomized Trees

ERT frequently takes into account homogenous weak learners and teaches them separately from one another in parallel. Then, homogenous weak learners are combined using some deterministic averaging procedure. The advantage of ERT compared to other types of bagging trees is the reduction in bias and variance. The reduction in bias relates to the tree's construction process, which involves sampling from the entire dataset not from a subset of the dataset. The reduction in variance is a result of the decision trees' nodes being split randomly, which means that the algorithm is not greatly influenced by specific dataset features or patterns (63).

Extreme Gradient Boosting

XGBoost frequently takes into account uniformly weak learners and trains them sequentially in a highly adaptive manner. Then, uniformly weak learners are combined using a deterministic approach. Compared to other gradient boosting methods, XGBoost has advantages, such as a more direct path to the least error, faster convergence with fewer steps, and simplified calculations to increase speed and reduce compute costs. One of the disadvantages of XGBoost is complexity. In order to simplify the model and produce predictions that are more reliable and precise, it combines L1 and L2, which are lasso regression and ridge regression regularization, respectively. Overfitting, a common issue in machine learning, can be avoided with the aid of regularization algorithms, which are part of XGBoost. The numerous hyper-parameters involved might be complicated to configure and tune, making the process time-consuming (40).

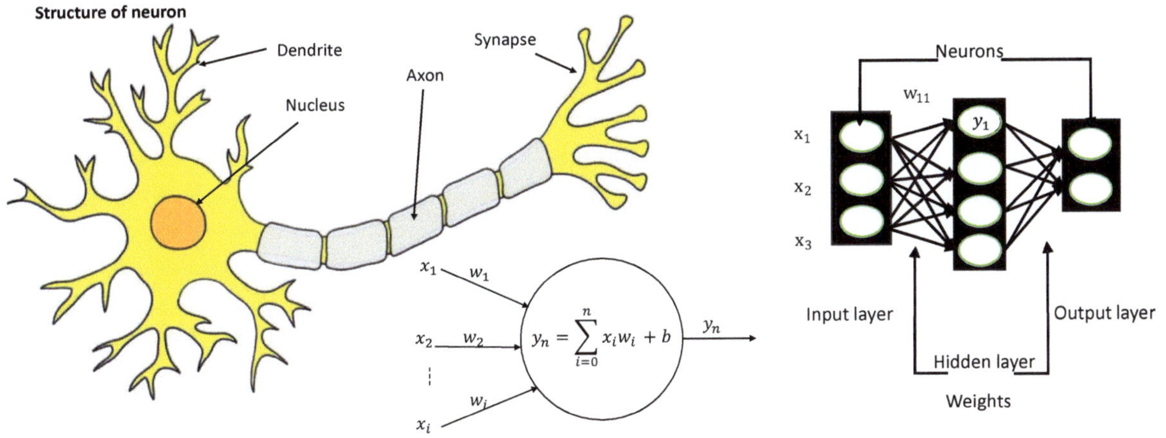

(a) (b)

Figure 2.4: The biology of a neuron in the brain (a - left) and the applied neuron in NNs (a - right). A simple NN architecture (b). This Photo was modified by the author based on (3; 189) and licensed under CCBY .

2.5.2 Neural Networks

Neural Network (NN)s are a branch of the machine learning field. These algorithms' fundamental concepts are mostly inspired by how the brain functions. The brain's primary computational component is the neuron. Dendrites and axons are the components entering and exiting the neuron, respectively, in the way that neurons are interconnected. An axon and a dendrite are connected by a synapse as shown in Figure 2.4 (a - left).

Both inputs and outputs are referred to as activations. The ability of a synapse is to scale x_i with a scaling factor of w_i, which is considered the most significant function as shown in Figure 2.4 (a - right). The brain's learning synapses are associated with weight changes. As a result, different weights produce various reactions to an input. The process of adjusting weights is known as learning (189).

There are three layers in a straightforward NN: input, hidden, and output as shown in Figure 2.4 (b). It is based on the perceptron theory. The perceptron is considered a neuron model (155). It is a non-linear function since this action only allowed the neuron to produce an output if the inputs exceeded a certain threshold.

Stochastic Gradient Descent

This straightforward network is being trained to optimize performance by modifying the weights of neurons. The objective of deep neural networks (DNNs) training is to reduce the discrepancy between the ideal perfect scores of weights and the scores determined by the DNNs based on the current weight as shown in equation 2.6. The procedure used to update the weights is gradient descent. When training the network, Stochastic Gradient Descent (SGD) can be used to change the NN model's parameters. The weight-updated definition of gradient descent is updated (w_{ij}^{t+1}) using a multiple of the gradient of the loss (L) with respect to each weight, which is the partial derivative of the loss with respect to the weight. The loss (L) is the cost function and α is the learning rate. The target is to minimize this cost function. To lessen the overall loss, the technique is continually repeated and the NN model's parameters are optimized (189).

$$w_{ij}^{t+1} = w_{ij}^{t} - \alpha \frac{dL}{dw_{ij}} \tag{2.6}$$

Backpropagation is the method used to compute the gradient where the influence of loss is propagated backward through the network to compute how each weight has an impact on loss. One such training example is x_1, which is trained for these parameters. Throughout the training, the NN is active in both directions. In the feed-forward algorithm, the prediction is compared to the target connected to x_1 in addition to forecasting the output. It keeps running until it reaches the loss layer which is responsible for updating the parameters. In mathematical optimization, the partial derivative vector represents the gradient. The gradient's decision to move w_i in a particular direction causes an increase in loss. w_i should be updated in the opposite direction of the gradient in order to minimize the loss function. Although gradient descent appears straightforward in math, it is very difficult. An unstable loss function is shown, for instance, if the parameters are updated using only the gradient updated from one training example. Since the gradient is computed using only one training example rather than the entire dataset, the average loss of all training examples will oscillate up and down with a relatively high frequency. For decreasing these oscillations, mini batches of SGD are recommended which is splitting the dataset into small batches and for each batch, the gradients are computed (207).
In order to summarize how a NN functions, it takes a single vector and turns it into a number of hidden layers. Neurons with learnable weights and biases make up each hidden layer. Each neuron performs a dot product, receives some inputs, and is fully connected to every neuron in the layer behind it. The output layer is the final fully-connected layer and it categorizes the class scores (6; 189).

Activation Functions

All activation functions are non-linear. There are different activation functions, for example, a threshold function which behaves like a step function. The most frequently used in the literature on deep learning and NNs are the logistic sigmoid function as shown in equation 2.7 (left) and the hyperbolic tangent as shown in equation 2.7 (middle). In addition, Rectified Linear Unit (ReLU) as shown in equation 2.7 (right) has been recently used in multiple architectures (6). Furthermore, the idea of sparsity has come to be a fascinating topic in the field of machine learning. ReLU is considered to be the activation function that applies the concept of sparsity since it provides rising to real zeros. In ReLU, all of the negative values are set not to be activated, which indicates that only certain pieces of information are activated. In addition, the gradient is back propagated which is considered to be an advantage for SGD. For instance, after the weights are uniformly initialized, approximately 50% of continuous output values from hidden units are real zeros; this percentage can go up due to regularization that causes sparsity.

Using ReLU is recommended because it increases the non-linearity of NN and helps identify complex patterns and objects. It has been realized from the point of view of computation that ReLU is simpler and easier than the other two activation functions since it does not include an exponential function as shown in the equations 2.7. ReLU is considered the most widely utilized in hidden layers of NNs (6).

$$y = \frac{1}{1+e^{-x}}, \quad y = \frac{e^{x}-e^{-x}}{e^{x}+e^{-x}}, \quad y = max(0,x) \tag{2.7}$$

Deep Neural Network

For real-time systems, deep learning is a rapidly expanding discipline that requires a lot of data to be effective (98). Deep learning is considered to be a black box that has only one module doing two operations together; extracting features and classification, while a machine learning algorithm has two modules; a fixed feature extractor and a trainable classifier. Deep learning outperforms any other machine learning method (189). One of the most famous algorithms in deep learning is Convolutional Neural Network (CNN).

Convolution Neural Network CNN was first introduced by Yann LeCun, 1998 (98). CNN is quite similar to ordinary NNs. However, it assumes that its inputs are images, which gives us the chance to comprehend certain architectural characteristics. Consequently, it is easier to implement and there are fewer parameters in the network (6). For instance, a single fully-connected neuron in the first hidden layer of a regular NN would contain 3072 weights if an input image of size 32x32x3 was used. The number of parameters would rapidly expand as the size of an input image increased (6). While CNN is different, it depends on sharing weights. Hence, the number of weights used for the same input image is 5x5x3 = 75 weights instead of 3072 weights in addition to a bias value. Three architectural concepts are combined. The first concept is the local receptive field, the second is the replication of weights, and the third, when the pooling layer is applied, is spatial subsampling (98; 189). These concepts ensure some degree of distortion invariance, shift, and scale.
Convolution Layer The convolution layer is the main layer in CNN. It consumes most of the operation time in the architecture. Each layer generates a feature map that preserves important distinctive information. Each convolution layer has a set of 2D input feature maps, each of which is called a channel. Each channel is convolved with a different 2D filter from the filter stack, one for each channel. The results of the convolution at every point are summed over all the channels as shown in equation 2.8. A 1D bias can also be added to the filtering result (189). These calculations produce output activations, which provide one channel of an output feature map. To provide more output channels and enhance the reuse of filter weights, multiple input feature maps can be employed as a batch. The convolution layer is crucial because it finds all of the object's edges, including those discovered by SGD and those that are horizontally or vertically oriented. Convolution produces good results in a particular direction if bias is included. The representation of more complicated properties and the acquisition of significant objects,

such as the classification of tissues, increases with the depth of the convolution layer (207; 189). Padding is another term that is crucial to understand in the context of the convolution layer. It is used to adjust the width and height such that they are the same length. Although it often has zero values, it can have other values (207). The size of the output volume is controlled by three hyperparameters: depth, stride, and zero-padding as shown in equation 2.8 (6).

$$O[z][u][x][y] = B[u] + \sum_{k=0}^{C-1} \sum_{i=0}^{R-1} \sum_{j=0}^{R-1} I[z][k][U_x + i][U_y + j] \times W[u][k][i][j] \tag{2.8}$$

$$0 < y < E, \quad 0 < x < F, \quad 0 \leq z < N, \quad 0 \leq u < M$$

where: O, I, W, and B are the matrices of output feature maps, input feature maps, filters, and biases, respectively. E and F are the output feature maps of plane height and width, respectively. N is the batch size of 3D feature maps. M is the number of output feature maps.

$$E = \frac{H - K + U + 2P}{U}, \quad F = \frac{W - K + U + 2P}{U} \tag{2.9}$$

where H and W are the input feature maps of plane height and width, respectively. K is the filter plane height and width. U is the step size or stride and P is the padding size.

Dropout It is a way to enhance the generalizability of deep learning methods. It makes all weights connected to a certain percentage of neurons equal to 0. For example, if the percentage of neurons is 0.5, these neurons are set to zero (207).

Normalization It controls the input distribution across layers in order to accelerate training and enhance accuracy. The distribution of layer input activations; mean μ and standard deviation σ are normalized such that it has a zero mean and a unit standard deviation. There are numerous varieties of normalization; for example batch normalization and local response normalization. Batch normalization is performed between the convolution layer and the non-linear activation function while local response normalization is performed after the non-linear activation function (189).

Loss Layer The last layer is the softmax layer or loss layer. It gets the exponential of each neuron and divides it by the sum of the exponential of each neuron. It is employed to determine whether the predicted value and the target differ from one another. The softmax layer is useful for training by gradient descent. It is considered a cross entropy loss. It eliminates several issues that would be challenging to be trained again, such as when one class is equal to one while all the others are zero. For more illustration, the probability of the maximum value of the previous layer will be increased by exponential functions compared to other values and the sum-up of the output will always equal one. The softmax is not needed for prediction, it is only useful for training CNN parameters using a set of training examples (207).

Recurrent Neural Networks Recurrent neural network (RNN) are deep learning algorithms but the main difference between recurrent and NNs is that they use data from earlier inputs to modify the current input and output. In contrast, conventional NNs consider inputs and outputs to be independent of one another.

Bidirectional Long Short-Term Memory with an Attention Layer One of the architectures that are famous in RNN is Long short-term memory (LSTM). Each LSTM consists of three main gates: forget gate, input gate, and output gate which is called a cell state. The forget gate is the gate responsible for adding or removing information to the cell state, for which the sigmoid function is used (214). Bidirectional long short-term memory (BLSTM) processes data in two directions. As a result, this model can use both past and future data. Additionally, there is a layer called the attention layer, which is added as a cascaded layer to the BLSTM. The idea of the attention layer is to apply an activation function to the output to obtain a weighted output vector instead. It consists of two cascaded steps of activation functions. For example, those activation functions can be tanh, sigmoid, or ReLU. In the next equations, tanh is used as an activation function in the two cascaded steps as shown in the following equations 2.10, 2.11, 2.12, and 2.13 where H is the output vector, T is the length of the output vector and w is the trained parameter vector. Based on (214), bidirectional long short-term memory with an attention layer (Att-BLSTM) has a better performance compared to BLSTM without an attention layer.

$$M = tanh(H) \tag{2.10}$$

$$\alpha = softmax(w^T M) \tag{2.11}$$

$$r = H\alpha^T \tag{2.12}$$

$$h^* = tanh(r) \tag{2.13}$$

3 Biomechanical Model Based Registration

For registration of MRI to full X-ray mammograms, a biomechanical model based registration is used (160). The method originates from an earlier work in our group (156; 158; 76). It has been applied in an automated registration workflow and tested with clinical datasets (77)(78).
In this chapter, a literature review of biomechanical model based registration between MRI and X-ray mammogram will be discussed and followed by an explanation of the methods proposed for achieving registration between the two modalities.

3.1 Biomechanical Models

One of the main factors from the point of view of registration between MRI and X-ray is the change in intensities and deformation of tissue structures in the breast (75).
There are several ways to register between MRI and X-ray mammograms in literature. The most frequently used method for this purpose is using Finite Element Method (FEM). Alternative methods are to register using transformation models such as affine transformation. In the case of 3D, this includes 12 degrees of freedom. Although this method is very simple compared to non-rigid registration or biomechanical models, registration between MRI to X-ray mammography could achieve errors of $13.1\,mm$ for 49 patients (121).
Free form deformation using a B-spline model, where a set of control points' coordinates were tuned to generate a deformation that varied smoothly (75), shows an error of $5.6\,mm$ for ten patients of large deformation (141).
In addition, thin plate splines (24) have been implemented for modeling breast compression deformation during an X-ray (19). Though these mentioned methods achieved considerably good results, there are still some limitations. Mertzanidou et al. requires a user interface step to select landmarks on the pectoral muscle boundary. Additionally, these cases contain limited rigid deformation which is far away from real mammographic compression (121). Also, Pereira et al. does not include patients with high dense breasts (141). Biomechanical models are the state of the art, as they can respect highly nonlinear deformation and model it on a physical basis.
The challenges in breast image registration are mainly three factors: a large deformation, a non-rigid problem, and soft tissue deformation (75). That is why researchers tend to use soft tissue simulations instead (75).

The main goal for using those simulations is to be applicable to each patient specifically regarding size, shape, and geometry. The target is also to represent the internal structures of different tissues and predict the deformation of each tissue according to the simulation. Additionally, it could be used in clinical use from a computation point of view (75).

For that purpose, many biomechanical models, either linear or non-linear, have been suggested for the simulation of the deformation of the breast such as mass spring method (153; 138), mass tensor method (154), and most prominently in the last years FEM.

3.1.1 Finite Element Methods

Finite Element Methods allow having finite discrete representations of an object for solving the differential equations describing the physical conditions approximately (137). The boundary conditions are considered as the input and out of which, an algorithm for approximating the corresponding solution comes (137). The continuum problem is approximated by a method where the continuum is partitioned into a finite number of elements and a finite set of parameters determines the performance of these elements. The entire system's solution as an assembly of its elements pursues the same principles that apply to standard discrete problems (215).

Biomechanical modeling (BM) of the behavior of anatomical structures under different loads is a necessary step for numerous academic and clinical applications. A set of Partial Differential Equations (PDEs) and the accompanying boundary conditions must be satisfied in order to derive appropriate BM. Corresponding PDEs control the physical phenomenon being modeled such as the deformation of organs like liver, prostate, stomach, breast, and other virtual organs in augmented reality applications (143).

FEM specifically for modeling breast deformation has been used for several applications with different complexities of biomechanical models and materials (75; 59). More specifically, the mammographic compression of the breast has been modeled using FEM with different models using diverse tools for computing the FEM simulation. The main differences are in the complexity of modeling such as homogenous or heterogenous tissues, type of materials, and geometry of the breast. For example, it could be modeled patient-specific or by a simplified geometry. In addition, it was solved using different solvers such as dynamic or static solvers. In this thesis, only an overview of software packages, computation time, and type of studies are presented in Table 3.1 (referred to as "Optim." for optimization).

For the registration between MRI and X-ray using biomechanical models, three main factors need to be defined, as shown in Figure 3.1. First, the geometry of the breast includes typically tetrahedral or hexahedral meshes. Second, the material properties include linear or non-linear materials. Hyperelastic is one example of non-linear elastic. Third, compression simulation and boundary conditions include forces, displacements, and contact problems. Some models, mainly who used FEM for registration purposes have been collected and sorted ascendingly based on the number of nodes as shown in Table 3.2.

Table 3.1: Literature review of finite element methods for breast deformation

Reference	Application	Computation Time	Tools	Studies
Azar et al. (15)	MR image-guided biopsy	< 30 mins	ABAQUS	Clincial
Samani et al. (165)	Image registration	-	ABAQUS	Phantom
Ruiter et al. (157)	Cancer diagnosis	-	ANSYS	Clinical
Ruiter et al. (158)	Image registration	-	ANSYS	Clinical
Tanner et al. (191)	Breast deformation	-	ANSYS	Clinical
Chung et al. (41)	Image registration	-	-	Phantom
Han et al. (70)	Breast deformation	312 mins (Explicit)	ABAQUS	Clinical
Hopp et al. (78)	Image registration	20 mins (Optim.:120)	ABAQUS	Clinical
Lee et al. (99)	Image registration	-	CMISS	Clinical
Mertzanidou et al. (122)	Image registration	120 mins	NiftySim	Clinical
Liu et al. (105)	Simulation compression	-	ABAQUS	Clinical
García et al. (60)	Image registration	61 mins (Optim.)	NiftySim	Clinical

Table 3.2: Literature of biomechanical models

Biomechanical Model			Reference
Geometry Definition (see details in Table 3.3)		Material Definition	
#nodes	#elements	Model	
-	64	Neo-Hooken	Lee et al. (99)
800	2500	Hyperelastic Neo-Hookean	Mertzanidou et al. (122)
3712	2793	Wellman non-linear elastic	Azar et al. (15)
15939	16841	Hyperelastic	Samani et al. (165)
29307	161997	Hyperelastic Neo-Hookean	Han et al. (70)
50000	200000	Hyperelastic Neo-Hookean	García et al. (60)
54843	37755	Neo-Hookean / Mooney–Rivlin	Tanner et al. (191)
61000	313000	Hyperelastic	Solves-Llorens et al. (177)

Before creating the meshes, the MR images need to be segmented first. From the literature, manual segmentation or interactive approaches have been often applied (165; 192; 70; 122), while some methods tend to apply automatic segmentation **(158; 78)** (99; 177). The bold ones are from this group's research.

3.1.2 Biomechanical Model and FEM Simulation

To model the geometry of the breast, some methods used in literature have been collected in Table 3.3 to show the variance of geometries of the breast. Extraction of a volume or a surface mesh based on segmentation is done by these methods. The volumetric meshing is done using tetrahedral or hexahedral meshes. It was claimed that hexahedral meshes in Finite Element (FE) models save more time and are more accurate (165). One disadvantage of hexahedral meshes, it is hard to model curved surface geometries, which in turn leads to some inaccuracies in surface strain. Therefore, two methods have been suggested to solve this specific problem for improving the mesh. First, a smoothing technique was

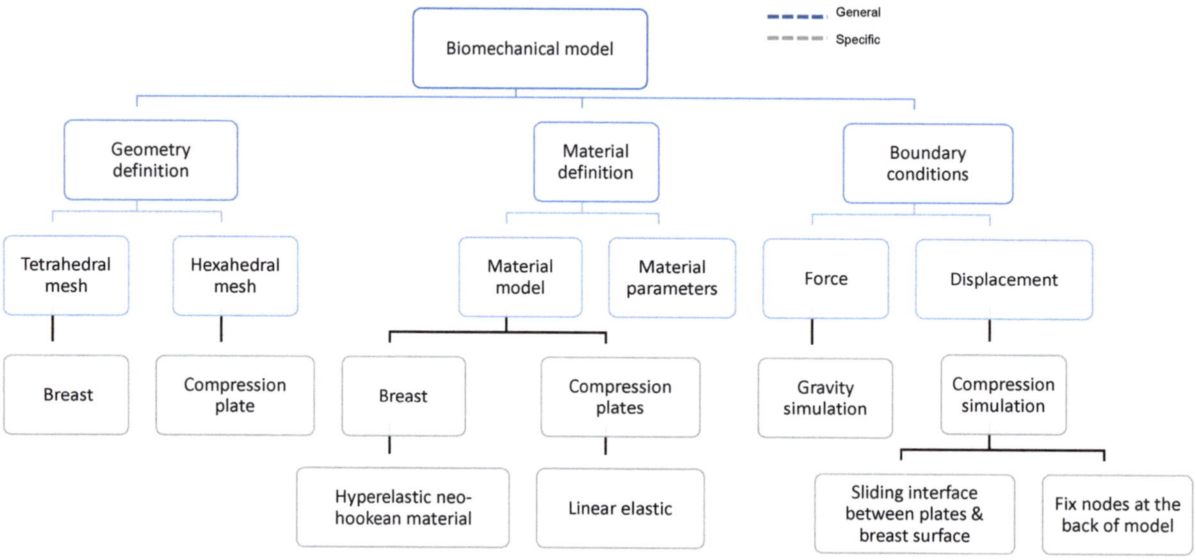

Figure 3.1: The components of a biomechanical model and the specific components to build a biomechanical model of breast deformation.

used which smooths irregular boundaries at the model surfaces and material interfaces (32). Second, a method based on the transfinite interpolation technique is considered to be more efficient and accurate than applying the first method (165). One disadvantage often occurs for the transfinite interpolation: it requires a large number of elements and this large number of deformed elements makes it harder to converge. This leads in most cases to the use of tetrahedral meshes. Tetrahedral meshes can also have undesirable characteristics such as slow convergence with mesh refinement in between the steps (165). But it is more common to be used for breast geometry based on literature (75). Hexahedral meshes have been used by (158; 42; 99). The mesh granularity is another dimension of difference proposed in literature. The breast anatomy and shape are accurately represented by a fine mesh. However, this requires a lot of time to be solved by a FE solver. In addition, it might cause folding problems (75). For modeling breast compression, the number of elements varies (75). It could be a few hundred such as (158; 42; 99), or several thousand such as in (122), or up to ten thousand such as in (78; 165; 192; 70; 177).

To have a realistic biomechanical modeling of the breast, comprehending anatomical characteristics is necessary. The elasticity of cancer or benign tissues varies from one another. Infiltrating ductal carcinoma varies from 1% to 15% strain. At 1% strain, stiffness is ten times as fatty tissue, and two and a half times as glandular tissue. While at 15% strain, it is 70 times stiffer than fatty tissue and five times as glandular tissue (75). The elasticity of malignant tumors is stiffer than benign lesions (75). Values for malignant tumors were reported as 15.9 KPa (106), or 146.6 KPa (14) while for benign lesions 7 KPa (106), or 45.53 KPa (14) was given.

Table 3.3: Literature of biomechanical models - geometry definitions

Nodes	Fat	Gland.	Muscle	Skin	Tumor	Tool	Reference
8-node trilinear	✓	✓	×	×	✓	custom in C	Azar et al. (15)
3-node triangular (only surface)	×	×	×	✓	×	custom in C	Azar et al. (15)
8- hexahedral	✓	✓	×	×	×	Matlab	Samani et al. (165)
4-node quadrilateral (only surface)	×	×	×	✓	×	Matlab	Samani et al. (165)
tetrahedral	✓	✓	×	✓	×	Simpleware	Solves-Llorens et al. (177)
3-cubic hexahedral	✓	✓	✓	×	×	-	Lee et al. (99)
tetrahedral	×	×	×	×	×	TetGen	Mertzanidou et al. (122)
tetrahedral	✓	✓	✓	×	✓	TetGen	Han et al. (70)
10-node tetrahedral	✓	✓	×	×	×	ANSYS	Tanner et al. (191)
tetrahedral	✓	✓	×	×	×	TetGen	García et al. (60)

Krouskop et al. (95) reported that the elastic modulus of fatty tissue is almost constant over a wide range of applied strain (up to 30%) while glandular tissue is strain dependent. It remains constant across a small range (less than 10%) and then afterward it increases with respect to the applied strain. The elasticity of glandular and fatty tissues are nearly the same at low strain levels while at higher strain levels, glandular tissue is stiffer than fatty tissue with one order magnitude higher (75). The values of fatty tissue range from 0.5 - 25 KPa while for glandular tissue range from 2 - 66 KPa (62).

To model the mechanical response of breast deformation, various materials have been used such as linear elastic, non-linear elastic, hyperelastic, and exponential models. Linear elastic models are treated to be less accurate models for large deformation which is the case for breast deformation (75). In literature two groups used linear elastic models (192; 9). One limitation for Tanner et al. (192) is that surface displacements are known. For modeling the breast, hyperelastic material is the most commonly used (75; 17; 191; 42; 70; 78; 99; 122; 158; 165; 177). Hyperelastic material (112) which is considered a non-linear elastic model and does not rely on the strain rate, responds elastically and quickly to high strain. More details about the hyperelastic material will be given in Section 3.2.3.

During MRI acquisition, women lie in a prone position while during X-ray mammography acquisition women stand in an upright position (59). In this case, MRI is considered gravity loaded, therefore, an unloaded state is required before mammographic compression. Several approaches have been proposed to estimate the unloaded state. A simple inversion of gravity was employed by Lee et al. (99). However, obtaining the unloaded reference state by this simple method is considered to be not accurate (146).

Another technique is an iterative approach which was first proposed by Carter et al. (36). It assumes internal stresses equal to zeros, applying gravity in the anterior direction while computing the reference point. The method was expanded later by Eiben et al. (52), relaxing the stresses of the breast model and inverting gravity.
Inverse FE approaches have been proposed by Govindjee and Mihalic (65) for obtaining the unloaded state configuration. The strategy entails solving the inverse motion by reparameterizing the equilibrium equation with a numerical approach. Eiben et al. examined these methods. It was observed that the iterative and the inverse methods show similar zero gravity estimation while the simple inversion method only works for small or highly constrained deformations (59).
Boundary conditions for the compression simulation were defined by displacements in the compression direction to the surface of breast nodes (158; 74). Another technique is modelling a contact problem between plates and breast using friction model (9; 78; 99), or frictionless model (42; 122; 70).
It has been realized that muscular tissue affects simulation compression (75). One approach was to constrain nodes close to the muscular tissue. In literature, several methods followed this approach (165; 42; 78). The second approach was to allow these nodes to slide along the chest wall (9; 192; 70; 99; 177; 122).
There are several models used in literature to model glandular tissue, for example, an exponential function of strain was used by Azar et al., (16) to fit the curve of Wellmann et al., (206). Also, the two parameters Mooney Rivlin model were used by Yin et al., (211) which was first introduced by Krouskop (95). In addition, a second order polynomial hyperelastic model was applied in (61). The last two models are also used for fatty tissue (75). If the skin is modeled, it is considered to be an incompressible tissue like fatty, glandular, and muscular tissues. An exponential hyperelastic model (200), a linear model (165) or a non-linear model has been applied to model the skin (16).
Experiments measuring the stress-strain relationship by ex-vivo tissue samples are usually carried out by recording the forces and displacements of the samples undergoing strain. The measured sample points are used to determine the stress-strain relationship of the material by an interpolation function. The applied mathematical formulation of the model is a simplification. The stress-strain relationship observed in the experiment is given with respect to the reference configuration. To describe the material model, linear elastic, hyperelastic, and exponential models are mostly used in the literature.
Several methods for MRI to X-ray mammogram registration with clinical data have been proposed (see Table 3.4). There is no fair comparison between these methods with respect to accuracy since the clinical datasets are different and features used vary. Also, differences in views such as MLO and CC may cause unfair comparisons. Moreover, the effect of muscular tissue in the compression simulation and the huge uncertainty of positioning the breast before compression is challenging and varies a lot (75).
After collecting an overview of what has been done in literature regarding FEMs for allowing the possibility of registration between MRI and X-ray mammography, the method used in this thesis will be discussed in detail in the upcoming section.

Table 3.4: Evaluation of clinical datasets available from literature

Reference	Views	#datasets	mean error
Ruiter et al. (158)	CC / MLO	6	4.3*mm*
Hopp et al. (77)	CC	11	11.8*mm*
Hopp et al. (78)	CC	79	13.2*mm*
Hopp (76)	MLO	52	15.3*mm*
Lee et al. (99)	CC / MLO	6	19*mm*
Solves-Llorens et al. (177)	NA	14	4.5*mm*
Mertzanidou et al. (122)	CC / MLO	10	11.3*mm*
Liu et al. (105)	CC	10	-

Table 3.5: Literature of skin model

Skin	Linear / Non -Linear
Azar et al. (15)	×/✓
Samani et al. (165)	✓/×

3.2 Proposed Method

The proposed registration method between MRI to full X-ray mammography is based on the work originated by Hopp et. al, (78). It estimates a configuration of the breast that is comparable to its shape in X-ray mammography using the breast geometry as observed within 3D MRI. A biomechanical model is used in combination with a FEM simulation, which mimics virtually the deformation applied to the MRI.

The biomechanical model considers four tissue types: fatty, glandular, and muscular tissues, as well as skin. It uses an estimation of the unloaded state of the breast before applying compression using a contact problem between the breast and the rigid compression plates.

It has been evaluated in several clinical use cases and obtained clinically relevant results, e.g. for multimodal diagnosis of MRI and X-ray mammography with an average localization error of about 13 to 18*mm* (78; 46), image fusion of MRI contrast kinetics with X-ray mammograms (77; 47), and for multimodal computer-aided diagnosis (79; 80).

For obtaining the patient-specific breast geometry, first, an image segmentation for three tissues is applied to the MRI followed by interpolating images to make them isotropic, see in Appendix B.1 for more details for the interpolation method. For simplicity, the shape of the breast is only modeled until a small distance behind the sternum position in the posterior direction. Based on the segmented MRI, the model geometry is generated, dividing the breast anatomy into approximately 2000 to 2500 4-node elements using a tetrahedral meshing algorithm, see in Section 3.2.2. Afterwards, an isotropic hyperelastic Neo-Hookean material model is applied (206) to account for non-linear and incompressible tissue behaviour

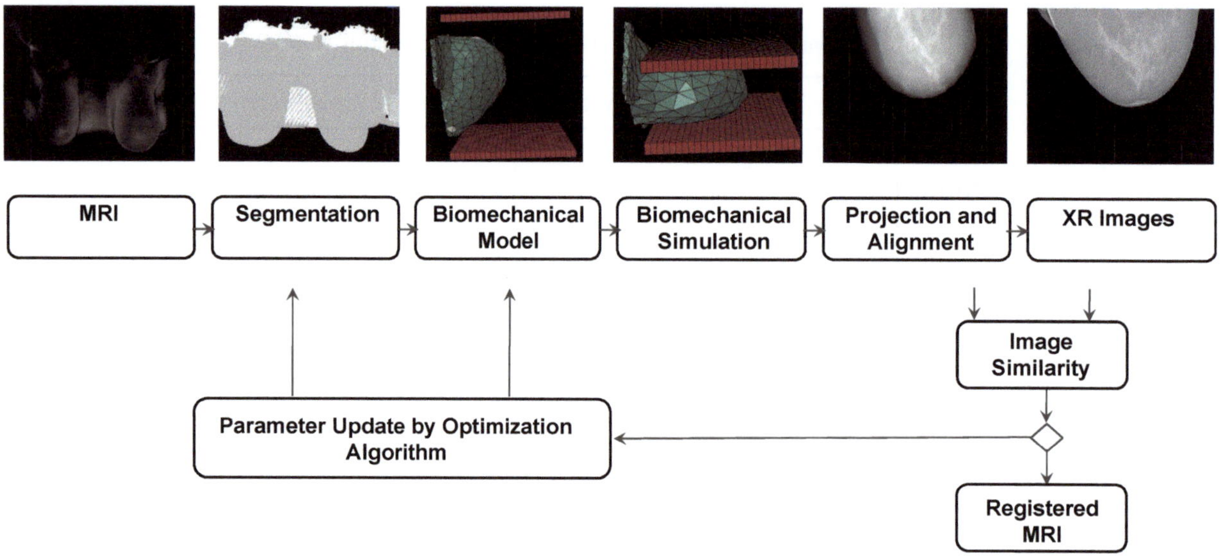

Figure 3.2: The block diagram of the biomechanical based registration between MRI and full X-ray mammogram.

during the deformation process of the breast which will be discussed in detail in Section 3.2.3. An FEM is used to solve the mechanical deformation problem. The mechanical deformations are computed using the dynamic FE solver in the commercial FEM software ABAQUS / Explicit (175).
"After the deformation simulation, the deformation field is applied to the segmented MR image, resulting in a volume containing the deformed configuration of the breast. The deformed MRI volume is calculated by linear interpolation of the deformation field at all voxels and trilinear interpolation of the voxel intensities. This volume serves as a basis to carry out an X-ray-like projection. A raycasting algorithm is applied for this purpose. The geometry of the X-ray acquisition process using the available information from the metadata is taken into consideration such as source to detector distance, source to patient distance, and compression thickness.
The biomechanical model based registration is optimized to adapt to patient-specific characteristics and uncertainties in the metadata. For this purpose, several parameters are optimized such as the compression thickness from the mammogram's metadata which might have an error in the range of $5\,mm$ (201), the Young's modulus of fatty and glandular tissue, the positioning of the compression plate and a possible patient rotation around the three main axes. Optimization is done using a simulated annealing scheme optimizing all the parameters at once. The employed optimization criterion is to measure the shape similarity of the deformed MRI and the full X-ray image" (Said et al. (160)). For this purpose, the DICE Similarity Coefficient (DSC) (see Chapter 6) is used to calculate the overlap of the full X-ray mammogram with the projection of the deformed MRI.
The workflow mainly consists of the following points as shown in Figure 3.2:

1. segmenting MR volume images
2. interpolating MR images and segmented images to derive an isotropic resolution

3. generating meshes for the patient-specific breast geometry
4. carrying out the deformation simulation
5. applying X-ray projection
6. optimizing patient-specific characteristics in steps 4 and 5 in order to align with the original X-ray mammography for getting the best deformed MR image

In this thesis, several processing blocks have been modified to generalize robustness, stability, accuracy, and acceleration of computation time. Those blocks are mainly: segmentation, meshing, and deformation simulation. In the first block, a novel segmentation algorithm is proposed using unsupervised NNs. This work has been published in (162). It will be discussed in detail in Section 3.2.1. In the second block, the mesh quality based on specific criteria is validated and it will be discussed in detail in Section 3.2.2. The third block is the deformation simulation and it will be discussed in detail in Section 3.2.4. The goal was mainly to robustify the model regarding some parameters: position of plates, rotation angles, boundary conditions, etc. Additionally, the stability of the deformation simulation without losing accuracy was investigated. The model was accelerated by removing some steps in the simulation that were previously used. In addition to that, a novel method using machine learning and deep learning algorithms was proposed for replacing FEM. It has been published in (164) and it will be detailed in Section 3.2.5. Afterward, the model was used as part of the MRI to spot mammogram registration.

3.2.1 Segmentation

One of the critical steps in image registration is segmenting images into specific tissues, mainly fatty, glandular, and muscle tissues. Hence, MRI segmentation plays an essential role in the automatic creation of biomechanical models in such applications. It extracts the Region of Interest (ROI) through a semiautomatic or automatic process (45; 193). For breast MR images, different image segmentation methods have been suggested. Among the methods is clustering. It was applied by distinguishing the global characteristics of the ROI from the foreground in images (135; 90). Intensity-based clustering such as K-means and Fuzzy C-means (FCM) are most commonly used in breast MRI segmentation (89). K-means is often employed due to the uncertainty, irregular, and fuzzy borders in breast MR images (90; 89). To enhance segmentation techniques, FCM was used (135). Other approaches of segmentation based on region-based methods, such as active contour, region growing, and region merging, were used (212; 202). However, breast MR images include artifacts in patient-related, signal processing-dependent, and hardware-related which might cause noise, intensity inhomogeneity, and weak boundaries which require complex procedures (45; 53).
Most of the automatic breast segmentation techniques proposed in literature only divide tissue into two categories: fatty and glandular tissues (34; 67; 208). There are also techniques for segmenting muscular tissue such as a Bayesian model. However, it is time-consuming. To obtain a segmentation using this approach, it might need up to 7 hours to compute (66). Recently, deep learning algorithms have been

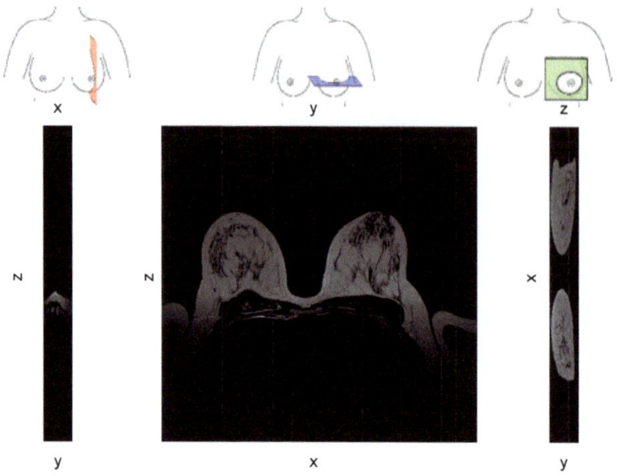

Figure 3.3: One of the clinical cases from the Medical University of Vienna for MR image in three views without any preprocessing: sagittal (left), transverse (middle), and coronal (right).

applied to medical image segmentation, enhancing the performance of segmentation techniques (100; 124). U-Net is a famous architecture that applies convolution to extract high-level semantic information and low-level detail information (152). U-Net has been adapted to segment MRI data into fatty and glandular tissue (213; 68). In addition to the lack of muscle segmentation, these networks require labeled training data for robust performance, which is costly and time-consuming to obtain for 3D breast MRI data.

Here, a novel method is proposed using an unsupervised NN for segmenting three types of tissue simultaneously (162). The method was developed for segmenting T2-weighted images, which causes fatty tissue to be brighter than glandular tissue (no fat saturation). It consists of three parts, which are preprocessing, NN classification, and postprocessing as shown in Figure 3.4. First, preprocessing is intended to make the training of the network and the segmentation easier. It prepares the MR volume for the subsequent processing step. Second, an iterative approach is used for NNs to generate a set of different segmentations of the same MR volume. The network is randomly reinitialized and retrained for each iteration. Third, postprocessing improves segmentation by eliminating common mistakes and selecting the most reasonable segmentation. With some minor adaptions in the postprocessing which will be illustrated, it could also segment T1-weighted images (no fat saturation).

Preprocessing

Preprocessing is responsible for improving segmentation results for training NN. It deals with two problems that might appear. First, the intensities of muscle are enhanced such that they differ from the background. Second, it corrects the bias field which affects MRI's intensities considerably and may lead to the misclassification of different tissue types if not adapted. Preprocessing consists of four steps: improving contrast, masking, correcting bias field, and cropping data, see Figure 3.5.

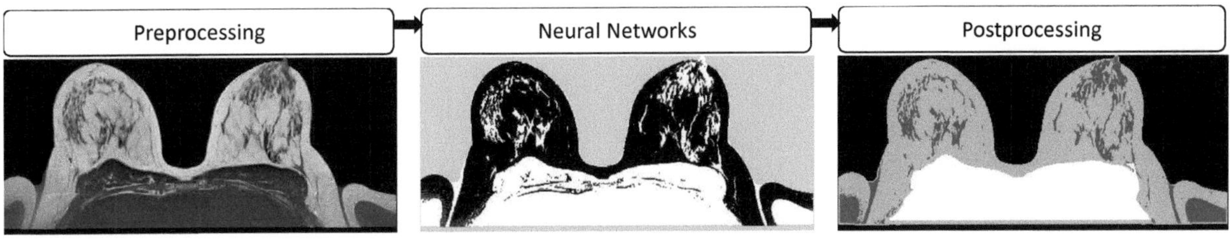

Figure 3.4: The block diagram of the segmentation process indicates the output from each step: preprocessing (left), NNs (middle), and postprocessing (right).

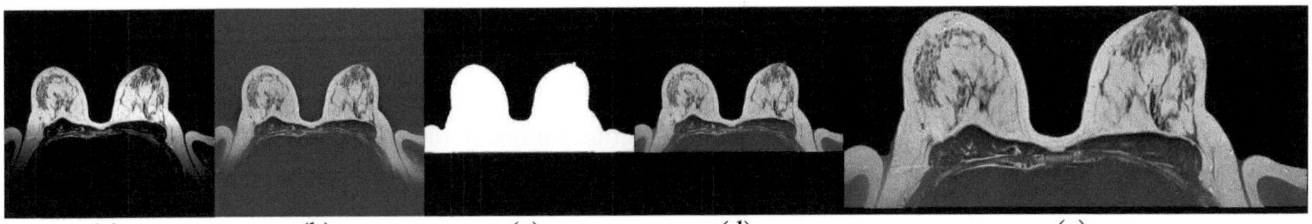

Figure 3.5: The steps of preprocessing in one of the slices: The enhanced image (a). The corrected bias field image (b). The binary mask using K-means (c). The corrected bias field image is multiplied by the K-means mask (d). The cropped image (e).

Improving Contrast The first step is stretching the histogram intensities of the MRI which improves the contrast. The intensity distribution in various datasets contains nearly 99% of voxels in less than 30% of the intensity range. The maximum intensity of the other 99% of the voxels is assigned to the remaining 1% of the voxels as shown in Figure 3.6. As a result, the contrast is enhanced as shown in Figure 3.5 (a) in comparison to the original image (Figure 3.3 (middle)).

In order to avoid the problem of missing parts in muscular and glandular tissue as they are similar to the background in intensities, a certain position is estimated from the left and right breast in each slice based on the position of the sternum position. The holes that occur after this position in the muscular and glandular tissue are filled.

Bias Field Correction Due to the inhomogeneity in the coils of the magnetic field, a bias field often exists in MRI data. This results in intensity inhomogeneities of the same tissue type. It has a higher effect mostly on the superior and inferior slices in MRI. Song et al. (178) concluded that bias field correction is not completely a solved problem, however, a few practical approaches have been proposed in literature. In this thesis, the bias field correction based on Fuzzy C-means (Bias Field Correction based on Fuzzy C-means (BCFCM)) is used (7). It performs segmentation by Fuzzy C-means clustering and a bias field correction simultaneously. The Fuzzy C-means clustering is the extension of K-means clustering. In parallel, the method adapts the intensity with respect to the found classes to eliminate the bias field. Then the class centers are recalculated and the process is repeated iteratively. The best solution for the T2-weighted images based on the clinical data is currently initializing only two centroids. By that,

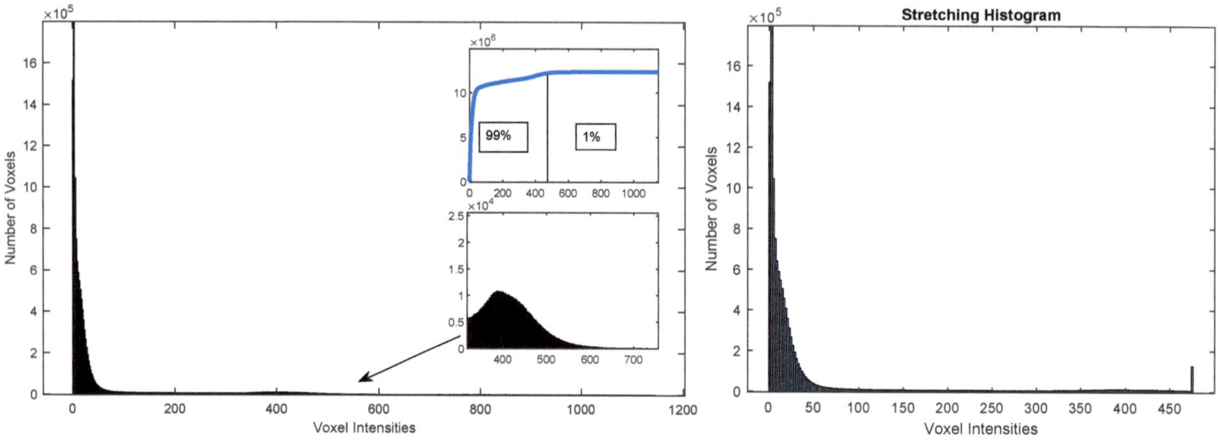

Figure 3.6: The distribution of voxels intensities using histogram for one of the cases for T2-weighted MRI volume (left). Improving the contrast as seen in Figure 3.5 (a) using stretching histogram (right).

homogeneity within classes have been increased and the bias field is corrected. However, one of the side effects is that the background becomes brighter.

Masking Since the breast muscle and the background have nearly the same intensity values in T2-weighted MR images, an air mask is implemented using K-means to get a binary mask for the breast volume excluding the background as shown in Figure 3.5 (c). Then, an element-wise multiplication between this mask and the corrected bias field data is applied as shown in Figure 3.5 (d).

Cropping Data For decreasing the volume of MRI in order to reduce the computation burden, one step is to get only the bounding box of the breast. This is done by cropping all the voxels of the background from posterior, inferior, left, and right as shown in Figure 3.5 (e). Additionally, the positions of the 100 brightest pixels of the middle slice are stored. They will be used further in the postprocessing for the evaluation of segmentation as the brightest pixels are considered fatty tissue with respect to the T2-weighted characteristics.

Neural Networks

The NN architecture is based on the work of Kanezaki et al. (93; 88). It is an unsupervised image segmentation method but has not yet been utilized for breast MRI segmentation and has only been used for purposes in non-medical imaging applications. This network consists of convolution layers, ReLU as an activation function, batch normalization layer, and softmax layer. They are cascaded together such that the output of a specific layer depends on the input of the previous layer in a feedforward scheme.

This network's loss function consists of two criteria. First, voxels with similar features are assigned to the same class. Cross entropy loss is applied for this purpose as shown in Figure 3.7 (first row). Second, spatially continuous voxels are also assigned to the same class. The Mean Absolute Error (MAE) is

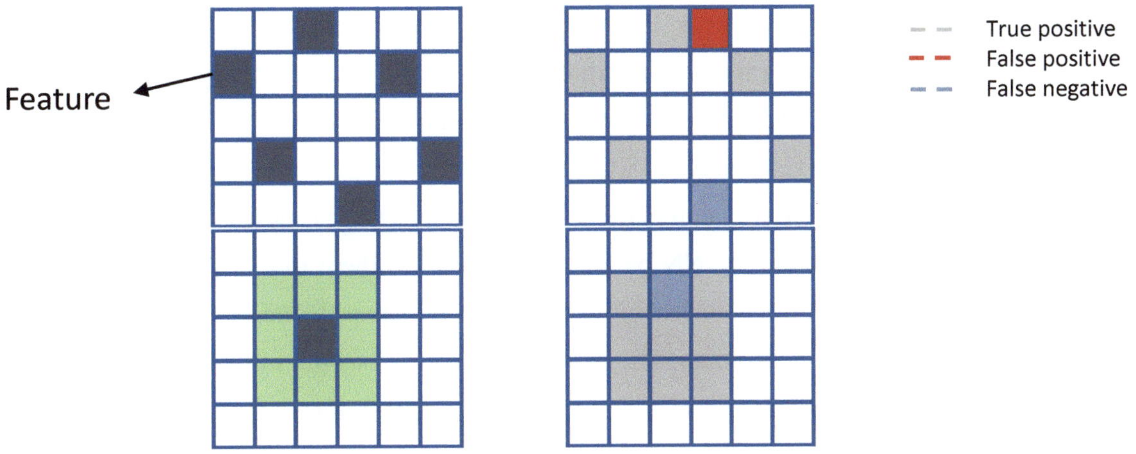

Figure 3.7: Concepts of similar features to be assigned to the same class (first row). Spatial continuous voxels in 3D dimensions to be assigned to the same class (second row).

calculated for this purpose between each element in the input x and target y, where x is the difference between the voxel-wise feature to its neighbouring voxels and y is a matrix full of zeroes as shown in Figure 3.7 (second row). The step size that has been empirically used for both loss functions based on the available datasets is one and three, respectively.

This network is modified to be adapted for 3D datasets with anisotropic volume dimensions. The loss function of spatial continuity is expanded into the third dimension. An iterative approach is used in NNs that generates different segmentations of the same MR volume because of the random initialization of the NN. For each iteration, the network is randomly reinitialized and retrained.

Architecture In this thesis, three approaches have been proposed for training the model. One approach was training the segmentation on the whole volume 3D. The second approach was training 2D slices separately and combining them to a 3D volume (denoted as 2D stacked). The third approach was to train a subset of the 3D volume of a dataset and adapt the same trained parameters to the whole volume. The advantages and disadvantages of each approach are trade-offs between accuracy and computation time which will be discussed in Section 7.1. With respect to accuracy, the first approach is considered to be the best compared to the other two approaches. From the perspective of computation time, the third approach is faster than any of the other approaches.

3D Volume and 2D Stacked Volume For the first two approaches, the architecture consists of two sub-networks. The first network with a filter size of [1x3x3] pixels is responsible for classifying the muscular tissue and the second network with a filter size of [1x1x1] pixels is responsible for detecting glandular tissue. The number of convolution layers is three and two for the first and the second network, respectively. After each convolution layer, there is an activation function in which ReLU is used in addition to a batch normalization layer. At the end of each sub-network, there is a fully connected layer. The learning rate (α) is empirically set to 0.2 and 0.05 for the first and second networks, respectively.

The number of channels in both networks is 25. A SGD optimizer is used. The number of iterations is 300 for training while the number of randomly reinitialized and retrained different segmentation is 10 for each network. The number of labels reached for stopping the training is four labels. These values have been selected based on empirical tests on the available data.

Subset from the Volume For the third approach, the idea is to select a subset from the volume, which is used for training, and then extract the NN parameters and assign them to the whole volume. Only the middle slice was selected to train the network. The idea of selecting the middle slice was to ensure that all the features were included in training the parameters since it has the full shape of the breast as shown in Figure 3.3. The architecture consists of only one network similar to the network architecture with a filter size of [1x1x1]. All of the parameters are the same except for two: the learning rate (α) and the number of random reinitialization and retrained segmentations. (α) is 0.2. The number of randomly reinitialized and retrained different segmentation is 20. It uses only one network, not two sub-networks as before. These values were selected based on empirical tests on the available data.

In the postprocessing, the final segmentation is selected from this series of segmentations. First, different numbers of iterations have been tried: the mean DSC of the available ground truth (eight cases) was calculated in each iteration starting from one to ten as shown in Figure 3.9 (left). The figure shows that starting from the third iteration, it varies. Hence, another experiment was done, in which each iteration 40 random times for each patient was simulated to determine if ten iterations were sufficient. In each iteration, the average of all the patients in the same specific random simulation is taken. Then the average of all the random simulations for this iteration is taken as shown in Figure 3.9 (right). In iterations starting from one to five, the breasts with small sizes could not identify the muscle as a class in some random simulations. It was obvious starting from the sixth iteration, that it could be considered a determined result for each patient. Hence, re-initialization 10 times for each sub-network is sufficient for the first two approaches. So, 20 iterations in total are used. For the third approach, 20 iterations for only one network are applied. By that, it would guarantee general robustness for new data.

Postprocessing

The postprocessing improves segmentation by removing common errors and choosing the most reasonable segmentation from the random re-initialization. It consists of three main steps: assigning class numbers to tissue types, splitting the combined class of glandular and muscular tissue, and finally an evaluation metric for selecting the most reasonable segmentation for both tissues glandular and muscular tissues.

Assigning Deterministic Class Labels There is no distinct class label for each tissue because the learning is done unsupervised. It varies between 0 and 24 based on the number of channels that have been selected in the architecture. Based on heuristic observations, the four segmented areas are sorted by the NN to deterministic tissue labels: 0 (background), 1 (glandular tissue), 2 (fatty tissue), and 3

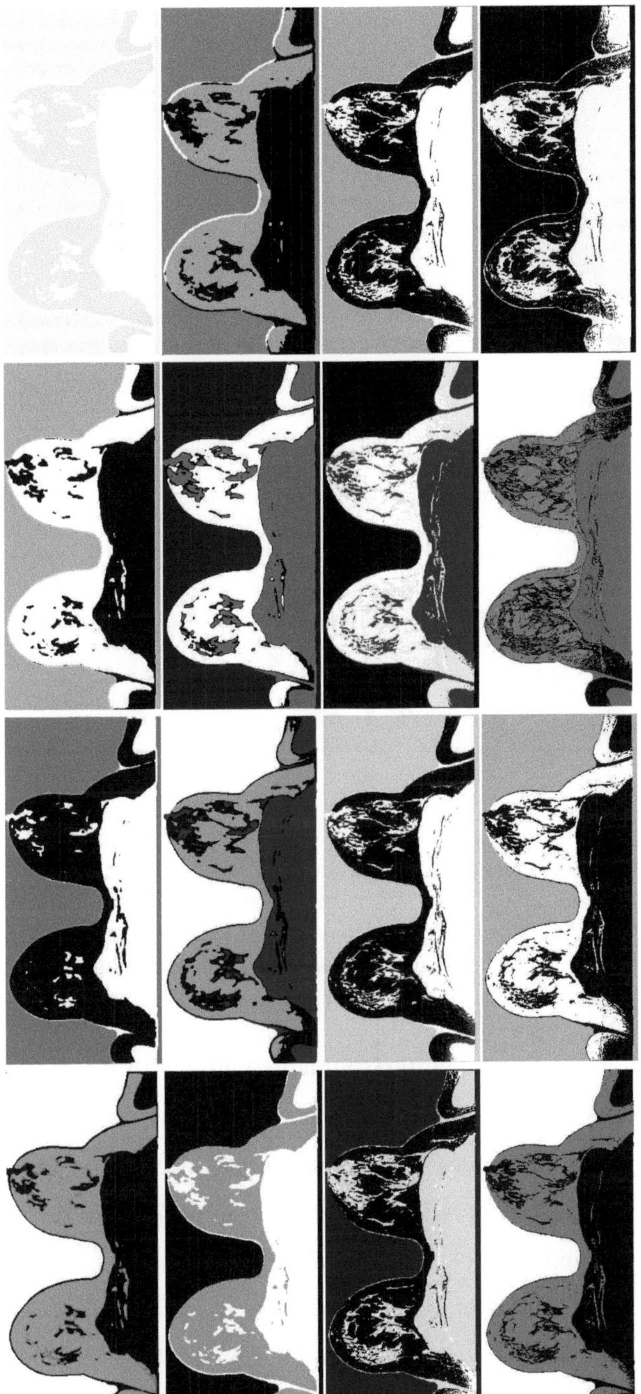

Figure 3.8: The output of the same MRI volume in the middle slice (transverse view) from NN of ten iterations for each sub-network using the first approach. The first two rows are from the first sub-network of the size filter [1x3x3]. The second two rows are from the second sub-network of the size filter [1x1x1].

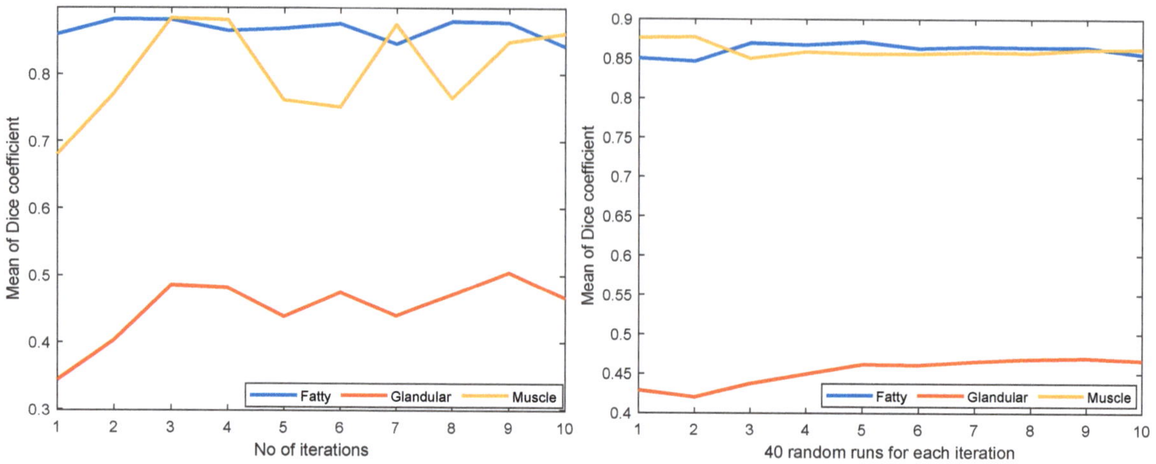

Figure 3.9: The mean DSC of a single run for the eight patients in each iteration (left) starting from one until ten iterations using the first approach. The mean DSC of 40 runs for the eight patients in each iteration starting from one until ten iterations (right).

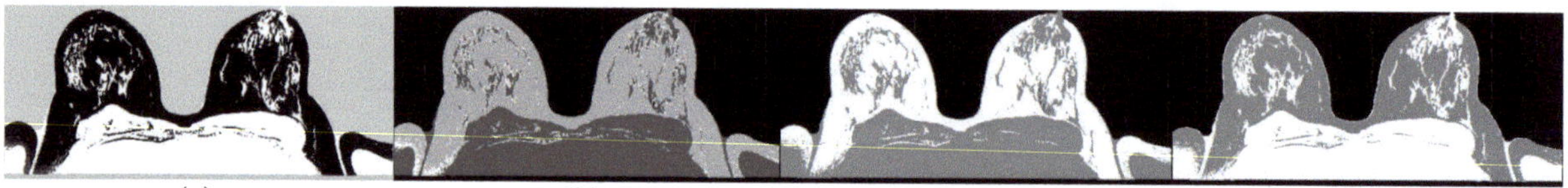

Figure 3.10: Assigning deterministic class labels in the postprocessing in the middle slice of one of the cases from Medical University of Vienna. The output image from NN in one of the iterations (a). Applying determined labels based on the four segmented labels (b). Compensating the loss of segmenting glandular tissue (c). The assigned classes for the three main labels (background, fatty, mixed class of glandular and muscular tissues) (d).

(muscle). It was observed using a histogram that the segmented area with the most voxels is given the label 0 as shown in Figure 3.10 (b) since the background includes the majority of the voxels as shown in Figure 3.11 (i.e. label 9 in this example). One common error might arise after this step, a small amount of voxels is assigned to i.e. label 7 (unassigned labels) as shown in Figure 3.11. Those voxels are considered to be members of the mixed class of glandular and muscular tissue. Hence, they are assigned to label 1 as shown in Figure 3.10 (c). However, other tissues can not be assigned unfortunately based on the histogram of a dataset. The ratio between fatty and glandular tissue depends on the age of the patient (85). Young women can have more glandular tissue than fatty tissue and vice versa (18) (85). Hence, another approach is needed for assigning the remaining tissues.

Due to the characteristics of the T2-weighted MR images, fatty tissue is considered the brightest tissue. Therefore, the 100 brightest pixels are examined in the middle slice in the original MRI dataset and the NN tissue class that these pixels belong to is assigned to label 2 for fatty tissue shown in Figure 3.10 (d). For now, glandular and muscular tissues will be assigned to the same class label 1. In the upcoming paragraph, it will be discussed how to differentiate between them.

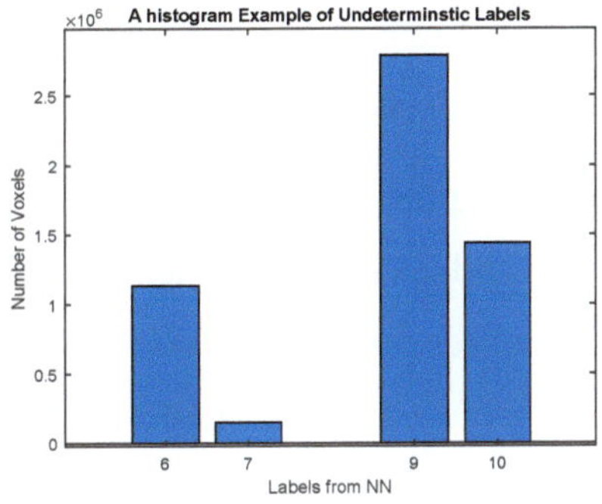

Figure 3.11: The histogram of the unassigned segmented four labels from NN for one example. The absolute label from the NN is not consistent for different random re-initialization.

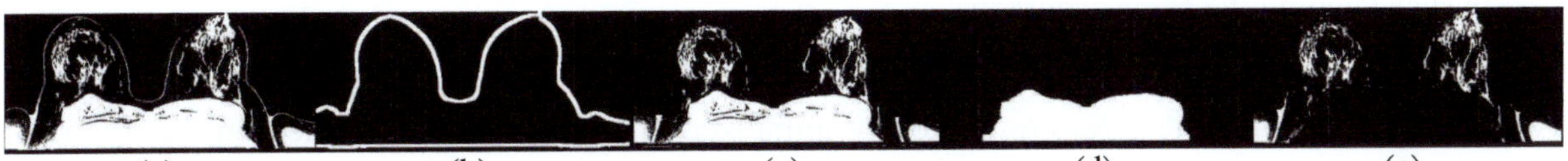

Figure 3.12: Differentiating between glandular and muscular tissues in the postprocessing in the middle slice of one of the cases from the medical university of Vienna: The mixed class of skin, glandular, and muscular tissues (a). The borders of the breast with a large area (b). The subtracted image from the first two images (mixed class without skin) (c). The binary mask of muscular tissue only (d). The binary mask of glandular tissue only (e).

Differentiating between Glandular and Muscular Tissues

(1) **3D Volume and 2D Stacked Volume** Before separating glandular and muscular tissue into two different classes, one needs to take into consideration that some errors could occur and need to be resolved.

First, the boundary of the breast is classified as glandular tissue. However, this misassignment is caused by the partial volume effect which leads to a gray value of pixel close to the breast surface similar to glandular tissue. It is handled by having two masks of the breast volume with different sizes. Those masks are subtracted from each other so that the borders of the breast can be extracted as shown in Figure 3.12 (b). Finally, these borders are assigned to fatty tissue label 2.

Second, some fatty tissues could appear inside the muscle area. This is because some intensities of muscle are similar to the fatty tissue. It is handled by filling these holes in the muscle area. Those two common errors are only done for the network of filter size [1x3x3].

Due to their similar intensity values in T2-weighted images, muscular and glandular tissues are frequently classified as the same class. From the sub-network of filter size [1x3x3], a binary mask of the mixed class of glandular and muscular tissues is extracted. Based on the available datasets, it can be empirically determined that glandular tissue consists of numerous small areas while muscular tissue consists of one single object with a large area as shown in Figure 3.12 (d and e). Thus, the largest connected object in each slice is identified and given the label 3 for muscular tissue. Label 1 is given for glandular tissue to the rest of the objects.

The final result is combined by taking the shape of the breast, which is set to fatty tissue, from the sub-network of filter size [1x1x1] and adding the mixed class of the same segmentation on top for glandular tissue and overlying the breast muscle from the sub-network of filter size [1x3x3]. By that, an accurate shape could be achieved, good muscular tissue, and detailed glandular tissue as shown in Figure 3.13.

(2) **Subset from the Volume** Instead of using two sub-networks, it is sufficient to use only one network of filter size[1x1x1] with the same procedures of assigning classes to glandular and muscular tissues keeping the same accuracy.

Evaluation Metrics The NNs are randomly initialized 10 or 20 iterations based on the approach. A selection metric is suggested to assess the quality of the muscular and glandular tissue segmentation in order to choose the most reasonable segmentation out of the ten different segmentation for each sub-network or out of 20 different segmentation for one network only. By combining several heuristic metrics, the muscle quality score (MQS) and glandular quality score (GQS) are determined.
For muscle tissue, first, the muscle occurrence (denoted as $Occurr_m$ in equation 3.1) is evaluated in each segmentation from the sub-network of filter size [1x3x3]. The number of muscle voxels is compared to the total breast volume. If the ratio is too small, it will be given a bad score. Second, the symmetry of the muscle (denoted as $SymErr_m$ in equation 3.1) is evaluated. The muscle symmetry error is the mean square error between muscle segmentation and its flipped version. If a one-sided muscle is only detected, then it will be given a bad score. Third, in the case of including extra tissues in the same mask of muscular tissue, the muscle needs to be corrected. Therefore erosion is applied to this binary mask, which causes a loss of details. The morphological process of erosion causes an object's boundaries in an image to get smaller. Hence, the number of erosion steps (denoted as $Eros_m$ in equation 3.1) that were done for the inaccurate detection of the muscular tissue in the anterior part of the breast is counted. As the number of erosion steps increases, the quality of segmentation decreases. Fourth, it is taken into consideration that the muscle should not occur in the anterior two third of each slice (denoted as $OccurAnterior_m$ in equation 3.1). It is only available in the posterior of the breast (136). Fifth, in the human body, muscular tissue should not be in contact with air (denoted as Air_m in equation 3.1) but rather with the chest wall. Therefore, it was evaluated how many voxels are in contact with air instead, which is considered in this case the background. These metrics are subtracted or added together based on the decision of whether they have a negative or positive effect on selecting a good muscle shape, as shown in equation 3.1. The values of these metrics have been weighted based on empirical tests of the available data. The weighted values are the same for both series. For $Occurr_m$, the constant value (a) currently is 500. For $SymErr_m$, the constant value (b) is 100. For $Eros_m$, the constant value (c) is $\frac{1}{15}$. For $OccurAnterior_m$, the constant

value (d) is 50 while for Air_m, the constant value (e) is 5×10^{-4}. These constants could be updated in the future.

$$MQS = a \cdot Occurr_m - b \cdot SymErr_m - c \cdot Eros_m - d \cdot OccurAnterior_m - e \cdot Air_m \tag{3.1}$$

For glandular tissue, first, the glandular occurrence (denoted as $Occurr_g$ in 3.2) is evaluated in each segmentation from the sub-network of filter size [1x1x1]. It is evaluated with the same concept as muscle occurrence: comparing the number of glandular voxels to the total breast volume. Second, the overlap of the muscle area detected in each sub-network (denoted as $Overlap_m$ in 3.2) is calculated in the case of the first two approaches. In the case of the third approach, the overlap of the muscle area between the selected muscle (based on evaluation metrics) and the current muscle of the specific iteration. If overlapping increases, the quality of segmentation of glandular tissue increases. Third, for the position of the 100 brightest pixels of the middle slice of the MRI volume (denoted as $PixPos_f$ in 3.2), a ratio of those pixels is calculated between the default labels (fatty tissue) and the actual segmented labels. If the ratio increases, it is likely that those pixels are segmented correctly. Fourth, it could be determined that different MRI datasets do not have glandular tissue in the lateral parts or the area posterior to the sternum position. This observation can be validated that the glandular tissue reaches from the axillary to the inguinal regions (136). A mask for this region has been created as shown in Figure 3.12 (b) and multiplied by the segmented data. By that, the percentage of glandular tissue in this area (denoted as $Borders_g$ in 3.2) could be calculated. If the percentage increases, it is considered bad segmentation. These metrics have been similarly subtracted or added together based on the decision of having a negative or positive effect on selecting a good glandular shape, as shown in equation 3.2. The values of these metrics have been weighted based on empirical tests of the available data. These weighted values are different for both series. For the T2-weighted series, the constant value (a) of $Occurr_g$ is currently 1.5. For $Overlap_m$, the constant value (b) is 0.3. For $PixPos_f$, the constant value (c) is 0.05, while for $Borders_g$, the constant value (d) is 30. For the T1-weighted series, the constant value of $Occurr_g$ (a) is currently 1.1. For $Overlap_m$, the constant value (b) is 0.5. For $PixPos_f$, the constant value (c) is 0.05, while for $Borders_g$, the constant value (d) is 300. These constants could be updated in the future for both series.

$$GQS = a \cdot Occurr_g + b \cdot Overlap_m + c \cdot PixPos_f - d \cdot Borders_g \tag{3.2}$$

3.2.2 Meshing

Based on the segmented MRI, the model geometry is generated, dividing the breast anatomy into approximately 2000 to 2500 4-node elements using a tetrahedral meshing algorithm. The iso2mesh library (54), which is based on Computational Geometry Algorithms Library (CGAL) (39) and tetgen (169), is used for creating the tetrahedral meshes used. The meshing algorithm first divides the outer and inner surfaces of the breast tissues into a triangular mesh based on the Delaunay triangulation (55). Afterward,

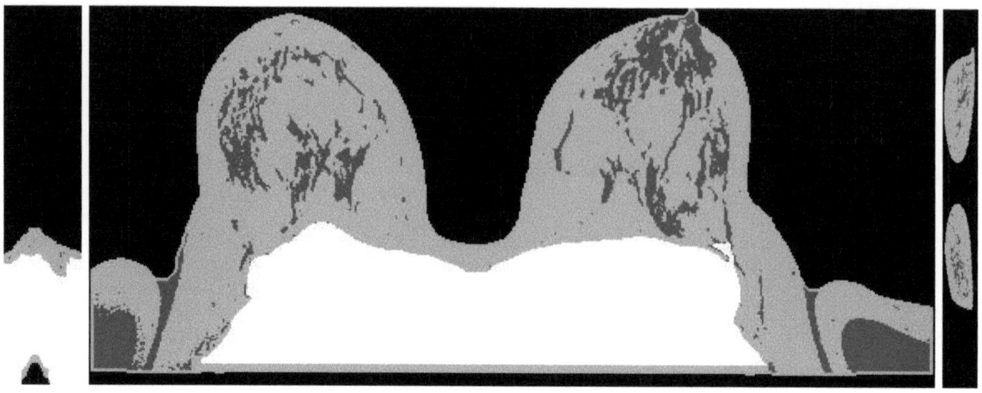

Figure 3.13: One of the segmented clinical cases for three tissues simultaneously: glandular (dark grey), fatty (bright grey), muscular tissue (white) in three views: sagittal (left), transverse (middle), and coronal (right).

the volumetric mesh is filled with tetrahedrons in between. Each element of the mesh is assigned to tissue-specific material properties according to the labels of regions produced, which in turn are based on the MRI segmentation. On top of the tetrahedral mesh, a skin layer of membrane elements with a thickness of $2mm$ is modeled. There are five parameters controlling the mesh generation, taking into consideration the shape, the size, and local refinement.

- maximum element volume (size - element)
- maximum radius-edge ratio: a ratio between circumRadius (R) and length edge (shape - element)
- deviation from isosurfaces (local refinement)
- maximum triangle delaunay circle radius (size - surface)
- minimum angle of a surface triangle (shape - surface)

An analysis is done to investigate if there is a correlation between breast segmentation in images and mesh coordinates. Two aspects were analyzed: the ratio between glandular and fatty tissues and the amount of glandular tissues compared to the breast. It is found that there is a strong correlation with an Pearson Correlation Coefficient (r) of 0.88 and 0.85, respectively using a polynomial fitting curve (first degree). This analysis verifies that the ratio of dense breasts which will be used further in the results is calculated correctly either based on images or meshes as shown in Figure 3.14.

3.2.3 Material Model and Boundary Conditions

After creating the mesh, a material model for describing the mechanical response of breasts during deformation needs to be implemented. Several parameters can be used to characterize the mechanical properties of the breast:

- Poisson's ratio (ν) defines the relative change of volume, either thickness or length when applying an external force

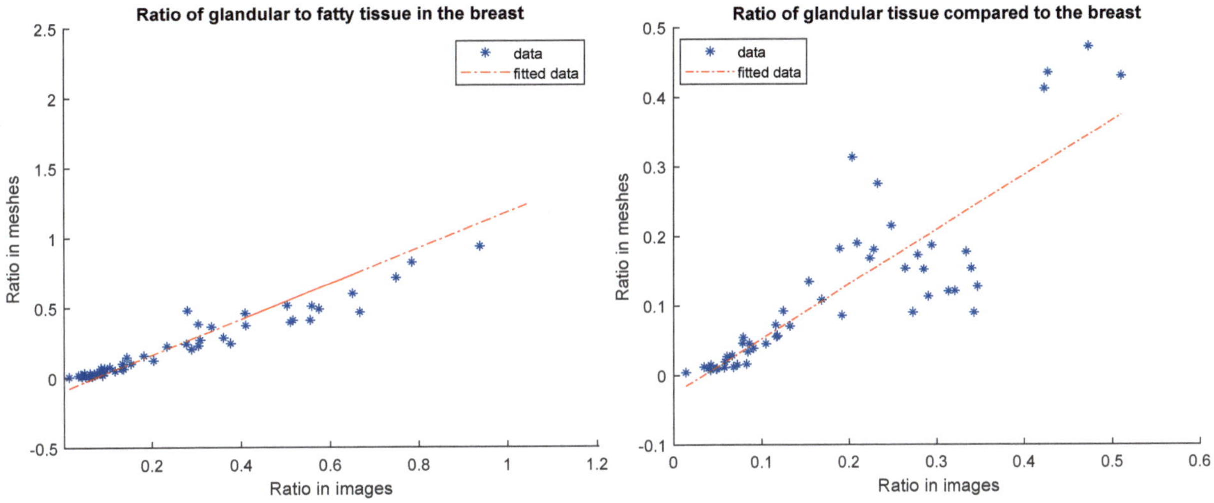

Figure 3.14: Comparison of the ratios calculated once by image coordinates and once by mesh coordinates. These ratios have been calculated based on the proposed segmentation methods. A strong correlation is found between the ratio of glandular and fatty tissue calculated in images and meshes of an r of 0.88 (left). A strong correlation is found between the amount of glandular tissue compared to the breast of an r of 0.85 (right).

- Young's modulus (E) describes the stress (σ) - strain (ε) relationship of the breast when exposed to deformation in linear elastic cases.
- Bulk modulus (κ) explains the magnitude of pressure change required to cause a specific volume change
- Shear modulus (G) or (μ) describes the deformation due to shear forces

Since breast tissue is considered to be an incompressible material (147), Poisson's ratio is assumed to be nearly $v = 0.5$, which leads the bulk modulus K to be infinity as shown in equation 3.8.

The hyperelastic material models are considered to be a solution for breast tissues. Since they are isotropic, non-linear, incompressible, and depend on the strain rate which leads to respond instantly elastically up to large strains. In addition, they could take into account the geometric non-linearity during the analysis step. That is why hyperelasticity was selected for modeling the stress-strain behaviour of such material (rubber-like material) (27; 195). Some typical hyperelastic models are:

- Polynomial
- Mooney-Rivlin
- Neo-Hookean

The strain energy density function for the polynomial model is defined as (27)

$$U = \sum_{i+j=1}^{N} C_{ij}(\overline{I}_1 - 3)^i(\overline{I}_2 - 3)^j + \sum_{i=1}^{N} \frac{1}{D_i}(J^{el} - 1)^{2i} \tag{3.3}$$

where U is the strain energy per unit of the reference volume, C_{ij} are the empirically determined material constants, and $(\overline{I}_1 3)$ and $(\overline{I}_2 3)$ are the first and second invariants of the deviatoric component of the left

Cauchy-Green deformation tensor. A correction factor for compressible material based on the material constant D_i is represented by the last term on the right-hand side. The D_i values determine the compressibility of the material: if all the D_i are zero, the material is taken as fully incompressible. If $D_1 = 0$, all D_i must be zero. J^{el} is the elastic volume ratio. Since J becomes 1 in the case of incompressibility, the last term on the right side can be omitted. Regardless of the value of N, the initial shear modulus and the bulk modulus depend only on the polynomial coefficients of order as shown in equation 3.4 (76):

$$\mu_0 = 2(C_{10} + C_{01}), \quad \kappa_0 = \frac{2}{D_1} \tag{3.4}$$

The Mooney-Rivlin model is a simplified version of the polynomial model with $N = 1$ and $C_{11} = 0$. The Neo-Hookean model is a specific case of the Mooney-Rivlin model when $C_{01} = 0$ as shown in equation 3.5 and for this case, the initial shear modulus and the bulk modulus are as shown in equation 3.6 (76). An isotropic hyperelastic Neo-Hookean material model was implemented to account for the large deformation, non-linear, and incompressible tissue behavior during the deformation process (206). It depends on the material parameters C_{10} and D_1. The C_{10} and D_1 are derived from the equations 3.7 and 3.8 as functions of Young's modulus E and Poisson's ratio ν as shown in equation 3.9 to describe the stress-strain relationship of the breast tissue to fit the reduced polynomial form (187). Poisson's ratio ν was set to 0.495 for all three tissues while Young's modulus differed from one tissue to another. E_{fat} was set to 1100 Pa, E_{Gland} was set to 2500 Pa, and E_{Muscle} was set to 6000 Pa (176).

$$U = C_{10}(\overline{I_1} - 3) + \frac{1}{D_1}(J^{el} - 1)^2 \tag{3.5}$$

$$\mu_0 = 2C_{10}, \quad \kappa_0 = \frac{2}{D_1} \tag{3.6}$$

$$G = \mu_0 = \frac{E}{2(1+\nu)} \tag{3.7}$$

$$\kappa_0 = \frac{E}{3(1-2\nu)} \tag{3.8}$$

$$C_{10} = \frac{E}{4(1+\nu)}, \quad D_1 = \frac{2}{3E(1-2\nu)} \tag{3.9}$$

$$\nu = \frac{3(\frac{\kappa_0}{\mu_0}) - 2}{6(\frac{\kappa_0}{\mu_0}) + 2} \tag{3.10}$$

The biomechanical model that was previously proposed by Hopp et al. mainly consists of four steps simulating the deformation of mammographic compression. First, since the MRI is taken in prone position as discussed in Section 2.1. The initial model is not stress-free and the breast is first simulated in an unloaded state, i.e., in a configuration in which no mechanical deformation and no gravity are applied to the breast. Then, an upright gravity simulation mimics the upright standing patient, followed by applying compression to the breast by moving the lower compression plate first to its desired position and then the upper plate to perform the actual compression. "In the posterior direction, the fixation of the breast at the chest wall is modeled. In terms of boundary conditions, the breast is attached to the body by modeling the muscle as an undeformable body. Nodes at the interface between the muscle and other tissue are restricted in moving in the anteroposterior direction. The boundary condition to perform the compression is thereby defined as a displacement of the compression plates. Acrylic glass plates modeled by 8-node hexahedrons are used. The compression of the breast model is achieved by defining a contact between the breast surface and the plate surface. Since the material is considered nearly incompressible, the applied deformation results mostly in a change of shape and not in a change of volume" Said et al. (160). The four steps are named:

step 1 unloaded state - removal of gravity
step 2 applying upright gravity
step 3 moving lower plate
step 4 moving upper plate

The compression simulation mimics two views of X-ray mammography, CC and MLO. For CC, compression plates and breast are in the same position at angle 0°, while for MLO there are two ways for implementing the simulation of compression either rotating the breast only by 45° or -45° based on the side of the breast or similarly rotating the compression plates only. Different scenarios have been analyzed to robustify and decrease the number of steps of the biomechanical model and simulation which will be discussed in detail in Section 3.2.4.

Based on the results of Section 3.2.4, mimicking the deformation simulation of the CC mammographic compression consists of two steps. First, the unloaded state of the breast is simulated in which no gravity is applied to the breast. This is simplified by applying a body force with a gravitational force in the anteroposterior direction. In the second simulation step, compression plates are added to compress the breast until the required compression thickness is achieved. The compression thickness is read from the metadata of the corresponding mammogram. The lower compression plate is assumed to be fixed at the inframammary fold and only the upper plate moved downwards toward the breast. In terms of boundary conditions, a displacement of the upper compression plate is defined and the lower plate is kept in position. 8-node hexahedrons are used to model the plates and a linear elastic material with material parameters mimicking acrylic glass is applied. The deformation simulation of the breast is considered a contact problem between the breast surface and the plate surfaces. It defines a small sliding interface with an adjustable friction coefficient which has been set to 0.25 taken from literature (210).

There are common problems that might cause non-converging deformation of the breast, such as the quality of the mesh, titling because of large breasts, or highly dense breasts. In order to avoid non-convergence problems, several investigations will be discussed in the next Section 3.2.4. For the tilting problem, it has been observed that it might be caused by not having a concave shape or symmetry of

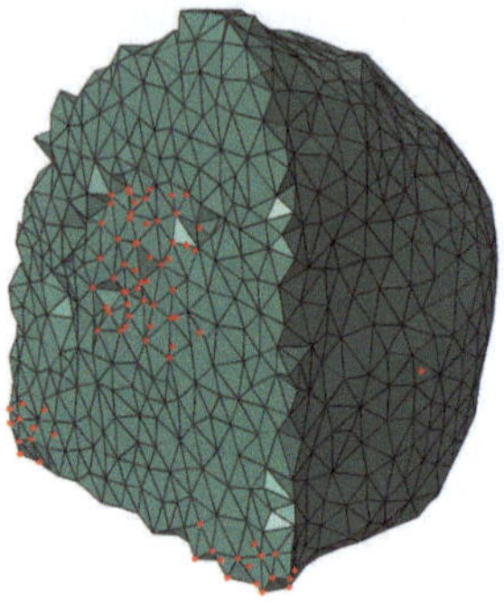

Figure 3.15: Example of the additional restricted nodes at the back of the breast in case of a very large breast to avoid titling. Nodes are restricted at the corners of the breast from the bottom back and one node is restricted in the middle of the breast.

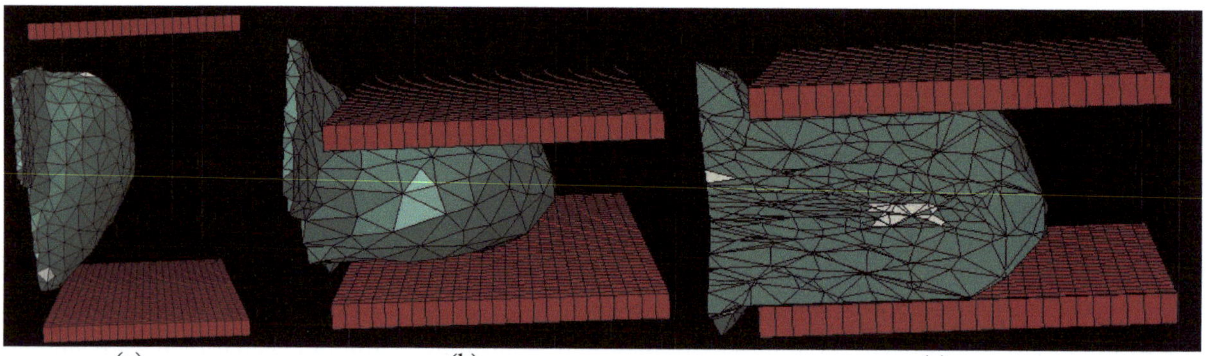

Figure 3.16: Example of CC case: the breast geometry in the biomechanical simulation before deformation (a), the deformed breast after applying the compression plates (b), and example of the inner distribution of tissues showing fatty (green) and glandular tissue (white) in the compressed breast (c) (160).

the breast, or by very large breast size, or by very glandular i.e. very dense breasts. One way to solve this problem is to restrict more nodes at the corners of the breast from the back bottom in addition to the interface nodes that have been previously explained in the anteroposterior direction. Also, a node in the middle of the breast with a ratio of one to a third of the breast size is restricted as shown in Figure 3.15. For mimicking the deformation simulation of the MLO mammographic compression, three steps are applied. First, the unloaded state of the breast is simulated in combination with applying the upright gravity. So in total using Pythagoras' law, a force of 13.81 m/s is applied in the negative direction. Then the second and third steps are moving first the lower plate toward the breast followed by moving the upper plate toward the breast. The boundary conditions are similar to that of CC view but rotating those restricted nodes with the same rotation value of MLO that was read from the metadata. The reason why three steps are selected instead of two in the case of MLO cases is that it increases the convergence and solving the deformation simulation more smoothly. It is necessary to solve this deformation simulation because 58.8% of the available data are MLO cases. Two factors are affecting the robustness of the

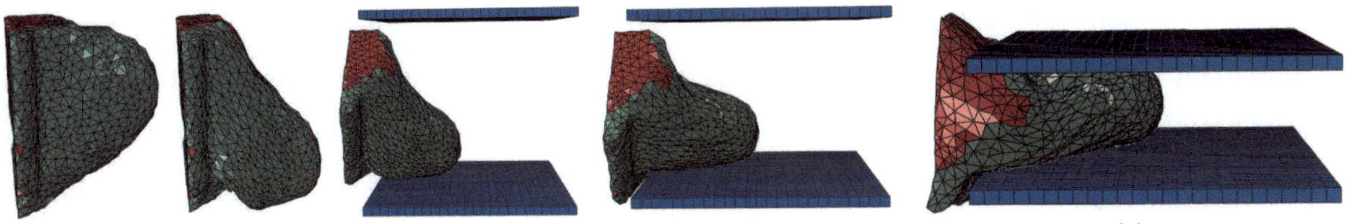

Figure 3.17: Example of MLO case: the breast geometry in the biomechanical simulation before applying deformation (a), the deformed breast after removal of gravity and applying upright gravity in one step (b), the breast geometry before applying deformation of the plates (c), the breast geometry after applying deformation of the lower plate (d), and the breast geometry after applying deformation of the upper plate (e).

deformation simulation for the MLO. First, the precision of the rotation angle for the muscle area in the breast is required. The technician is asked to rotate the patient so that as much of the axilla or muscle appears in the view. This causes the rotated view in the image, which needs to be detected to apply these findings to the biomechanical model as well. Thus, this value has been extracted based on the proposed method that will be explained in Chapter 4 in Section 4.2.1. Second, the position of the plates in the posterior direction. This value has been extracted like the explained algorithm for CC in Section 3.2.4, in addition to 2 cm more for compensating the posterior nipple line (PNL) for MLO cases. The general rule is that if the PNL is less than 1 cm, the MLO view is taken correctly and corresponds to the CC view (134). Since the available data varies a lot, 2 cm was added instead of 1 cm.

The simulation is started using these empiric standard parameters. The FEM is employed to compute the mechanical response of the breast tissue for the modeled compression. The mechanical deformations were estimated using the dynamic FE solver in the commercial FEM software ABAQUS / Explicit (175), selected for its robust convergence for highly non-linear deformations and contact problems. Although, ABAQUS / Explicit could require a lot of time or give a noisy response at the end of the last step as shown in Figure 3.27. Applying mass scaling can help in reducing the computation time required for the explicit solver (185). The non-linear system of equations is solved in ABAQUS / Explicit using a small-step incremental solution method. For each small time step, the geometric non-linearity of the breast is taken into account. ABAQUS / Explicit is set to automatically determine the step size and convergence criterion of each increment. To have a stable time increment size, it is based on two factors as shown in equation 3.11; the mesh's smallest element dimension L_{min} and the wave speed dilatation c_d in terms of Lame's constant and shear modulus as shown in equation 3.12. The effective hyperelastic material moduli are computed from the constitutive response of the material to provide the current dilatational wave speed c_d. The effective Lame's constants in an isotropic elastic material are defined as follows: in terms of Poisson's ratio ν and Young's modulus E as shown in equation 3.13. Since the mesh has multiple materials, the initial time increment is estimated by the element with the highest wave speed. Large deformations in non-linear problems cause the model's highest frequency to fluctuate over time, which in turn affects the stability limit. There are two methodologies for time incrementation control in ABAQUS/Explicit: fully automatic time incrementation and fixed time incrementation. In this thesis, fully automatic time incrementation is used and as a consequence, user intervention is not required. The analysis initially makes use of a stability limit determined by the highest element frequency over the entire model. It is calculated for every element based on its current dilatational wave speed. Compared to the true stability limit, which is based on the highest frequency of the whole model, it will offer a smaller stable time increment.

Finding the stable time needed for a dilatational wave to propagate across the smallest element dimension is crucial. That makes it simple to see how the explicit approach determines the time increment when the time increase is controlled by element-by-element stability estimation. The time increment will become less sensitive to the element size as the step progresses if the global stability estimate is applied.

$$\triangle t \approx \frac{L_{min}}{c_d} \tag{3.11}$$

In equation 3.11, L_{min} is the smallest element dimension in the mesh and c_d is the dilatational wave speed in terms of λ_0 and μ_0.

$$c_d = \sqrt{\frac{\lambda_0 + 2\mu_0}{\rho}} \tag{3.12}$$

with ρ is the density of the material,

$$\lambda_0 = \frac{E\nu}{(1+\nu)(1-2\nu)} \tag{3.13}$$

A smooth step displacement curve is used to approximate the model to a quasi-static configuration. The amplitudes are calculated based on the smoothing step to estimate the ramp step as shown in equation 3.14. The amplitudes are considered to be the displacements of moving the upper plate towards the breast. The simulation of the breast compression is divided into small time steps to calculate the displacements of the moving upper plate towards the breast.

The deformed configuration of the breast is extracted by requesting from ABAQUS the node positions at a certain time step (referred to as "step number"). This information is used as a feature for training the machine learning models, as described in Section 3.2.5.

$$a = A_i + (A_{i+1} - A_i)\xi^3(10 - 15\xi + 6\xi^2) \quad for \quad t_i \leq t \leq t_{i+1} \tag{3.14}$$

$$\text{where: } \xi = \frac{t - t_i}{t_{i+1} - t_i}$$

In this case, t ranges from zero until one second. The initial state represents the highest position of the upper plate before starting the simulation. The last state represents where the upper plate has stopped when it reaches the compression thickness from the metadata.

In the next section, several experiments will be discussed in order to show why these proposed steps, and specific parameters regarding stability, and convergence are selected.

3.2.4 Experiments for Optimizing Parameters

This section discusses four experiments that have been tested by a small subset of eleven patients that were available at that time. First, the evaluation of the quality of meshes is included. Second, how many steps are needed compared to the reality of mammographic compression was investigated. This was done using the evaluation methods of quality meshes. Third, some factors that could affect the simulation of models, such as the position of the plate in the Y direction and the thickness of the skin are examined. Fourth, an analysis of the stability of the model when being solved by FEM is investigated.

Quality of meshes

Solving the mechanical deformation problem using FEMs requires a good mesh quality, i.e., no degenerated tetrahedrons, for a stable solution, to achieve convergence and avoid re-meshing in between the steps of the simulation. Poor quality meshes could lead to non-convergence. These problems of poor quality meshes are listed below, as shown in Figure 3.18:

1. An element that is touched by the upper plate's corner degenerates (see Figure 3.18 (a))
2. An element that is fixed at the back (breast-muscle interface) is pulled down by the plate at single nodes (see Figure 3.18 (b and c))
3. An element gets squeezed excessively inside the breast (see Figure 3.18 (b))
4. The breast flows down the lower plate for finer meshes (see Figure 3.18 (c and d))

There are a lot of methods by which mesh quality can be analyzed. Three methods have been implemented and tested. All meshes were done using tetrahedrons. First, quality measures of meshes are calculated in MATLAB using a specific equation, which will be illustrated in detail below. Second, the quality of meshes is calculated using ABAQUS functions. Third, the quality of meshes is calculated by metrics provided in the iso2mesh toolbox. The goal of this analysis was to select which metric from these is the most effective in identifying the initial parameters for generating meshes, without the need for re-meshing in between the simulations based on the available data.

(1) **Ratio radius of tetrahedron meshes** The ideal shape for tetrahedron meshes is an equilateral mesh. However, it is usually difficult to generate tetrahedrons with edges of the same size for the meshes. In general, the tetrahedral elements can be described as a region bounded by polygon faces. An evaluation of the shape of a tetrahedron and how much it deviates from the equilateral ideal can be done using the ratio of inradius (r) and circumradius (R) (2; 31) as shown in Figure 3.19 and equations 3.15 and 3.16. The inradius is the length measured between a face's center and the tetrahedron center. Circumradius is defined as the radius of the sphere passing through

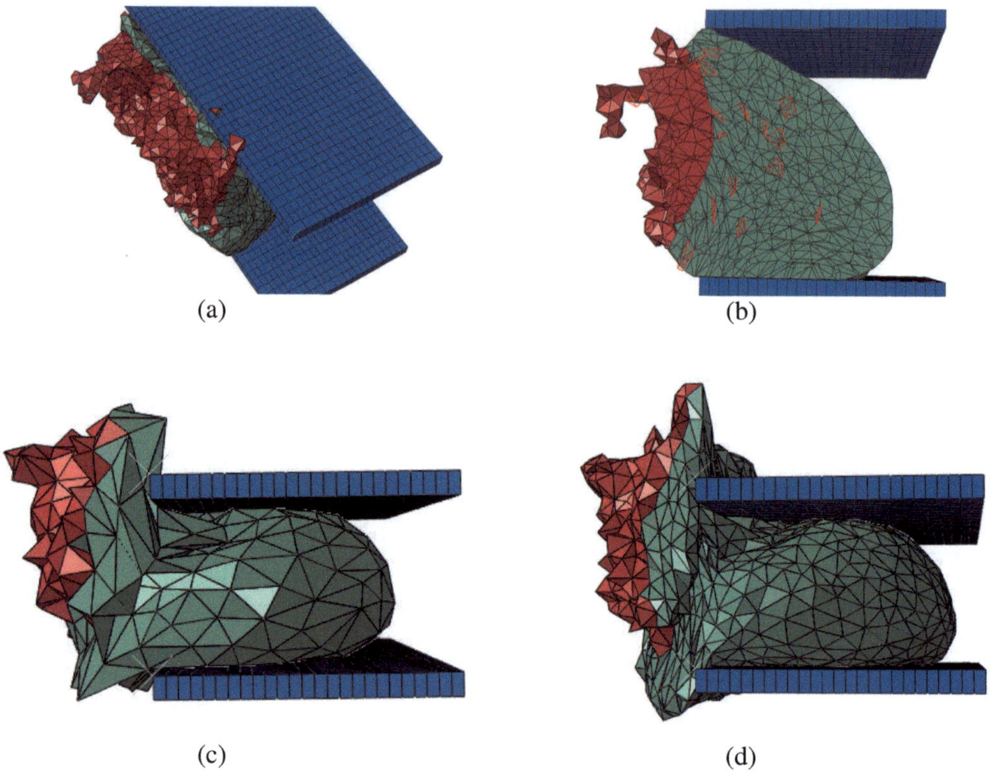

Figure 3.18: Problems of bad quality meshes that could lead to non-convergence as listed earlier. Figure a shows the first problem (1). Figure b shows the second and third problems (2 and 3). Figure c shows the second problem (2). Figures c and d show the fourth problem the finer (c) and the coarse (d) mesh using these parameters, the maximum element volume of 18 and 9, respectively, the maximum radius ratio edge of 3 in each, the deviation from isosurfaces of 5 and 4, and the maximum triangle delaunay circle radius of 35 in each.

the four vertices of the tetrahedron as shown in equation 3.16. The radiuses of these two spheres are $\frac{1}{3}$ for an equilateral tetrahedron. The more a tetrahedron deviates from the ideal shape, the lower this ratio will be. It is therefore an indication for the detection and rejection of bad shapes of tetrahedrons. Any tetrahedron with a ratio (see equation 3.17) less than 0.25 should be rejected and considered a distorted element (2) (31).

$$r = \frac{abs(\alpha)}{||N_{(a,b,c)}|| + ||N_{(a,b,d)}|| + ||N_{(a,c,d)}|| + ||N_{(b,c,d)}||} \tag{3.15}$$

$$\text{where: } \alpha = \overrightarrow{ab} \cdot (\overrightarrow{ac} \times \overrightarrow{ad}), \quad N_{a,b,c} = \overrightarrow{ab} \times \overrightarrow{ac}$$

$$R = \frac{\sqrt{(a+b+c)(a+b-c)(a+c-b)(b+c-a)}}{24 * v} \tag{3.16}$$

$$Quality\ measure = \rho = \frac{3 \cdot r}{R} < 0.25 \tag{3.17}$$

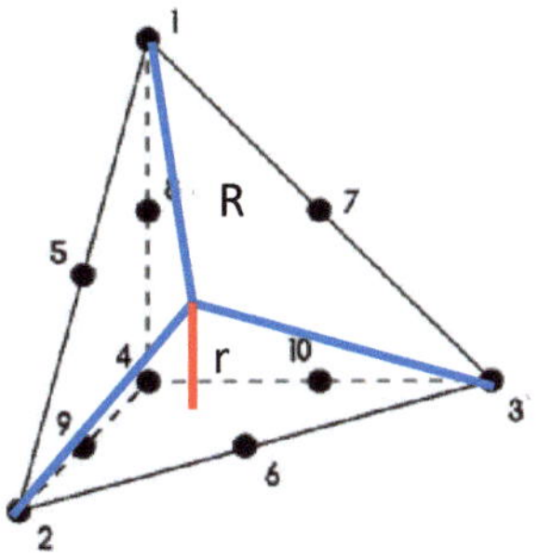

Figure 3.19: Diagram of a tetrahedron mesh showing inradius (r) and circumradius (R). This diagram was modified by the author based on (48) and is licensed under CCBY.

where: r is inradius, R is circumradius, v is the volume of an element, and a, b, c are the products of the lengths of opposite edges.

(2) **Quality metrics from ABAQUS** Based on the documentation of ABAQUS (188) 2017, there are three quality metrics for judging the quality of the mesh.

a) shape factor
b) aspect ratio
c) short edge

According to equation 3.18, the shape factor is the ratio of the element volume to the ideal element volume. The volume of an equilateral tetrahedron with the same circumradius of the element is the ideal (optimal) element volume (70). The aspect ratio is the ratio between the longest and shortest edge of an element, as shown in equation 3.19. The short edge is considered to be the edge with the shortest length in a tetrahedron mesh. There are certain limits in ABAQUS for tetrahedral meshes regarding these parameters for achieving convergence. The aspect ratio should not exceed 10, the shape factor should not exceed 0.0001, and the short edge should not exceed 0.01 (188).

$$Shape\ Factor = \frac{element\ volume}{optimal\ element\ volume} \tag{3.18}$$

$$Aspect\ Ratio = \frac{h_{max}}{2 \cdot \sqrt{6}r} \tag{3.19}$$

(3) **Quality metrics from iso2mesh** Based on the documentation of iso2mesh (54), a metric to judge the quality of a mesh is the JoeLiu metric (103) which takes into consideration radius ratio (ρ) as shown in equation 3.17 and it could be called as well the aspect ratio, mean ratio (η) as shown in equation 3.21, and σ_{min} as shown in equation 3.20.

$$\sigma_{min} = \sin(\frac{\theta_{min}}{2}) \tag{3.20}$$

where: θ_{min} is minimum solid angle

$$\eta = \frac{12 * (3v)^{\frac{2}{3}}}{\sum\limits_{0 \le i < j \le 3} l_{i,j}^2} \tag{3.21}$$

where: v is volume of an element and $l_{i,j}$ is length of edges of an element

After measuring the quality of the meshes, the most effective metric for avoiding the need for re-meshing in between the steps was the first method. If the ratio radius of tetrahedron meshes (see equation 3.17) is less than 0.25, then it is considered a distorted element. The best parameters found for generating meshes resulting in the least distorted elements were found as shown in Table 3.6 (second column). It is obvious from the table that the values selected are the upper bound of the analysis which could be a local minimum or not fully optimized. The reason why this range is selected is to preserve the shape and the anatomy of the breast and at the same time achieve convergence. For example, if the maximum element volume is greater than 15, the geometry of the breast is not correct and it does not converge. By selecting these parameters, It was found that 42 patients out of 51 patients available have only 1 to 2 % distorted elements using the initial parameters found in this analysis (see Section 7.3.1). By that, it is possible to avoid non-convergence and re-meshing between the steps of simulation.
For the other two methods of quality measures, it was more meaningful to use them in the next experiments because the investigations were done in between simulations to evaluate the distortion elements at each step in order to decrease the complexity of biomechanical deformation.

Simulation of steps of the model using the evaluation methods of quality metrics

The complexity of the biomechanical deformations was investigated using the percentage of distorted elements that could appear at each step in between the four mentioned steps (see Section 3.2.3). The percentage of distorted elements was calculated using the quality mesh metrics of the second and third methods for the same subset of patients. As explained, those four mentioned steps are needed if the goal is to apply real and physical mammographic compression regarding forces and displacements. However, the goal is to solve the mechanical deformation problem using FEM simulations which lead to convergence and at the same time decrease the number of steps. Thus, different scenarios have been tried by

Table 3.6: The range and the step size that have been tried for creating a tetrahedron mesh (first column). The best parameters were found with the least distorted elements (second column)

Parameters	Range / step size	Value
maximum element volume	7-15 / 1	15
maximum radius-edge ratio	2-3 / 1	3
deviation from isosurfaces	2-5 / 1	5
maximum triangle delaunay circle radiu	20-35 /5	35
minimum angle of a surface triangle	-	30

changing the order of these steps to investigate which step leads to the most distorted elements as shown in Figure 3.20, 3.21, 3.22, and 3.23.
The first step is defined from 0 to 1 second, the second step is defined from 1 to 2 seconds, the third step is defined from 2 to 3 seconds, and the fourth step is defined from 3 to 4 seconds. The first scenario (denoted as the full scenario) consists of the four mentioned steps. Through all the next scenarios, the gravity (unloaded state step) is removed and the order of upright gravity has been changed. The second scenario (denoted as upright+plates) consists of three steps applying an upright gravity followed by applying compression plates with the same sequence as explained before. The third scenario (denoted as lower plate+upright+upper plate) consists of three steps as well, but adding the upright gravity in between moving the lower plate and moving the upper plate. In the fourth scenario (denoted as plates only), the upright gravity is removed and the plates are moved in the same sequence as before.
Figure 3.20 shows that the percentage of distorted elements either from ABAQUS or from iso2mesh in the last two time steps (3 and 4) between 2 to 3 and 3 to 4 seconds lead to the most distorted elements because of the large deformations that happen in these two steps, additionally, the jump of distorted elements between the upright gravity step (2) and moving lower plate step (3) is high from 0.8% to 2.5% (the absolute difference is 1.8%). Figure 3.21 shows that not only distorted elements exist in the last two steps but also in the first step from 14.8% to 33.7%. The jump from the upright gravity step (1) to the next steps of compression plates (2 and 3) is higher than the Figure 3.20 in descending order from 14.8 % to 11.5% (the absolute difference is 3.3%). This shows how the upright gravity affects the distorted elements in the next steps. In Figure 3.22, the percentage of distorted elements increased compared to Figure 3.21 in the three steps from 14.9% to nearly 34.4% when the upright gravity step was added (2) in between. Additionally, there is nearly no jump in the upright gravity step from 34.3% to 34.5%. Figure 3.23 compared to Figure 3.22 shows that the percentage of distorted elements with or without the upright gravity, is nearly the same effect. The percentage of distorted elements is from 14.9% to 34.4% in the "lower plate + upright + upper plate" scenario. The percentage of distorted elements is from 14.8% to 34.1% in the plates only scenario.
From this analysis, adding the upright gravity step before the compression of the lower plate was found to have a negative effect, although it is a more realistic scenario for mammographic compression. In general, the upright gravity step has a negative effect on the model causing excessive distortion of elements. So in this thesis, removing the upright gravity step is proposed for CC views for the purpose of convergence and decreasing the complexity of deformation simulation.
By having less distorted elements and decreasing the number of steps of the biomechanical model, the probability of having the simulation fail and the probability of losing precision in the simulations would be decreased. From an optimization point of view, the computation time will be decreased as well. From this analysis, it was concluded that the proposed scenario would be having first the unloaded state, fixing the lower compression plate at the inframammary fold, and then moving the upper plate downwards

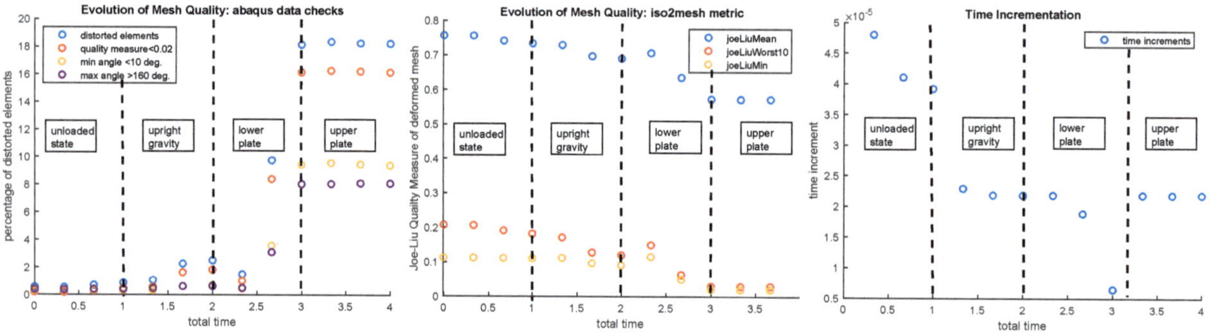

Figure 3.20: Monitoring of mesh quality of one of the cases of datasets from ABAQUS and iso2mesh in different scenarios for the FEM model. First scenario (full scenario): four steps (unloaded state, upright gravity, moving lower plate, and moving upper plate).

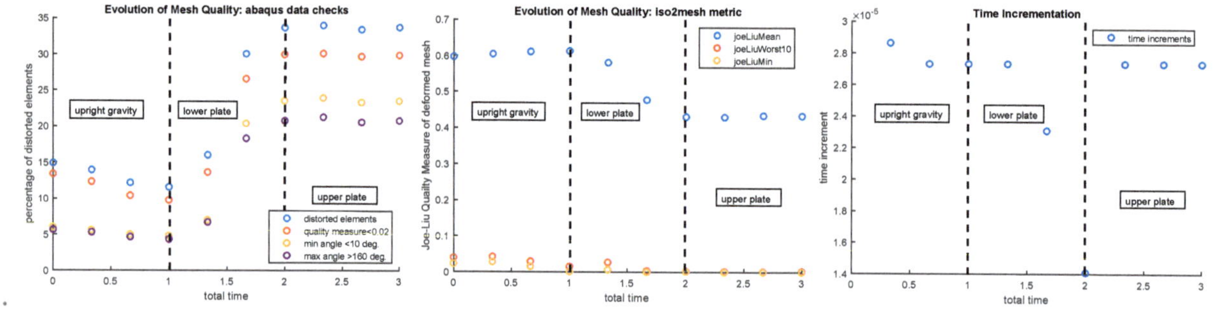

Figure 3.21: Monitoring of mesh quality of one of the cases of datasets from ABAQUS and iso2mesh in different scenarios for the FEM model. Second scenario (upright+plates): three steps (upright gravity, moving lower plate, and moving upper plate).

to compress the breast until the desired compression thickness which is considered the fifth scenario (denoted as gravity+plates) that will be used further. To sum up, the five scenarios that are investigated:

1. unloaded state, applying upright gravity, moving lower plate, and then upper plate "full scenario"
2. applying upright, moving lower plate, and then upper plate "upright+plates"
3. moving lower plate, applying upright, and moving upper plate "lower plate+upright+ upper plate"
4. moving lower plate and then upper plate "plates only"
5. unloaded state, fixing lower plate, and moving the upper plate "gravity+plates"

Simulation of steps of the model using TRE

The analysis of the four scenarios was continued from Section 3.2.4 in addition to the proposed fifth scenario. The goal of this experiment is to achieve the best accuracy with the least errors. This experiment has been evaluated by Target Registration Error (TRE) as shown in equation 3.22. It was observed that not only the scenarios are affecting TRE as shown in Figure 3.25. There are two more factors, which are the position of the lower plate in the inferior direction and the thickness of the skin.

$$TRE_{Biomechanical} = ||C_{XR} - C_{MRI}|| \tag{3.22}$$

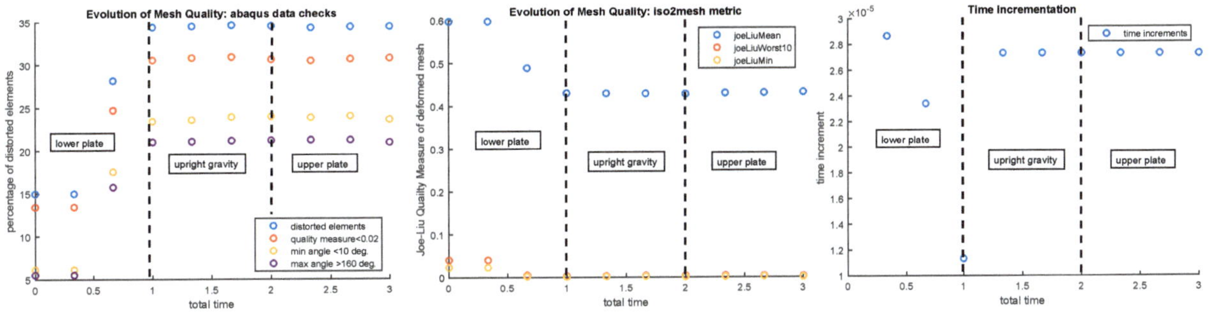

Figure 3.22: Monitoring of mesh quality of one of the cases of datasets from ABAQUS and iso2mesh in different scenarios for the FEM model. Third scenario (lower plate+upright+upper plates): three steps (moving lower plate, upright gravity, and moving upper plate).

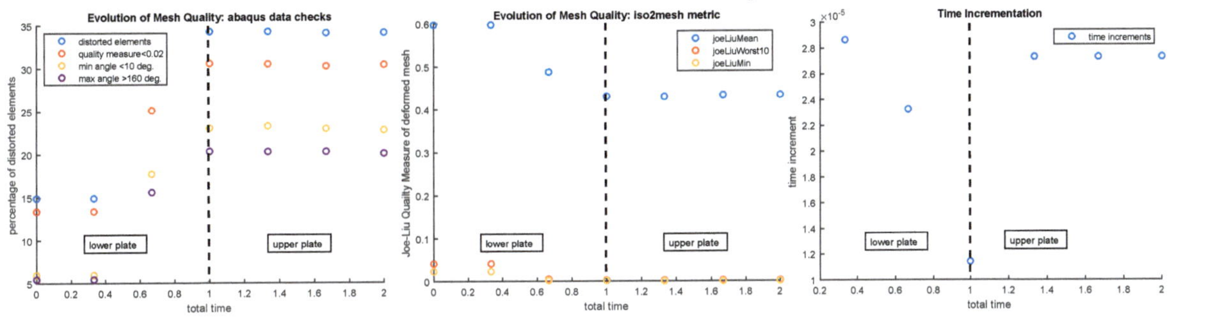

Figure 3.23: Monitoring of mesh quality of one of the cases of datasets from ABAQUS and iso2mesh in different scenarios for the FEM model. Fourth scenario (plates only) two steps: (moving lower plate and moving upper plate only).

First, the position of plates in the posterior direction at the back of the breast is the main parameter needs to be estimated robustly, as shown in Figure 3.16 (a). Not only the position of plates in the posterior direction is important but also the position of the lower plate in the inferior direction affects the TRE. There are four algorithms proposed for the estimation of the position of the lower plate in the inferior direction. First, the position of the inframammary fold is estimated by selecting the middle slice of MR images (transverse view) and calculating the bent shape of the muscle at this slice. Second, the first hit (non-background pixel) from the horizontal side of the middle slice of the MRI is calculated (see Figure 3.24 arrow indicates 2). The position that corresponds to the inferior position in this calculated location is extracted and considered to be the position of the lower plate in the inferior direction. Third, the first hit (non-background pixel) from the vertical side is calculated, as shown in Figure 3.24 (see arrow indicates 3), and similarly the position that corresponds to the inferior position is extracted and calculated. Fourth, the position of the lower plate is estimated to be equal to the position that corresponds to the inferior position in the calculated sternum position. From this analysis, it was observed that a simpler simulation with unloaded state and compression by moving one plate only (gravity+plates) is as good as a complex four-step scenario (full scenario) as shown in Figure 3.25. Also, it was observed earlier that the other scenarios including the upright step cause a lot of distorted elements, which leads to the variance of TRE in these scenarios from one algorithm to another. Additionally, it has been realized that TRE varies based

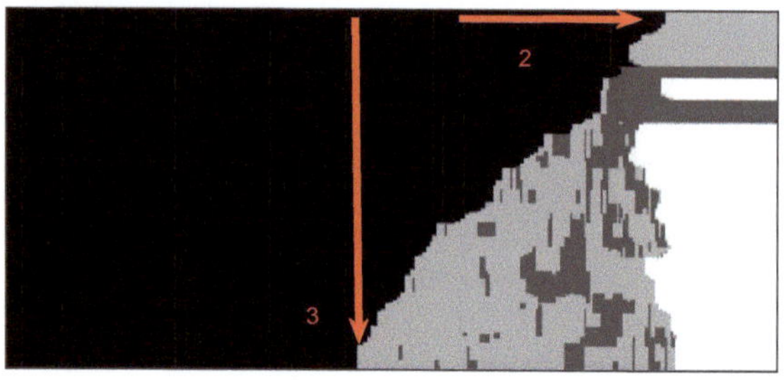

Figure 3.24: Two algorithms explained out of the four algorithms for the estimation of the position of lower plate in the inferior direction. (2) indicates the second estimation. (3) indicates the third estimation.

on these algorithms (1, 2, 3, and 4), and the first algorithm is the best of them which is currently used. In the first scenario (full scenario) and the fifth scenario (gravity+plates), the median of TRE is quite similar in the four algorithms.

$$mean\ Error = \left\| D_{specific\ massscaling} - D_{non\ massscaling} \right\| \tag{3.23}$$

Second, the thickness of skin from $0.5mm$ to $2.5mm$ with step size $0.5mm$, respectively, as stated in literature (87) is analyzed for the four algorithms of estimating the position of the lower compression plates in the inferior direction in the first and fifth scenarios as shown in Figure 3.26. This experiment is analyzed using the evaluation metric of minimum mean nodal displacement error. It is calculated between the nodal displacement without mass scaling as a reference (ground truth) and with the selected mass scaling of $5e^{-5}$ (based on the next experiment) at this given range of skin thickness as shown in equation 3.23. The goal of this experiment is to find the minimum error achieved concerning skin thickness and reserving the same performance of the full scenario at the same time.
From this analysis, it was observed that the magnitude of the minimum mean nodal displacement in the fifth scenario using algorithms 2 and 3 (see Figure 3.26 b and c) in comparison to the magnitude of the full scenario did not have the same performance. Therefore, algorithm 2 and algorithm 3 are not taken into consideration in the comparison. In the other two algorithms (1 and 4) (Figure 3.26 (a and d)), the magnitude of minimum mean nodal displacements achieved in the gravity+plates scenario compared to the full scenario could be nearly the same performance. It was also observed that the error achieved using the fourth algorithm, as shown in Figure 3.26 (d), is considerably the least error. Through all the different skin thicknesses in Figure 3.26 (a and d), skin thickness at $2mm$ is the minimum mean nodal displacement error. It was hard to select which algorithm to be used with a skin thickness of $2mm$. Since the goal in all the experiments is to predict the position of lesions correctly, therefore, the evaluation using TRE is more important. In the previous experiment, the first algorithm for estimating the position of the lower plate was the best concerning TRE. Hence, as a conclusion from these two experiments, the first algorithm is selected with a skin thickness of $2mm$ to be used in the model.

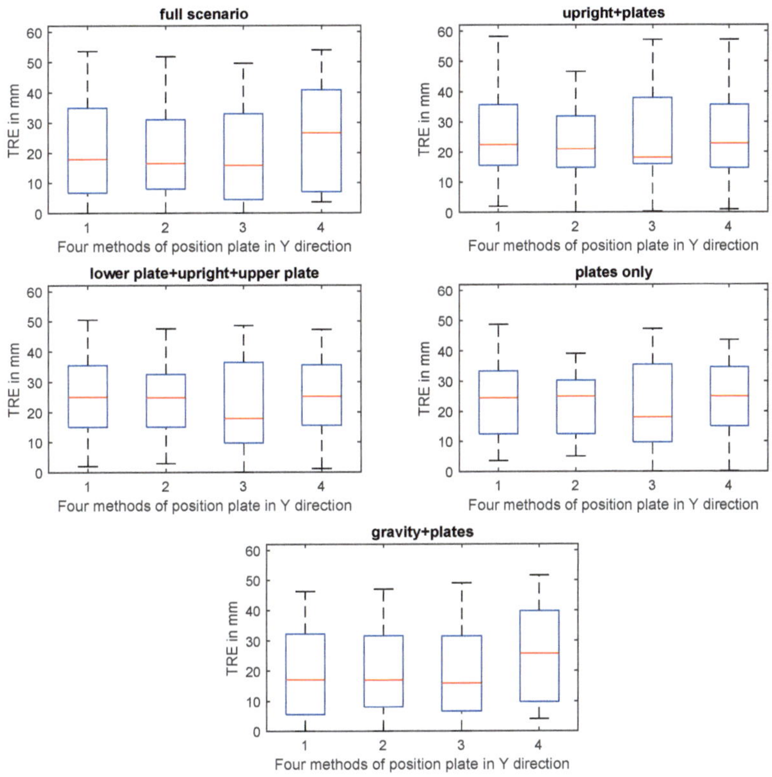

Figure 3.25: Analysis of TRE for different scenarios and different estimations for the position of lower plate in the inferior direction (Y direction). First scenario (full scenario): four steps (unloaded state, upright gravity, moving lower plate, and moving upper plate). Second scenario (upright+plates): three steps (upright gravity, moving lower plate, and moving upper plate). Third scenario (lower plate+upright+upper plates): three steps (moving lower plate, upright gravity, and moving upper plate). Fourth scenario (plates only): two steps (moving lower plate and moving upper plate only). Fifth scenario (gravity+plates): two steps (unloaded state, fixing the lower plate, and moving only the upper plate). For different estimations in Y direction for the position of lower plate Y, (1) indicates first algorithm, (2) indicates second algorithm, (3) indicates third algorithm and (4) indicates fourth algorithm.

Stability of the model

Having a stable model is important to verify that deformation is solved without any effect of noise. Damping oscillations are considered noisy behaviour that might appear randomly in between the simulation in the four steps (see Figure 3.30) or only in the last step (see Figure 3.27). The oscillations might also appear because of the dynamic solver or large deformations that happened to the breast. To analyze the damping oscillations that happen in the simulation and tackle this problem, specific nodes are extracted. The displacements of those nodes are tracked as a function of the simulation time. Those specific nodes were selected based on the geometry of the breast, i.e., the back, front, middle, surface, and bottom of the breast, in order to see the oscillations in all areas of the breast. It was visually seen that the tissue was oscillating in the simulation. It bounces back and forth, which might be because of a too coarse time step, i.e., too much force is applied, or because of the selected mass scaling, or because of the total time of the four steps together. Three experiments were done to validate this using three parameters.

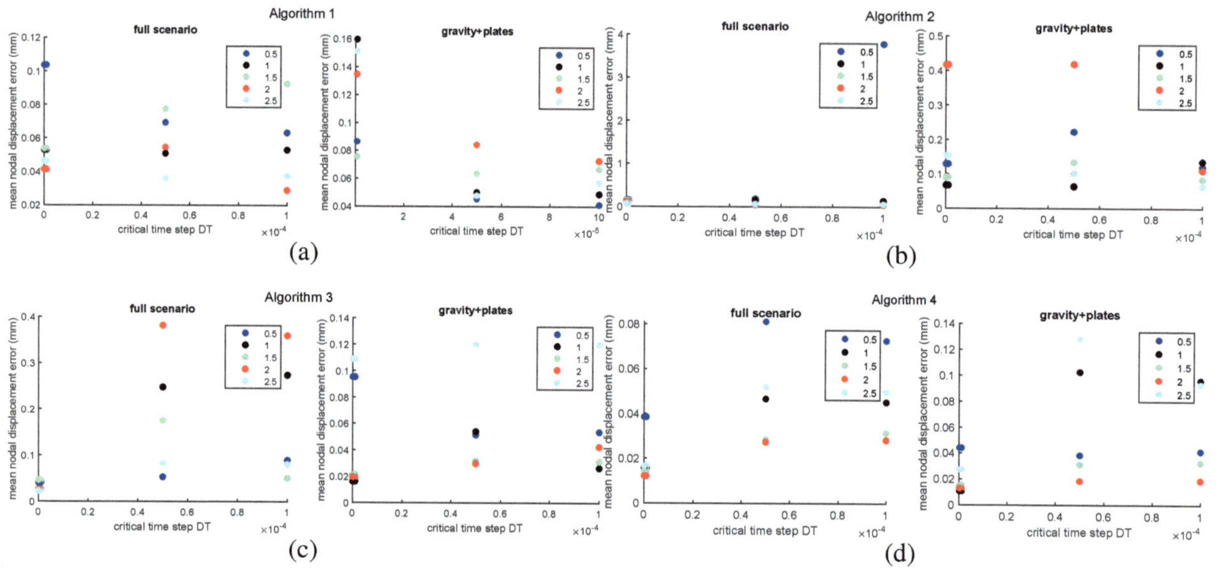

Figure 3.26: Effect of skin thickness $(0.5, 1, 1.5, 2, and 2.5)\, mm$ on mean nodal displacement error in the first and fifth scenarios using the four algorithms for the position of lower plate in the inferior direction (Y direction) for one of the datasets. Full scenario (left) and gravity+plates (right): first algorithm (a), second algorithm (b), third algorithm (c), and fourth algorithm (d).

- time increment
- mass scaling
- time step

The first experiment was done to vary between the initial time increment and mass scaling, as shown in Figure 3.27 for the full scenario. From this experiment, it was observed that initial time increments do not affect the stability of the system, as different values were tested from $1e^{-6}$ to 0.01. Hence, using the default initial time increment of 0.01 is sufficient for the model.

The second experiment was trying different mass scaling factors with values of $1e^{-5}$, $5e^{-5}$, $1e^{-6}$, $1e^{-7}$, $1e^{-8}$, $1e^{-9}$, $1e^{-10}$ and without mass scaling. This experiment has been tried for two scenarios: full scenario and the gravity+plates scenario, as shown in Figure 3.28 and 3.29. All of these trials for varying mass scaling were done with the same time increment of 0.01. This value was observed in the previous experiment. From this experiment, it could be concluded that mass scaling is not the reason for oscillations. Also, it shows how the upright gravity step causes most of the oscillation in all the steps that come afterward. Hence, a mass scaling of $5e^{-5}$, which is considered the default, was used.

The third experiment was increasing the total time of the four steps by two compared to the usual time in the following steps: upright gravity step, moving lower plate, and upper plate step, as shown in Figure 3.30. It was observed that increasing the total time does not prevent the oscillations, but it increases the oscillations more and more. As a conclusion, the time steps can be kept the same.

To have a stable FEM model based on the documentation of ABAQUS, two main factors were found to be affecting the stability of the system. One of them is the Kinetic Energy (KE) of the deforming material of the breast, which should not exceed a small fraction of 5% - 10% of its Internal Energy (IE) throughout most of the process (186). A KE and IE of the whole model for one of the cases is shown in Figure 3.32. The second factor that may cause this behavior is that the breast is incompressible, and assuming that the material is completely incompressible is not feasible in ABAQUS/Explicit (184). In theory, an incompressible material has an infinite wave speed, which will result in a time step of zero. So, at some point, some compressibility for the breast needs to be provided, knowing that this would make bulk behaviour slightly softer than that of the actual material. This is based on Poisson's ratio ν, bulk modulus κ and shear modulus μ_0. It is recommended that a ratio of $R = \frac{\kappa}{\mu_0}$ should be between 20 and 100 (184). Since the goal is to define the soft tissue as incompressible rather than using the default value defined in ABAQUS which is 20. Therefore, it is suggested from (183) to use the upper limit in order not to introduce high frequency noise into the dynamic solution. Currently, the upper limit, which is 100, is used by using values mentioned in Section 3.2.3 for Poisson's ratio and Young's modulus. An experiment was tried manually to change D_1, and C_{10} coefficients in ABAQUS which are interpreted for bulk modulus and shear modulus respectively to have a ratio less than 100. It was found out that for nodes in the middle bottom of the breast, the oscillation decreases significantly as shown in Figure 3.31, while for the other nodes, the oscillation decreases slightly compared to the other simulations. In which, it was concluded that the chosen Poisson ratio and Young's modulus needs to be further investigated in future.

Despite the fact that this model has been used to register between MRI and X-ray images with success, the FEM is time-consuming and thereby limits clinical applicability and integration into clinical workflows. Furthermore, the image registration considers an iterative scheme for optimizing meta parameters of each patient specifically such as the compression thickness, the material parameters, or rotations of the breast to cope with uncertainties in the clinical data. Therefore, the FEM simulation has to be computed numerous times, which increases the overall computation time further. Simulations with standard parameters take around 20 minutes on an Intel(R) Core i7-9700K Central Processing Unit (CPU) with 3.60 GHz. In contrast, optimization of some parameters for increasing accuracy takes around 2 hours or even more. Additionally, if the number of iterations is multiplied for only one parameter requires high computation time.

Numerous strategies have been proposed to reduce the computational effort of FEM in order to achieve real-time compliance. Some of them concentrated on improving linear solvers, which are the main FEM bottleneck (120). The Graphics Processing Units (GPU) implementation by (71) relies on a total Lagrangian explicit dynamics (TLED) formulation by (123), which is considered the most optimal method for modeling breast biomechanics (120). NiftySim is a GPU-based solver (70) used for breast modeling. The Simulation Open Framework Architecture (SOFA) also uses a GPU-based solver to speed up computations and has been used for prostate deformation simulation (127).

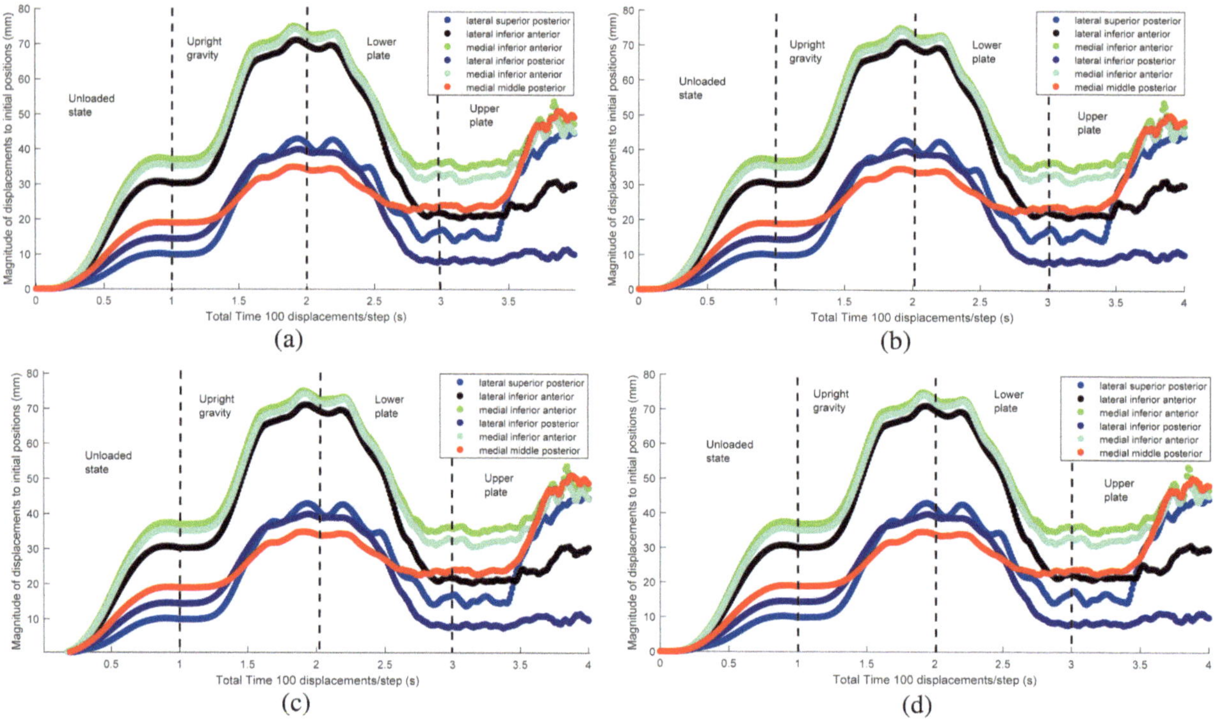

Figure 3.27: Magnitude of specific nodes through steps of full scenario in FEMs by different mass scaling factors and different time increment values. Mass scaling of $5e^{-5}$ and time increment of $1e^{-5}$ (a). Mass scaling of $5e^{-5}$ and time increment of $1e^{-6}$ (b). Mass scaling of $1e^{-5}$ and time increment of $1e^{-5}$ (c). Mass scaling of $1e^{-5}$ and time increment of $1e^{-6}$ (d).

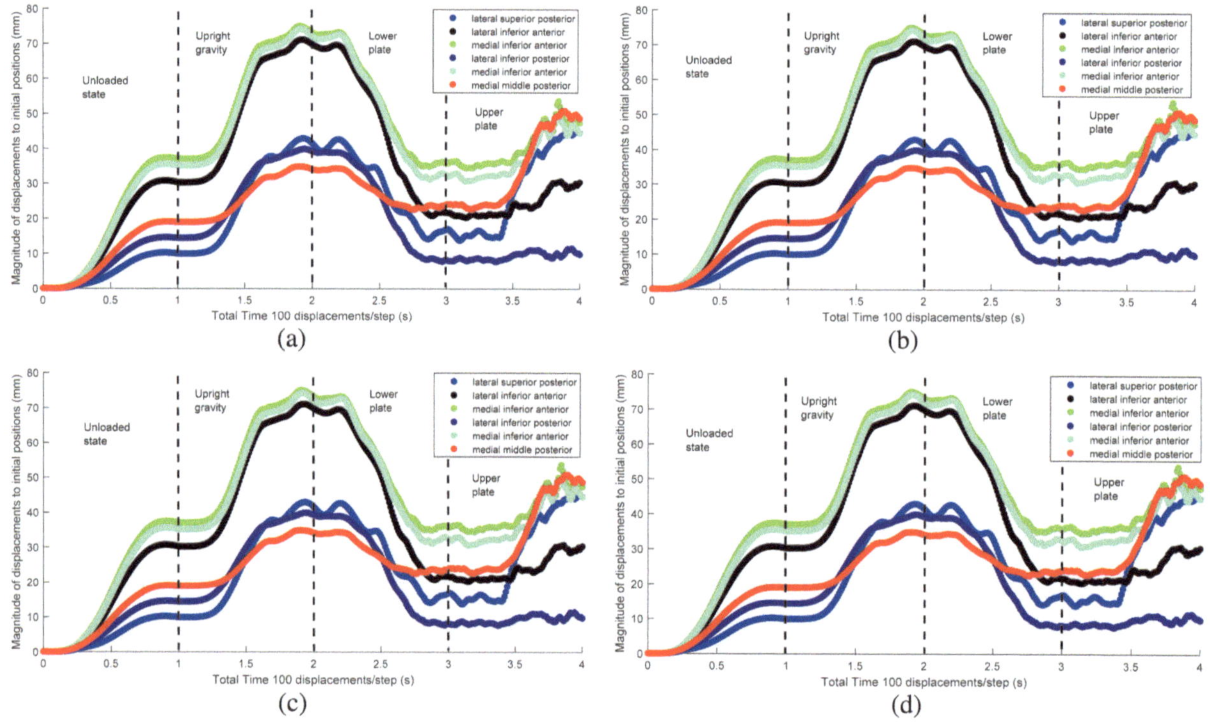

Figure 3.28: Magnitude of specific nodes through steps of full scenario in FEMs by different mass scaling factors. Mass scaling of $1e^{-6}$ (a). Mass scaling of $1e^{-7}$ (b). Mass scaling of $1e^{-8}$ (c). Mass scaling of $1e^{-9}$ (d).

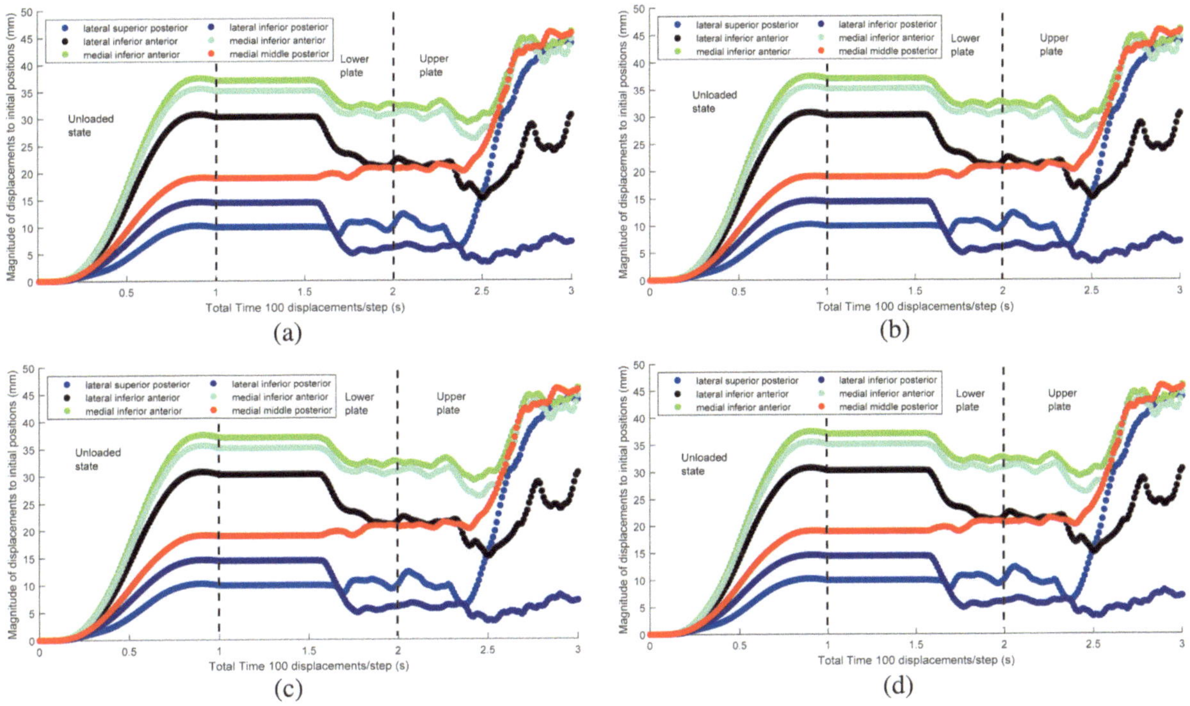

Figure 3.29: Magnitude of specific nodes through steps of unloaded state and moving plates in FEMs by different mass scaling factors. Mass scaling of $1e^{-6}$ (a). Mass scaling of $1e^{-7}$ (b). Mass scaling of $1e^{-8}$ (c). Mass scaling of $1e^{-9}$ (d).

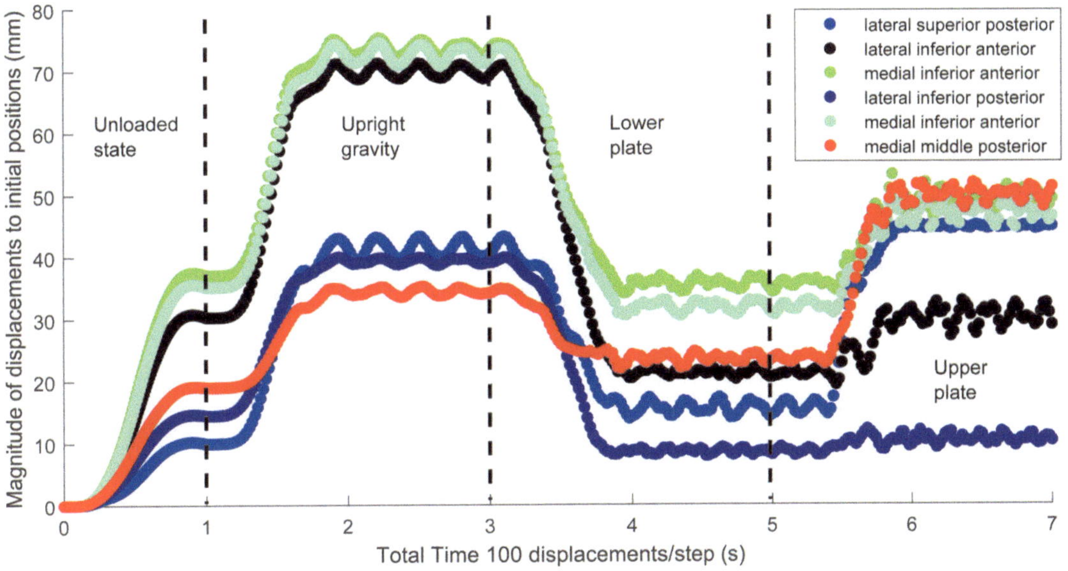

Figure 3.30: Increasing total time of the four steps of full scenario 2 times more in each of the following steps: upright gravity, moving lower plate, and moving upper plate step.

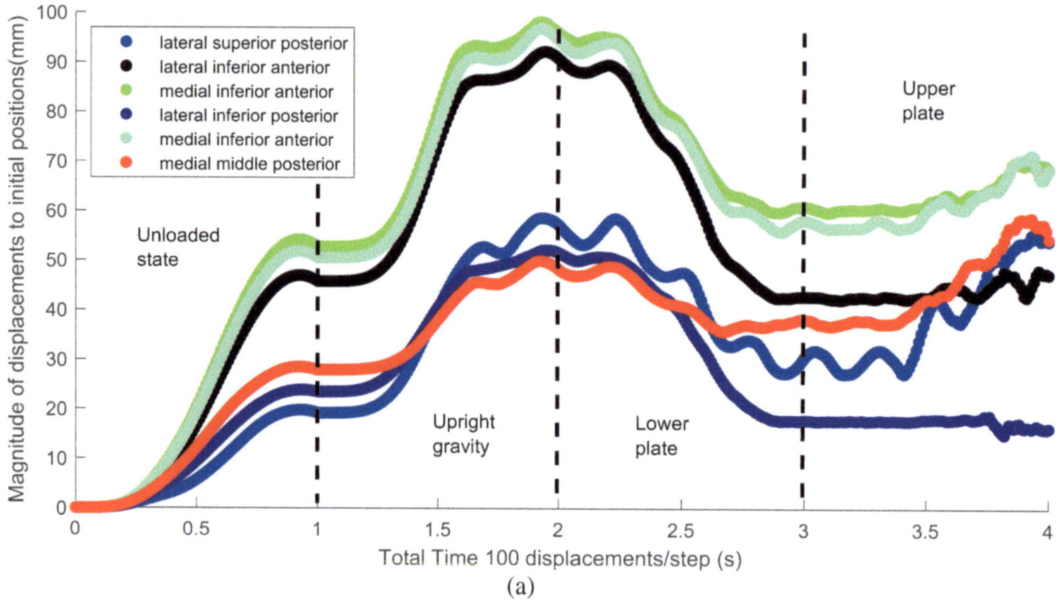

(a)

Figure 3.31: Manually changing D_1 and C_{10} coefficient to have a ratio less 100.

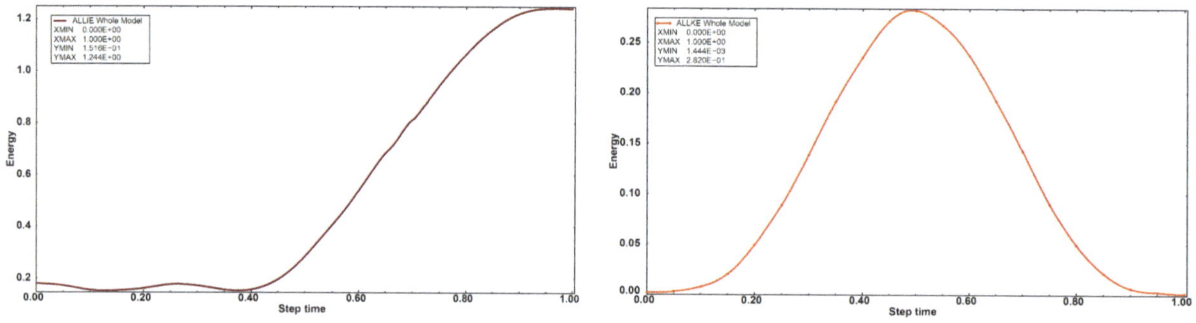

Figure 3.32: The total IE (left) and KE (right) of the FEM model to make sure the ratio is less than 10% in the compression step.

3.2.5 Machine Learning Algorithms for Simulation of Biomechanical Model Deformation

A lot of research has been done to predict the mechanical behavior of anatomical structures using different algorithms of machine learning algorithms (143). The fundamental benefit of machine learning models is their ability to forecast outcomes in real-time once the mapping function has been computed offline. The three main types of algorithms are NNs, tree-based algorithms, and support vector regression (SVR) machines. Although lots of anatomical structures using machine learning have been investigated (143), for the simulation of breast deformation, there is only limited research (see in Table 7.7 in Chapter 7).

Mendizabal et al., (120) was particularly concentrated on the simulation of deformations of US images for US guided breast biopsy. They used a U-net architecture that has been trained for a small synthetic dataset. Their main goal was to find a relationship between the partial surface deformation under the US probe and the deformation inside the breast. For that purpose, they were able to make an accurate prediction for the lesion displacement in real-time. Their model did not take into consideration the heterogeneity or complex boundary conditions that exist in clinical cases. Their model is insensitive to patient-specific elastic properties as its input is only the surface displacement.

For evaluation, they used the displacements that have been generated by FEMs as ground truth and then calculated the normalized mean error. In addition, since their goal was an accurate prediction, they calculated the TRE between their prediction and the annotated lesion in the phantom breast. They used different probe displacements from less than $12.5\,mm$ to more than $27.5\,mm$, achieving a mean TRE of $2.7\,mm$ to $5.8\,mm$. Their main limitation was that they had to train the model again for each new geometry, which is restricted to the types and number of compression tools such as different probe shapes (120).

Martínez-Martínez et al. (116) and Rupérez et al. (159) modeled the breast deformation of a biomechanical model derived from MRI under mammographic compression. Their main goal was to accelerate the multimodal registration and simulate the behavior of the breast tissues in image-guided interventions such as biopsies. They proposed three models which are decision tree (DT), ERT, and random forest. They first did their experiments based on phantoms and then expanded to clinical datasets.

They evaluated their methods by first calculating the mean 3D Euclidean distance between the nodes predicted by the models and those extracted from the FE simulation as a ground truth. Their experiments showed that ERT has the best performance with an average error of $0.62\,mm$. One limitation is certainly the number of datasets (10 phantoms, 10 clinical cases) that have been evaluated. This might cause overfitting problems and prevent generalization of the model. In addition, the model of FEM has been used to generate the data and used as a ground truth. It considers compression ratios of only 20%, which is considerably lower than the compression ratios applied in mammography. Their model was restricted to three tissue types fatty and glandular tissues and skin (116; 159).

In this thesis, three models from machine learning and deep learning algorithms are being proposed and compared to overcome the excessive computation time required by FEM; i.e. ERT and XGBoost based

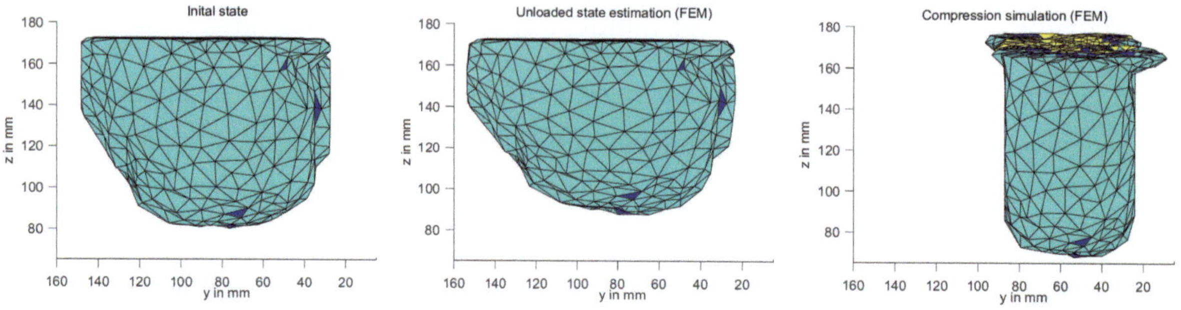

Figure 3.33: A clinical case of our biomechanical model in the sagittal plane (lateral view): The initial state, the unloaded state estimation followed by the compression step (164).

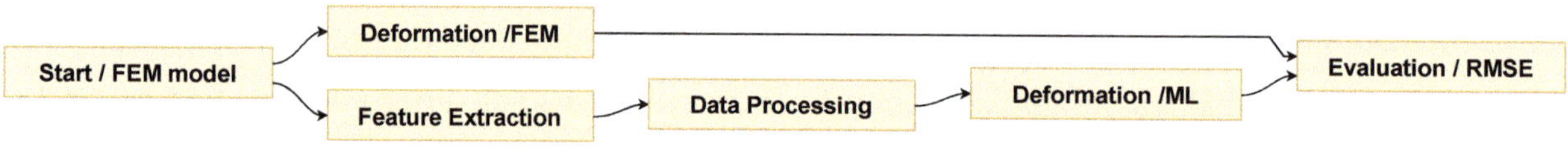

Figure 3.34: The workflow of the proposed method of machine learning algorithms for replacing FEMs

on machine learning, and one model consists of encoder and decoder models of a Att-BLSTM based on deep learning to predict the deformation of a biomechanical model for CC views. The biomechanical model can be considered complex as it considers four tissue types: fatty, glandular, and muscular tissues and skin. Also, compression ratios up to 76% were considered given the metadata of the mammograms that have been included in this study (164).

Data Cleaning

The deformed configuration of the breast is extracted by requesting from ABAQUS the node positions at a certain time step (referred to as "step number"). This information is used as a feature for training the machine learning models. For the unloaded state step and the compression step, 100 and 70 steps, respectively are extracted as intermediate steps between the initial state and the last step. For the unloaded state step, intermediate steps can not be removed as the forces that have been applied are the inverse of the gravity direction during the whole step. So deformation of the breast during the whole step is essential. For the compression step, since the focus of that step is the contact problem (deformation of the breast when it is compressed by the plates), some initial steps could be removed which is the time taken for the upper plate to move till it starts to contact the breast as shown in Figure 3.35 and 3.36. During this time, the breast hypothetically should remain the same but due to some remaining forces from the steps before, the breast is slightly moving and oscillating without any contact. Removing this duration of time helps in getting better results for the model. A threshold method is used for calculating the sum of displacements of all nodes of the breast for estimating the time taken to remove these steps. It has been empirically tested to set this threshold to $5mm$ which is around 25 $\sim$ 30 % out of 100 steps. This threshold is sufficient for removing these steps.

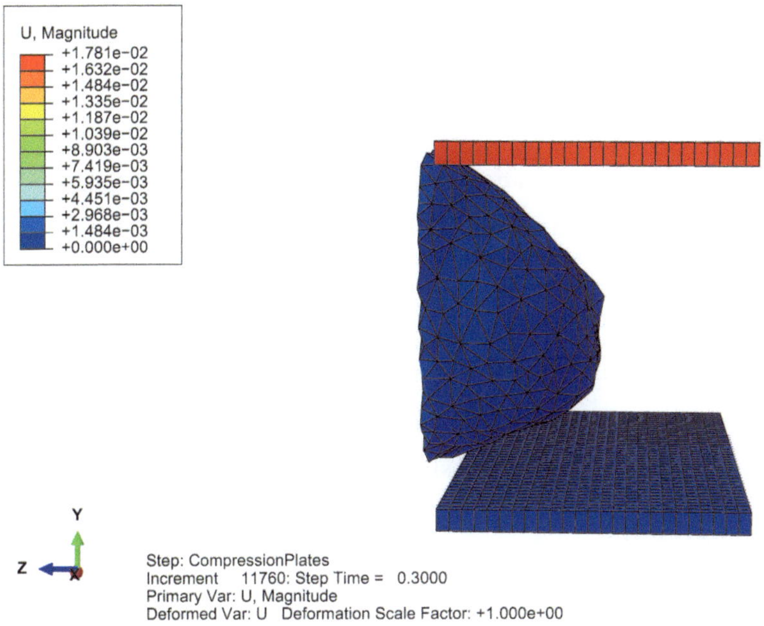

Figure 3.35: Simulation of the compression step in case of CC view in one of the clinical datasets, showing when the upper plate starts to contact the surface of the breast at a time step of 0.30.

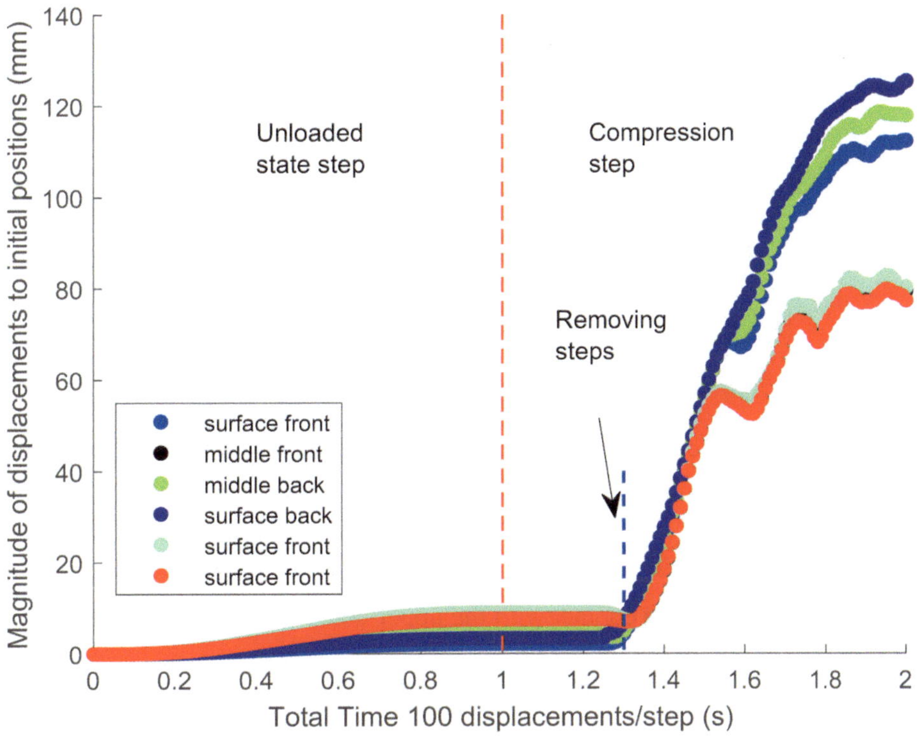

Figure 3.36: Some nodes were selected based on the geometry of the breast, validating that in the compression step around 25 $\sim$ 30% could be omitted, since the upper plate did not touch the surface of the breast.

Data Generation and Features Extraction

Extracted features mainly contain three different types of information. First, the geometry of the breast described by the initial coordinates of breast mesh is included. Second, the properties regarding tissues such as the breast volume and fraction volumes of tissues given by the proposed segmentation are included. Third, the external forces that cause the deformation such as the position of the compression plates and surface nodes of the breast mesh are also included. The biomechanical model for CC views consists of two steps: unloaded state estimation and compression simulation. For the unloaded state estimation, eleven features are extracted that are similar to the features proposed in (116; 159). For the compression simulation step, a total of 22 features are extracted: the same eleven features from (116; 159) and additionally, eleven features that have shown to empirically improve the results in the experiments, especially for the large deformations during the breast compression simulation. A description of the 22 features is summarized in the following list (164):

- Feature 1-3: The three components of the nodal coordinates (X, Y, and Z) of each node in the initial breast geometry
- Feature 4: The step number at which the nodal coordinate was acquired
- Feature 5-7: The volume fraction of each segmented tissue (fatty, glandular, and muscular tissues) compared to the total volume of the breast
- Feature 8: The volume of the breast mesh in mm^3
- Feature 9-11: The fraction of elements for each tissue type (fatty, glandular, and muscular tissues) compared to the total number of elements in the breast mesh
- Feature 12: The displacement of the upper compression plate towards the breast for the particular step
- Feature 13-14: Y and Z coordinates of the edge of the fixed lower compression plate
- Feature 15-16: Y and Z coordinates of the edge of the moving upper compression plate for the particular step
- Feature 17: a flag indicating if the node of the current instance is a surface node of the breast mesh which will potentially get into contact with the upper compression plate
- Feature 18: a flag indicating if the node of the current instance is a surface node of the breast
- Feature 19-21: 3-bit binary value representation of the tissue type to which the node belongs based on the breast mesh as shown in Table 3.7
- Feature 22: Laterality of the breast mesh (left or right breast)

Table 3.7: Tissue labels

bit 2	bit 1	bit 0	Class
0	0	1	muscle
0	1	0	fatty
0	1	1	interface (fatty, muscle)
1	0	0	glandular
1	0	1	interface (glandular, muscle)
1	1	0	interface (fatty, glandular)
1	1	1	interface (three tissues)

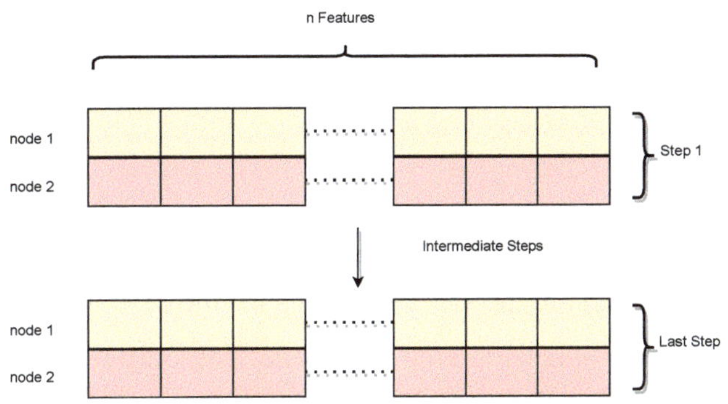

Figure 3.37: Data structure of the features, which are considered as the input for training the machine and deep learning models.

Finally, one instance of 22 features represents the information of one node in a breast.

After extracting the features, a 2D matrix was organized where the columns represent the 11 and 22 features for the unloaded state estimation and compression step, respectively, and the rows represent the nodes at one time step as an instance as shown in Figure 3.37.

Data Processing

Before feeding the instances into the machine learning algorithms, a preprocessing step is done, which consists mainly of normalization.

It is responsible for bringing different breast geometries into a similar reference space to increase robustness and training stability.

In this application, due to the large variance in the geometry of the breasts, such as position, shape, and size, normalization is an essential step. It brings the plates and the breast mesh into the same coordinate system. It has been empirically tested that the z-score normalization has the best performance for this problem: a pretest was performed comparing two common normalization methods: z-score and min-max scaling normalization (173). A small dataset of 100 breasts was used for training and 10 breasts were predicted using the machine learning and deep learning models, which will be described in the next subsection (models). The performance using the Root Mean Square Error (RMSE) was evaluated (see Chapter 6) and analyzed the percentages of nodes in the breast model with an RMSE less than $2mm$. For the z-score normalization on average 14.4% of the nodes had an RMSE less than $2mm$. For the min-max scaling normalization on average 12.7% of the nodes had an RMSE less than $2mm$. Since the z-score method showed better performance than the min-max scaling on average, it was selected for normalization. The main concept of the z-score is based on calculating the standard deviation of distances of data points from the mean. It aims to have all instances have a mean of zero and a standard deviation of one (150). For this problem, this means that the X, Y, and Z coordinates of the nodal positions V of one patient dataset are changed to V_{norm} in which they are centered around the origin (μ) and the standard deviation (σ) of all nodal positions from the origin is one, as shown in equation 3.24.

$$V_{norm} = \frac{(V - \mu)}{\sigma} \tag{3.24}$$

Models

The next step after generating the features and data processing is to build up machine learning models for training. The inputs of the model are the eleven features for the case of the estimation of the unloaded state while 22 features for the compression step are used. The output of the models is the predicted deformation of the instance sequentially starting from step number one till the last desired required step, i.e. the three coordinates X, Y, and Z of the nodal position belonging to one instance for a particular step. Since the problem is a supervised problem, there are three main approaches to follow: linear model, tree-based models, and artificial NNs. Two models are based on tree-based ensemble models: ERT similar to the model used in (159), XGBoost, and one model is based on RNN.

Extremely Randomized Tree and Extreme Gradient Boosting The main difference between bagging ERT (63) and boosting XGBoost (40) is that ERT learns trees independently from each other and in parallel. By that, each model is trained using a different sample from the same training dataset, and the prediction is made by averaging the weak predictions together. Contrary, XGBoost learns trees based on the previous one sequentially in an adaptive way. By that each new model is trained based on the performance of the previously trained model and the prediction is made by correcting the error of the previous model. The loss function used for ERT and XGBoost is the RMSE.
For XGBoost, some hyperparameters affect the model:

- number of estimations
- maximum depth
- minimum child weight

The number of estimators is equal to the number of decision trees in the model. The larger the number of trees is, the higher the ability to fit and solve complex problems is. However, one drawback of increasing the number of estimators is increasing the complexity of the model, which also could cause an overfitting problem (133).

The maximum depth is considered to be the depth of the decision trees. Increasing the depth will lead to fitting complex problems in a better way but it will also lead to increasing the complexity of the model and it might cause an overfitting problem (133).

The minimum child weight is to calculate the sum of the sample weights. The tree expansion is stopped if the sum of the sample weights is less than the selected threshold. Increasing the minimum child weight will lead to a decrease in the complexity of the model and will affect the number of branches of trees (133).

For the unloaded state and compression steps, the parameters are empirically tested on a small dataset of 100 datasets to validate the parameters. For the unloaded state step, the number of estimators is set to 55, maximum depth is set to 6, minimum child weight is set to 50, and learning rate (α) is set to 0.2. For the compression step, the number of estimators is set to 80, the maximum depth is set to 7, the minimum child weight is set to 6, and the learning rate (α) is set to 0.08.

Encoder and Decoder based on Att-BLSTM The third model is from the family of RNN. The architecture consists of an encoder and a decoder (108). The encoder consists of a Att-BLSTM (214). The decoder consists of an LSTM. The loss function for this architecture is the root mean square.

Similarly, the parameters are empirically tested on the same small dataset. For the unloaded state, the parameters are set as follows: the activation function in the attention layer in both stages is set to tanh and ReLU, and the learning rate (α) is set to 0.03. For the compression step, the activation function in the attention layer in both stages is set to sigmoid and the learning rate (α) is set to 0.03 as well.

3.3 Summary

First, the literature review of registering between MRI and X-ray mammograms specifically using biomechanical models has been presented. Second, the methodology of the proposed method has been illustrated: to register between MRI and X-ray mammograms, a patient-specific model has been implemented to mimic X-ray mammographic compression in the two views CC and MLO. This model was originated from the work of Hopp et al. The biomechanical model has been solved by FEMs using ABAQUS/Explicit. In this thesis, three main blocks have been investigated: segmentation, meshes, and the biomechanical model simulation. First, a novel approach has been proposed for segmenting three tissues simultaneously using unsupervised NNs for T2-weighted MRI series (162). Second, the ratio between circumradius and inradius was used to investigate the quality of the initial mesh in order to avoid re-meshing in between the simulation steps and achieve better convergence. If the ratio is less than 0.25, it is considered a distorted element. Additionally, other quality metrics read from iso2mesh and ABAQUS were used to investigate the complexity of the biomechanical model and to answer the question if the model could be simpler. By that, applying upright gravity shows a negative effect on the model. In addition, positioning the lower plate in the inferior direction and the thickness of the skin has an impact. Those two factors have been examined using TRE and minimum mean nodal error, respectively. It was observed that the first algorithm for estimating this position with a skin thickness of $2\,mm$ is the most robust. To analyze the stability of the model, different parameters have been investigated such as time increment, mass scaling, time step, the ratio between KE and IE, D_1, and C_{10}. It was found that time increment and mass scaling do not affect the stability. While increasing the time step leads to a more oscillating model. The latter two factors have a large impact on the stability of the model and the decrease in oscillations. Despite the fact that FEMs could solve the deformation, it requires around 20 minutes to solve using the standard parameters. Additionally, it might take up to several hours to optimize parameters even if it is only one parameter that needs to be optimized since it is based on the number of iterations. Thus, three models have been proposed based on machine and deep learning algorithms to replace the model solved by FEMs (164). The three models are ERT, XGBoost, and Att-BLSTM. By proposing this method, solving the deformation could be done in 5 seconds.

4 Image Based Registration

After aligning the MRI to the full X-ray mammogram, image based registration methods have been developed to align full X-ray with the spot mammograms.

4.1 Intensity-Based Methods

Intensity-based registration transforms two images by considering one image as a reference image and the other one as a moving image (21). The key here is that it only uses intensities like gray levels to estimate the transformation. Point-based, surface-based, and voxel-based registrations are three categories of intensity-based methods (110). It registers images with the same or different dimensions, by rigid transformations or deformable registration. Images can be intermodal and intramodal.

Registration using control points and surface-based control points minimizes the distances between corresponding anatomical landmarks and surfaces, respectively. For this thesis, it could be extracting e.g. glandular tissue features from a full X-ray mammogram and spot mammogram. There are two disadvantages to this method. First, the difficulty of extracting different landmarks automatically for different cases. This difficulty of extracting landmarks automatically is illustrated in Figure 4.1. Three different automatic methods have been used for extracting glandular tissues automatically from spot mammograms. These methods were first implemented to enhance the image contrast of spot mammograms which will be explained in the proposed method. Then afterward to extract the glandular structures, a threshold is selected. This specific threshold is calculated by adding the mean and standard deviation of the enhanced image. It is obvious that glandular structures in the three methods vary from one enhancement to another which makes it hard to rely on. This likely leads to a manual extraction which would require a lot of time. Second, in some cases, the identification of glandular tissue features is difficult for dense breasts since the dimensions of a spot mammogram image are considerably small with 50×50 mm only which is considered a subset of the breast not the full view of the breast image.

Pixel (2D) and voxel (3D)-based registrations optimize a functional calculation of the similarity of corresponding pixels pairs for a given feature. One advantage of pixel-based methods is that feature calculation is simple. The accuracy is independent of precise segmentation or extracting landmarks as in point or surface-based methods (110). Some challenges may arise with using only the pixel (2D) or voxel (3D) values. First, the algorithm might only function consistently if the similarity measure is determined using the voxels in a specific area of interest in the image for example, an ROI and not opposed to all voxels of the image. Second, to prevent aliasing in the subsampled images (21; 148), sometimes it is

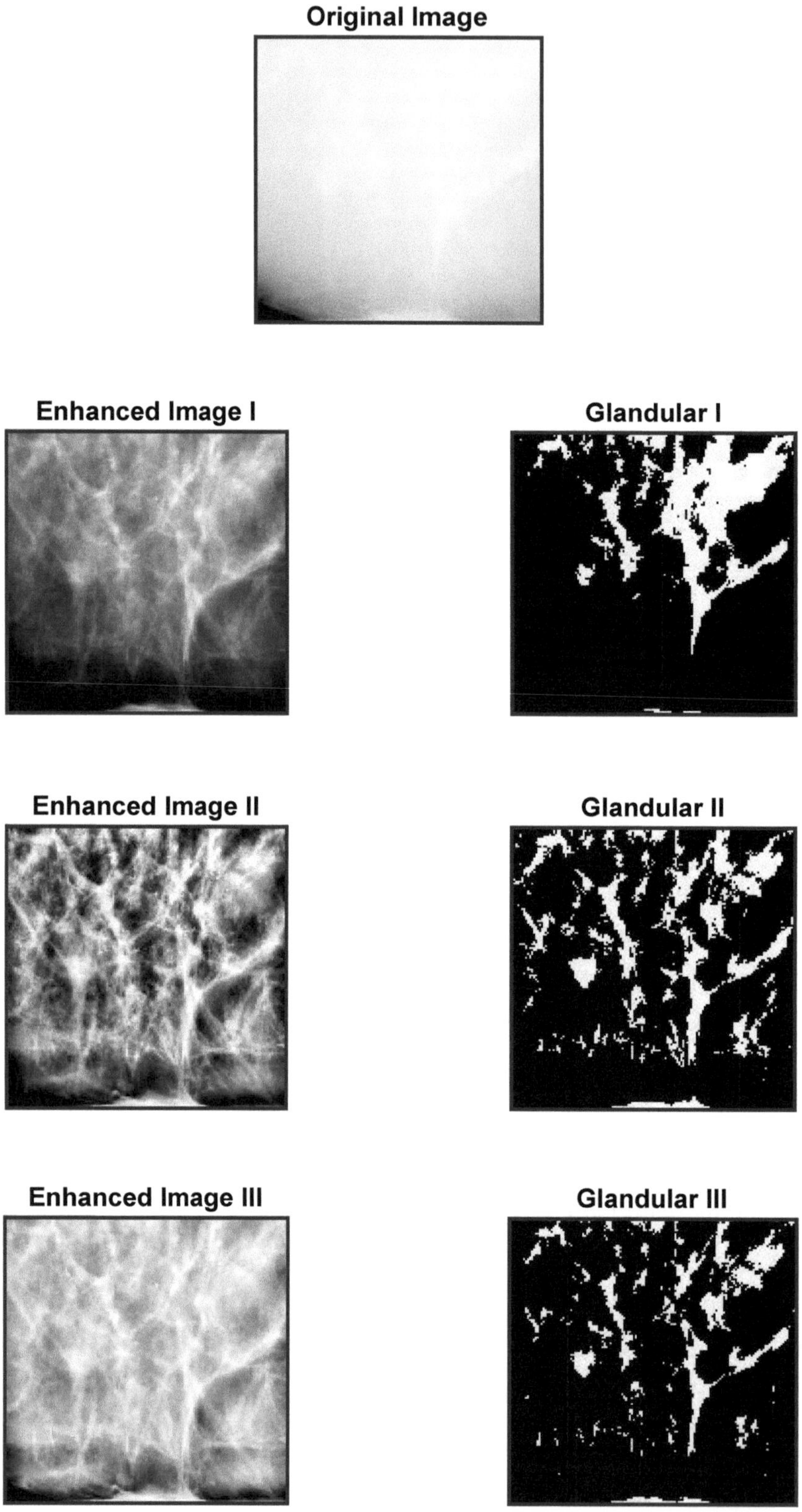

Figure 4.1: Variance in automatically extracted glandular tissue structures. The original image of the spot mammogram was received from the Medical University of Vienna. Three algorithms for enhancing the spot mammogram were applied (first column). Automatically extracting the glandular tissue was based on a threshold based on the mean and the standard deviation (second column).

helpful to blur the images before sampling. The amount of blurring utilized may vary depending on the application. Third, preprocessing is necessary and it will probably depend on the modalities being registered as well as the body part being analyzed. Fourth, for some modalities, segmentation of particular parts based on the body's anatomy might be helpful to avoid registering these parts. This is considered as pre-knowledge, e.g. in which areas no registration is allowed, e.g. muscle area. Fifth, robustifying the intensity-based methods is one of the most challenging tasks in order not to be constrained to a small set of image types but to respect patient variability (21). Different image similarity metrics can be selected based on the problem (119). The next paragraphs will outline the most frequent similarity metrics used in this thesis. In all the image similarity metrics, the reference image is denoted as X. The moving image is denoted as Y. Some symbols used in the following equations, are defined in this Section **??**.

(1) **Convolution (Conv)** It is a 2D convolution operation between the reference image and the moving image as shown in the equation 4.1. The moving image Y is considered the filter that slides over the reference image X, performing an element-wise multiplication. Then, it sums up results into a single output pixel (160; 33). The maximum convolution value is considered to have the highest probability that the two images are correctly aligned.

$$Conv(i,j) = \sum_{m}\sum_{n} X(i+m-1, j+n-1)Y(m,n) \tag{4.1}$$

where m and n refer to the number of rows and columns, respectively and i and j are the locations of pixels in the reference image X.

(2) **Correlation Coefficient (Corr)** It measures the correlation coefficient which is the difference between the intensity value at that pixel and the mean intensity of the whole image for every pixel location in both images: reference image X and moving image Y as shown in equation 4.2 (13). The maximum value of Corr is considered to be the highest value for a correct alignment.

$$Corr = \frac{\sum_m \sum_n (X_{(m,n)} - \overline{X})(Y_{(m,n)} - \overline{Y})}{\sqrt{(\sum_m \sum_n (X_{(m,n)} - \overline{X})^2)(\sum_m \sum_n (Y_{(m,n)} - \overline{Y})^2)}} \tag{4.2}$$

where m and n refer to the locations of pixels in either reference image X or moving image Y.

(3) **Mutual Information (MI)** It measures the degree of dependence of the reference image (X) and moving image (Y) by measuring the distance between the joint distribution $p_{X,Y}(x,y)$ and the distribution related to the instance of complete independence $p_X(x)p_Y(y)$ (109). MI is based on the definitions of the joint histogram, probability distribution function (PDF), entropy, joint entropy between two images, and conditional entropy of X given Y and of Y given X, respectively (110; 109; 180; 44; 102). The entropy of an image is defined as $E(X) = -\sum_x p_X(x) \log_b p_X(x)$ where p is the probability of a pixel having a certain gray level intensity (e.g. 256 for 8-bit images), and b is the base of the logarithm function, while the joint entropy $JE(X,Y)$ is the

measure of the uncertainty associated between the two images and is defined as $JE(X,Y) = -\sum_{x,y} p_{XY}(x,y) \log_b p_{XY}(x,y)$. The conditional entropy is the amount of uncertainty left in X when knowing Y and vice versa as shown in equation 4.3 (109; 180; 21). By measuring the information theory using mutual information instead of joint entropy or joint histogram, the problem of calculating only the PDF from the joint entropy (the region of overlap between the two images) is overcome. The higher the value of MI is, the higher the probability of correct alignment is.

$$MI(X,Y) = \sum_{y \varepsilon Y} \sum_{x \varepsilon X} p_{(X,Y)}(x,y) \cdot \log \frac{p_{(X,Y)}(x,y)}{p_X(x) p_Y(y)} \tag{4.3}$$

(4) **Normalized Mutual Information (NMI)** Similarily NMI is based on the definitions of entropy and joint entropy between the two images as shown in equation 4.4. It is considered the extension of MI and it has been proposed to counter the decisive effect of increasing MI while the registration quality is low (94). Similarly as well for the value of NMI, if the value of NMI is higher, then the probability of higher correct alignment is.

$$NMI(X,Y) = \frac{E(X)+E(Y)}{JE(X,Y)} = \frac{\sum_{x \varepsilon X} p_X(x) \cdot \log p_X(x) + \sum_{y \varepsilon Y} p_Y(y) \cdot \log p_Y(y)}{\sum_{x,y} p_{(X,Y)}(x,y) \cdot \log p_{(X,Y)}(x,y)} \tag{4.4}$$

(5) **Ratio Image Uniformity (RIU)** A ratio (R) is calculated by the division of each pixel in the reference image X by each pixel in the moving image Y for estimating the registration transformation. The uniformity is then determined by calculating the normalized standard deviation of R. Additionally, the roles of images X and Y are reversed and averaging the two calculated uniformity (21; 167). A minimal RIU value is considered to be the highest probability of correct alignment.

$$RIU = \frac{\sigma_R}{\mu_R} \tag{4.5}$$

where: $\mu_R = \frac{1}{N} \sum_{(i)} R(i)$, $\sigma_R = \frac{1}{N} \sum_{(i)} (R(i) - \mu_R)^2$, $R(i) = \frac{Y(i)}{X(i)}$, and N is the number of pixels.

(6) **Mean Square Error (MSE)** It is the mean square error between the reference X and the moving image Y as shown in equation 4.6. The optimum value is minimum which represents the correct alignment between the reference and moving image (166).

$$MSE = \frac{\sum (X_{(i,j)} - \hat{Y}_{(i,j)})^2}{n} \tag{4.6}$$

where: n is the number of pixels.

(7) **RMSE** It is the root mean square error between the reference and the moving image as shown in equation 4.7. The optimum value is minimum since it is based on MSE (38).

$$RMSE = \sqrt{MSE} \tag{4.7}$$

(8) **Erreur Relative Globale Adimensionnelle de Synthèse (ERGAS)** It evaluates the overall image quality in the reference image X as shown in the equation 4.8 (38). The lower the error is, the higher the probability that the moving image Y is located in the reference image X.

$$ERGAS = 100 * \frac{Res_X}{Res_Y}\sqrt{\frac{1}{N}\sum_{k=1}^{N}\frac{RMSE^2}{\mu_Y}} \tag{4.8}$$

Res_X and Res_Y are the resolution of the reference and moving images, respectively. N is the number of pixels.

(9) **Peak Signal to Noise Ratio (PSNR)** It is based on MSE and the peak value (denoted as peakval) of the reference image X is considered to be the maximum intensity. The optimum value is the maximum, although, it is based on MSE. The reason behind that is MSE is located in the denominator position as shown in equation 4.9 (166; 168).

$$PSNR = 10 * \log_{10}\frac{peakval^2}{MSE} \tag{4.9}$$

(10) **SNR** It is the signal to noise ratio between the reference X and moving Y images as shown in the equation in 4.10.

$$SNR_{x,y} = 10 * \log_{10}(\frac{\sum_{n=1}^{N}|X_n|^2}{\sum_{n=1}^{N}|Y_n|^2}) \tag{4.10}$$

(11) **Universal Objective Image Quality Index (UIQI)** This method is based on the Human Visual System (HVS) rather than the mathematical measurements like RMSE for example. Compared to the commonly used distortion measure such as MSE, it performs noticeably better (204). The main goal of this metric is to be suitable for different image processing applications and support fair comparison across various types of image distortions. It combines the multiplication of three factors: luminance distortion (l), contrast distortion (c), and loss of correlation (s) (203) as shown in equation 4.11. The optimum value that represents a good match is the maximum.

$$UIQI(x,y) = \frac{4\sigma_{xy}\mu_x\mu_y}{(\sigma_x^2+\sigma_y^2)(\mu_x^2+\mu_y^2)} \tag{4.11}$$

where: $l(x,y) = \frac{2\mu_x\mu_y}{\mu_x^2+\mu_y^2}$, $c(x,y) = \frac{2\sigma_x\sigma_y}{\sigma_x^2+\sigma_y^2}$, $s(x,y) = \frac{\sigma_{xy}}{\sigma_x\sigma_y}$

(12) **Structural Similarity Index Measure (SSIM)** It is the extension of UIQI of extracting three main features: luminance (l), image contrast (c), and structures (s) with additional regularization constants C_1, C_2, and C_3 for each feature, respectively as shown in equation 4.12 (204; 205; 166; 168). These regularization constants are used by SSIM function to prevent instability in image regions where the local mean or standard deviation is near zero. Consequently, these constants ought to have small non-zero values. Additionally, each feature has a parameter μ, σ, and γ that is used for adjusting the relative importance of each feature separately. For simplicity, these values are set to one (205). The SSIM is a perception-based model. It takes into account luminance and contrast masking terms as well as significant perceptual phenomena, such as picture degradation as a perceived change in structural information. The distinction between these methods and others is that they estimate absolute errors, unlike MSE or PSNR. The concept of structural information holds that pixels have high interdependencies, particularly when they are spatially close to one another. These dependencies carry crucial information about the hierarchy of the objects in the visual scene. Contrast masking is a phenomenon where distortions become less evident where there is significant activity in the image. While luminance masking is a phenomenon where visual distortions tend to be less visible in bright places (204; 205). Similarly, the optimum value for a correct match is the maximum.

$$SSIM(x,y) = [l(x,y)]^{\alpha}.[c(x,y)]^{\beta}.[s(x,y)]^{\gamma} \tag{4.12}$$

$$\text{where: } l(x,y) = \frac{2\mu_x\mu_y + C_1}{\mu_x^2 + \mu_y^2 + C_1},\ c(x,y) = \frac{2\sigma_x\sigma_y + C_2}{\sigma_x^2 + \sigma_y^2 + C_2},\ s(x,y) = \frac{\sigma_{xy} + C_3}{\sigma_x\sigma_y + C_3}$$

where $C_1 = (K_1L)^2$, $C_2 = (K_2L)^2$, $C_3 = \frac{C_2}{2}$ where $K_1 = 0.01$ and $K_2 = 0.03$ and L is the specified dynamic range value which in our case is 255 (205).

(13) **Multiscale Structural Similarity (MS-SSIM)** It is the extension of SSIM which supports more flexibility than single scale methods in incorporating the variations such as image resolution as shown in equation 4.13 (204). A low pass filter is applied iteratively for the reference X and moving Y images followed by downsampling the filtered images by a factor of 2. The original images are indexed as scale 1, and the highest scale as scale M which is obtained after M-1 iterations. The luminance (l) is only computed for scale M while the image contrast (c) and structures (s) are multiplied at each iteration (j). The overall MS-SSIM is calculated by combining the measurements at different scales. The optimum value for a correct match is the maximum.

$$MS\text{-}SSIM(x,y) = [l_M(x,y)]^{\alpha_M} \prod_{j=1}^{M} [c_j(x,y)]^{\beta_j}.[s_j(x,y)]^{\gamma_j} \tag{4.13}$$

where M is the highest scale (The default value is five).

(14) **BIAS** It is the difference between the mean of each image: reference X and moving Y image as shown in equation 4.14 (38). The ideal optimum value for a correct match is zero.

$$Bias = \overline{X} - \overline{Y} \tag{4.14}$$

(15) **Difference in Variance (DIV)** is the difference between the variance of each image as shown in equation 4.15 (38). The optimum value for a correct alignment is minimum.

$$Variance = V_X - V_Y \tag{4.15}$$

$$\text{where: } V_X = \frac{1}{N-1}\sum_{i=1}^{N}|X_i - \mu|^2, \ \mu = \frac{1}{N}\sum_{i=1}^{N}X_i$$

where N is the number of observations. If there is only one observation, then the weight is 1.

(16) **Standard Deviation Difference (SDD)** It is the difference between the standard deviation of the moving Y and reference X images divided by the mean of the reference image. In addition, the difference between the standard deviation of the moving and reference images divided by the mean of the moving image is calculated as shown in equation 4.16 and vice versa. The average of both values is calculated (38). The optimum value for a correct alignment is zero.

$$SDD = \frac{(\sigma_X - \sigma_Y)}{\mu_X} \tag{4.16}$$

Image similarity metrics have been categorized into classes based on how similar equations are. The first class of quality measures is the 2D convolution. The second class of quality measures is measuring the correlation coefficient. The third class of quality measures is MI and NMI. The fourth class of quality measures is RIU. The fifth class of quality measures is MSE, RMSE, ERGAS, and PSNR. The sixth class of quality measures is SNR. The seventh class is UIQI, SSIM, and MS-SSIM. SSIM is an extension of UIQI. Additionally, MS-SSIM supports more variation than SSIM such as image resolution (204). The eighth class is bias, DIV, and SDD. The optimal values of the calculated image similarity metrics are extracted. Using the optima of the image similarity metrics as a reference point with the best match of the moving image, the moving image is registered to the reference image.

4.2 Proposed Method

The proposed method consists of three main blocks: preprocessing, finding a ROI from the full mammogram which is considered the reference image as shown in Figure 4.2 (top), and then applying different image similarity metrics between the ROI (the reference image) and the spot mammogram (the moving image). The spot mammogram slides across ROI pixel by pixel computing the image similarity metric as shown in Figure 4.2 (bottom). Afterward, the optimum value either minimum or maximum is extracted

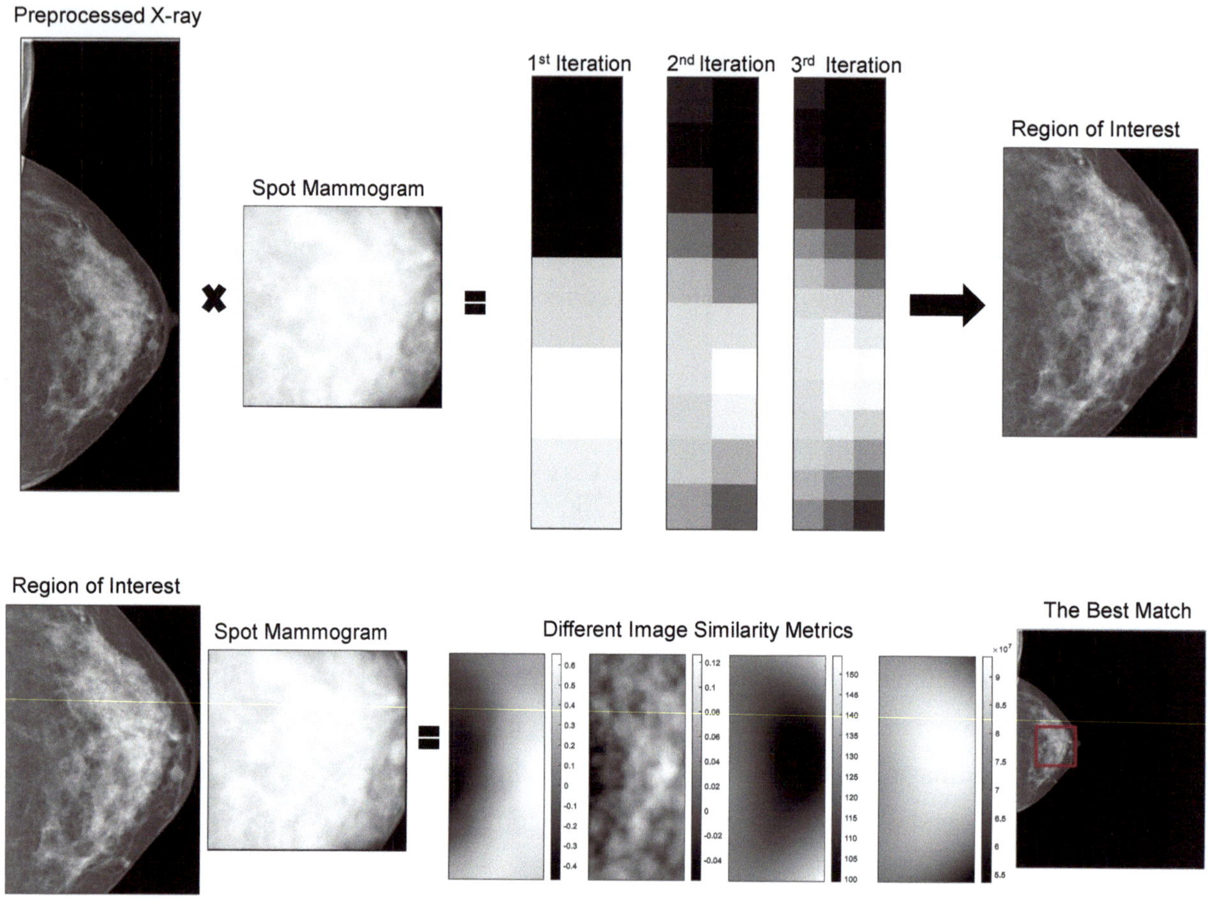

Figure 4.2: The block diagram of the image based registration between full X-ray mammogram and spot mammogram: finding ROI from the full X-ray mammogram using convolution operation (top) and applying different image similarity metrics between ROI and spot mammograms (bottom) to get the best match (red square).

based on the computed image similarity metric. The position of this optimum value is considered where the moving image is aligned with the reference image (161; 163). This matching process allows for resolving a translational transformation between both images. This proposed method is a rigid transformation since rotation is also applied in the preprocessing block. It is assumed from the clinical point of view that both images have the same forces applied although the gravity applied in cases of spot mammogram could be both in the prone and upright position. The most common here in this thesis is the prone position. For an X-ray mammogram, the gravity applied is only an upright position. Also, both images are considered to have the same compression thickness of the breast and the same position of the X-ray source. Moreover, both images are considered to have the same projection angle. While this is not the case, e.g., both images have CC views, there might be a slight difference in projection angles due to different imaging situations. Additionally, there is already a difference in views in mammography images and biopsy images, e.g. MLO for full X-ray mammogram and ML for spot mammogram. Despite all these differences, it is considered sufficient to solve the problem.

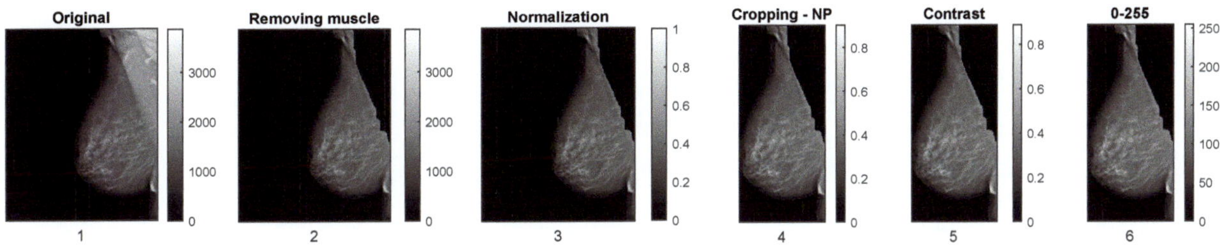

Figure 4.3: Steps of preprocessing block for X-ray mammogram images: the original image received from the Medical University of Vienna (1), removing muscle from the image using the proposed methods (2), normalizing the image (3), cropping until the nipple position (4), enhancing contrast using the proposed method (5), and scaling intensities into 0-255 levels (6).

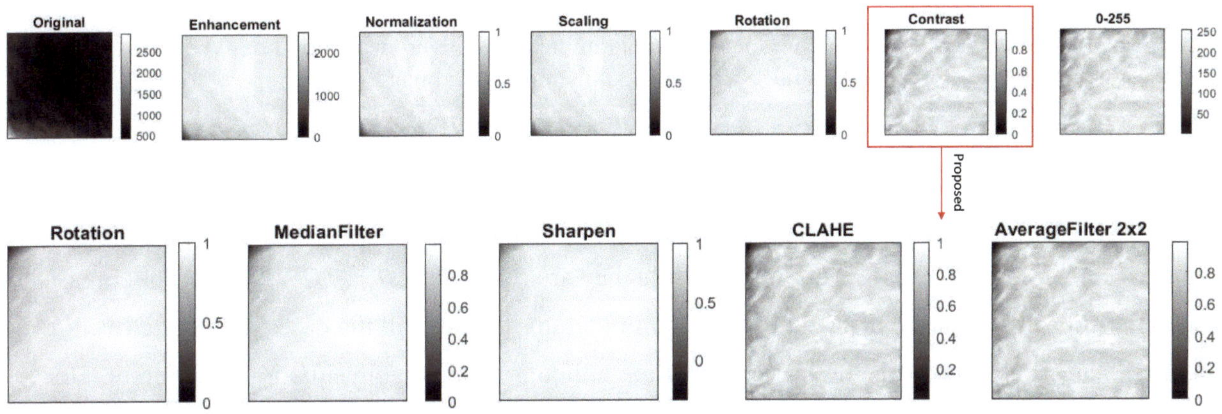

Figure 4.4: Steps of the preprocessing block for spot mammogram images (top): The original image has inverted intensities (first), image after enhanced intensities (second), normalization between 0 and 1 values (third), resampling to resolution of full mammogram (fourth), rotation (fifth), applying the proposed image contrast (sixth), and scaling the intensities into 0-255 levels (seventh). Steps of the proposed enhancement image contrast for spot mammograms (bottom): The rotated image is the input to the image (first), applying median filter (second), sharpen the image (third), applying CLAHE (fourth), applying an average filter with size 2x2 (fifth).

4.2.1 Preprocessing

Preprocessing aims to provide better quality for full X-ray and spot mammogram images regarding contrast, resolution, rotation, and removing unnecessary areas for the full mammograms as shown in Figure 4.3 and 4.4. First, the contrast of spot mammograms is adjusted to match that of the full X-ray mammogram by inverting contrast. Second, normalization is done for both images. Third, a resampling is applied for the spot mammogram using nearest neighbor interpolation since it has a different resolution (35).

Fourth, since spot mammograms could be taken from a prone or upright position taking into consideration three views (CC, ML, and LM as explained before) while the full mammogram could be taken only in the upright position taking into consideration two main views and two minor views (CC, MLO, ML, and Rolledlateral (RL), respectively), the spot mammogram needs to be rotated to compensate partly in the transformation matrix.

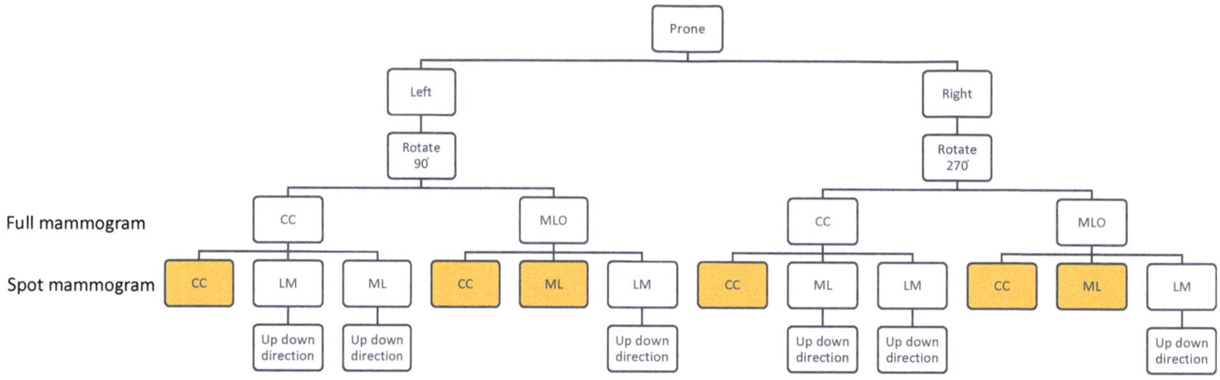

Figure 4.5: Twelve options of rotating the spot mammogram taken in the prone position based on the breast side and the combination of views between full X-ray and spot mammogram image.

The hierarchy of combinations of the standard views of full mammograms and views of spot mammograms taken from the prone position is shown in Figure 4.5. First, spot mammograms are rotated by 90° counterclockwise (CCW) for the left breasts while they are rotated by 270° CCW for the right breasts to match the coordinates of the breast in the full mammogram. Second, all the cases highlighted yellow in Figure 4.5 are considered to be in the same position although, the imaging situation is different. Hence, the deformation state is considered to be comparable. For the cases of CC (full X-ray mammograms) and CC (spot mammograms), the spot mammogram is considered to be in the same position at 0° rotation angle. Similarly, for the cases of MLO and CC, and MLO and ML, the variance of 45° between the full X-ray mammogram and the spot mammogram is considered to be acceptable to determine similarities. While the other cases (not highlighted) (CC and LM, CC and ML, MLO and LM) for full X-ray mammograms and spot mammograms, respectively, up-down direction rotation is applied for compensating partly the difference in the LM views as shown in Figure 4.5.

For the supplementary views ML and ML, ML and LM, and RL and ML for the full X-ray mammograms and spot mammograms, respectively, rotation of left and right breasts and up-down direction are applied. While for ML and CC for full X-ray and spot mammograms, respectively, rotation of left or right breasts is only applied.

For the upright position, spot mammograms are rotated by 270° CCW and 90° CCW for the left and right breasts, respectively in order to match the coordinates of the breast in the full mammogram.

Fifth, the background in some cases in the full mammograms is not assigned to zero intensity which might cause further problems in the next steps. Segmentation of the breast is applied to determine the shape of the breast and then multiplication of this segmented image to the original image to set all the background pixels to zero.

Sixth, the quality of the image for both images is necessary to be calculated since there are several methods to improve image contrast. The quality of images is calculated by measuring the properties of gray level co-occurrence matrix (GLCM), which are contrast, correlation, energy, and homogeneity as shown in equations 4.17, 4.18, 4.19, and 4.20 (219; 22; 49). Then the Euclidean distance (ED) is calculated between the ideal value (denoted as ref) and the actual value (denoted as act) for each property as shown in equation 4.21.

$$Contrast = \sum_{i,j} |i-j|^2 p(i,j) \tag{4.17}$$

$$Correlation = \sum_{i,j} \frac{(i-\mu i)(j-\mu j)p(i,j)}{\sigma_i \sigma_j} \tag{4.18}$$

$$Energy = \sum_{i,j} p(i,j)^2 \tag{4.19}$$

$$Homogeneity = \sum_{i,j} \frac{p(i,j)}{1+|i-j|} \tag{4.20}$$

$$ED = \sqrt{(Cont_{ref} - Cont_{act})^2 + (Corr_{ref} - Corr_{act})^2 + (Eng_{ref} - Eng_{act})^2 + (Homo_{ref} - Homo_{act})^2} \tag{4.21}$$

There are several techniques for improving the image contrast of medical images and removing noises such as Gaussian filter, median filter, and contrast limited adaptive histogram equalization (CLAHE) (196; 126). In this thesis, six techniques have been tested and evaluated using the properties of the GLCM matrix:

1. Adjusting image intensity values based on a ratio called gamma γ between the mean and standard deviation of the image (denoted as method (1))
2. According to a technique from Kumbhar et al. (97), the image contrast is improved by first applying a median filter, then sharpening, equating the histogram (CLAHE), and applying an average filter, as illustrated in Figure 4.4 (bottom) (denoted as method (2))
3. Using two cascaded steps of the same block of method 2 (denoted as method (3))
4. Using two cascaded steps as well: method (2) followed by method (1) (denoted as method (4))
5. Using only CLAHE (denoted as method (5))
6. Using wavelet transform for denoising the image (126) (denoted as method (6))

Based on this analysis, it was observed that method (2) is the best technique based on the available datasets which will be discussed in Section 7.4.1. An example of one clinical dataset is shown in Figure 4.6 and 4.7 (a). Additionally, the Euclidean distance between the ideal case and the measured parameters

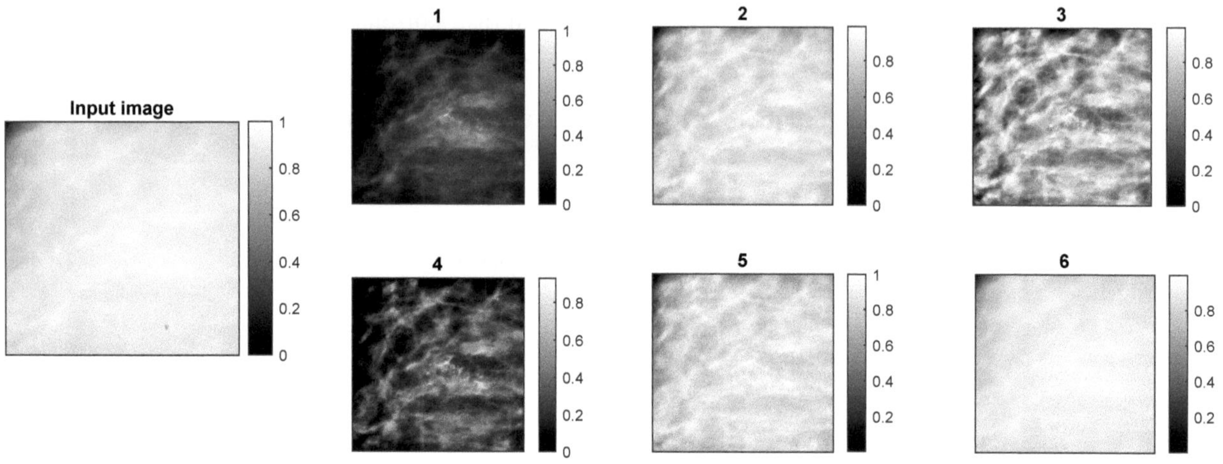

Figure 4.6: Six techniques have been analyzed for enhancing the image contrast of spot mammograms. The input image is the rotation image since the contrast enhancement is applied after rotation (see Figure 4.4). Based on the available dataset, method 2 is the best agreement for our problem.

of each property has been calculated as shown in Figure 4.7 (b). Last, the intensities are converted to 255 intensity levels for both images.

Since breast lesions are available in the breast shape not in the muscle area for MLO cases which is considered a pre-knowledge, the muscle is cut off to prevent the spot mammogram from aligning with it. Two methods have been evaluated to find the edge of the line dividing the muscle from the breast. First, the full X-ray mammogram is segmented using K-means clustering into three classes. Then classes background (0), glandular tissue (2), and fatty tissue (1) are assigned based on a histogram. One of the observations is that glandular and muscle are usually clustered in the same class. That specific class is extracted. Hence, separating between glandular and muscle tissue is applied using morphological operations as shown in Figure 4.8 (top). By that, the muscle is extracted. In some clinical cases, this method might not work. For the evaluation of this method, the rotation angle of Y is calculated. If the angle is larger than 40°, then this indicates that the method is not working because usually, this angle is in the range [0° - 35°] which leads to selecting the second method instead (28). In the second method, a Sobel filter is applied, which is an edge detection filter. Next, the gradient magnitude of the image is calculated, and finally, hough transform is used, which is a feature extraction technique used in image analysis (25) as shown in Figure 4.8 (bottom).

For speeding up the methods, the full mammogram is cropped to the nipple position in the anteroposterior direction as shown in Figure 4.3 (4).

4.2.2 Multi-Resolution Approach

After the pre-processing step, two cascaded stages of image similarity metrics are used. The first stage is responsible for finding a ROI which is used to speed up subsequent processing steps. Afterward, different image similarity metrics are applied to get the translational transformation parameters.

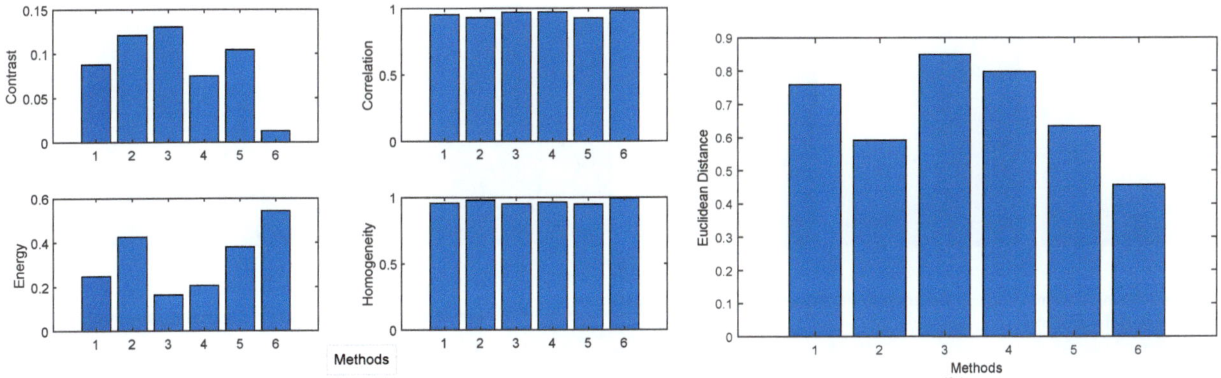

(a) (b)

Figure 4.7: Four properties have been extracted from GLCM: Contrast, Energy, Correlation, and Homogeneity. From 1 to 6 indicate the six techniques for enhancement of the image contrast. All six techniques have been evaluated using these four properties (a). Euclidean distance between the ideal values and the measure values for the four of them for each technique (b).

To get the ROI from the full mammogram where the spot mammogram is potentially located, a dynamic step size is applied using a 2D convolution operation as a constraint for image similarity. The convolution operation is measured between small squares of size 50×50 mm from the full mammogram and the spot mammogram. The dynamic step size depends on three factors. First, these squares are obtained from the full mammogram taking into consideration covering the full mammogram iteratively. These squares are selected by calculating the ratio between the X and Y coordinates of the input of full X-ray mammograms. The X coordinate is also split into a specific number of parts in each iteration. Then, based on the ratio and these partitions of X, partitions of Y will be calculated. Second, the maximum values of convolution are extracted based on the histogram of convolution values in all squares for all iterations for the last bin width. In this last bin width which is considered the bin that has the maximum convolution values, the number of areas is counted. Third, after a defined number of iterations combined, the input of the full mammogram is cropped to the region with the maximum convolution values. Based on empirical tests on the available data, three iterations are applied, splitting X coordinate into two, four, and six regions, respectively in each iteration. These regions could vary based on the size of the breast. It is illustrated for example in the first iteration as shown in Figure 4.2 (first row). The X coordinate is split into two regions. But, since it is a small breast with size $172.2 \times 81.2mm^2$, one region only will be processed in the first iteration and similarly for the next two iterations. Based on that, the ratio of X related to Y is 2.8. Hence, Y coordinates are split into five regions using rounding down. These five regions are convolved with the spot mammograms. The areas, that have maximum values of convolution values, are extracted i.e. the bright areas in Figure 4.2 (first row in the iterations).

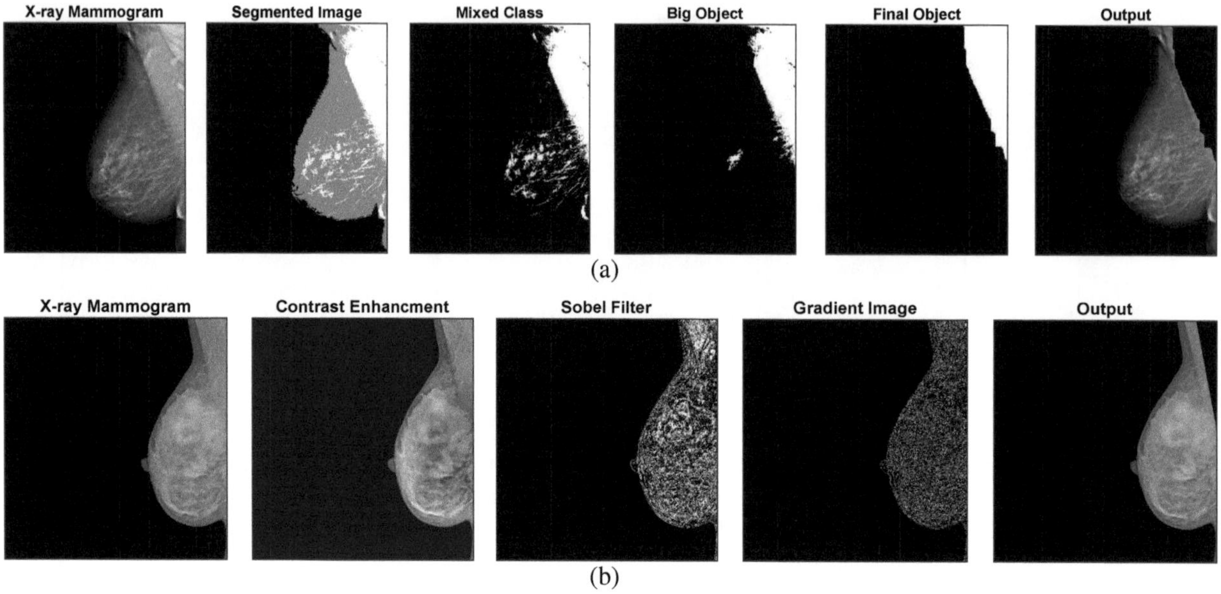

Figure 4.8: Two techniques proposed for removing the muscular tissue from the full X-ray mammograms. The steps of the first technique: original image (first), segmented image using K-means clustering (second), mixed class of glandular and muscular tissues (third), extracting the big object from the mixed class (fourth), apply erosion and dilation operations for getting the final object for the muscular tissue (fifth), and the final result after removing the muscular tissue (sixth) (a). The steps of the second technique: original image (first), applying the proposed contrast enhancement for the full X-ray mammogram (second), sobel filter (third), the gradient image (fourth), and the final result after removing the muscular tissue (fifth) (b).

4.2.3 Image Similarity Metrics

After estimating the ROI from the full mammogram a rescaling factor of 0.1 is applied for computational speed up for both ROI and spot mammogram images. All Image similarity metrics from Section 4.1, have been evaluated in order to select the best of them. For all metrics, the spot mammogram is the moving image that slides across the ROI as a reference image pixel by pixel. At each position the image similarity is calculated between the spot mammogram and the underlying area in the full mammogram (ROI).

4.3 Summary

This chapter first details the methodology of point-based, surface-based, and voxel-based registration, explaining why the latter category was the most convenient solution for the problem in this thesis. This is due to the difficulty in generating structures automatically from a subset of the breast. Then, it illustrates the equation of the image similarity metrics used. The proposed method consists of three steps. First, preprocessing allows for improving contrast and partially compensating the coordinates of full mammogram images. It also increases the probability of aligning the spot mammogram with the full X-ray mammogram by removing pre-knowledge information such as muscle area. For optimization purposes, it removes all the background pixels in the provided full X-ray mammogram. The second step is responsible for determining the ROI from the full mammograms to increase the chances of matching correctly with the spot mammogram. This is done using a multi-resolution approach which covers the full X-ray mammogram iteratively using a convolution operation. Third, different image metrics have been calculated by sliding the spot mammogram across ROI pixel by pixel. The optimum value of these solved metrics is extracted and considered to be the transformation parameter. This method is considered to solve a rigid transformation. Since it is rotating the spot mammogram, it then solves a translational transformation using image similarity metrics.

5 Clinical Data

In this chapter, the characteristics of the available clinical datasets, which have been used for evaluation from the Medical University of Vienna, will be discussed. For data exchange, they have been inserted in a Structured Query Language (SQL) server using a Matlab script, which has been adapted for the modalities used in this thesis. Additionally, the datasets could be visualized using a customized Digital Imaging and Communications in Medicine (DICOM) Viewer, which has been adapted also for the modalities used in this thesis. For evaluation purposes, lesions have been annotated in the three modalities using the free hand tool SLICER 3D (56) and have been saved in NIfTI data format. Thus, it could be processed again in Matlab. The full dataset consists of different types of MRI series, full X-ray mammograms, and spot mammograms. For MRI series, the series that has been used in the annotation, may be different than the series used in registration. The methods of segmentation are limited to T2-weighted series. Around 30% of the available clinical datasets (15 datasets out of 51) did not contain T2-weighted series. Hence, the method was adapted to be applied also for T1-weighted images. For annotation, either a subtraction series or a dynamic contrast enhanced is used. The subtraction series is easier to interpret the annotations by comparing them to the original sequences (194) and therefore, radiologists often annotate in this type of series. For full X-ray mammograms, there are two common views (182) CC and MLO. In some clinical datasets, those two views do not exist but rather the supplementary views such as ML or RL. For spot mammograms, the most important image is the scout image. The amount of clinical datasets received is 51 datasets. Although this available clinical dataset is considered a relatively small dataset, there are a lot of variances regarding:

- Patient age
- Combination of views for the two methods
- Resolutions
- Different machines
- Breast needle biopsies classified into categories according to histopathological appearances (10)
- Breast density
- Size of lesions
- Location of lesions

These variances were split into two categories. The first category is the variability which is read from metadata of each patient specifically such as patient ages, views, resolutions, and machines. The second category is the variability which is calculated based on the proposed methods in this thesis or information received from the expert radiologists. This category consists of breast needle biopsies according to histopathological appearances, breast density, size, and location of lesions.

5.1 Characteristics of Clinical Data from Metadata

5.1.1 Patient Age

The minimum patient age of the clinical dataset is 33, while the maximum age is 78 years. The mean and median ages are 55 and 56, respectively with a standard deviation of 11 years.

5.1.2 Combination of Views

The clinical datasets have differences regarding imaging situations and views. Those differences are important for the image based registration method between full X-ray and spot mammograms. Full X-ray mammograms are always in the upright standing position. There are different views such as CC, MLO, and also supplementary views such as ML and RL. Spot mammograms vary also in the imaging situation not only in the views. The imaging situation could be prone or upright position. The views could be CC, ML, and LM.

The biopsy of most of the available clinical datasets is done in the prone position. 51 datasets are split into 3 upright and 48 prone positions. For the three cases in the upright position, there is also a difference in the views of the X-ray mammogram and spot mammogram (see more details in Table 5.1). Additionally, one of these three cases does not have a T2-weighted MRI. Hence, an available T1-weighted MRI was used instead. For the 48 cases acquired in prone position, first, there are two standard views in the full X-ray mammograms: CC and MLO (the distribution is shown in Figures 5.1, 5.2, and 5.3). For the cases of full X-ray mammograms in CC view, six cases out of the 19 cases do not have a T2-weighted MRI series. While seven cases out of the 25 cases for MLO cases of full X-ray mammogram do not have T2-weighted MRI series. This leads to use T1-weighted MRI series instead. The most common combination in the available clinical datasets is CC for full X-ray mammograms and CC for spot mammograms (16 patients), MLO for full X-ray mammograms and ML for spot mammograms (8 patients), and MLO for full X-ray mammograms and LM for spot mammograms (14 patients).

Additionally, some cases have only the supplementary views such as ML and RL (190) while for the spot mammograms, there are three views: CC, ML, LM (the distribution for these cases are shown in Table 5.2. Several reasons might cause different views (82): small or thin breasts, superficial lesions close to the skin, deep lesions close to the chest wall, and lesions in the very posterior part (inner or outer quadrant). One of the four cases from this combination of views does not have T2-weighted MRI series.

5.1.3 Resolutions

For the biomechanical model based registration method, registration is done between T2-weighted MRI and full X-ray mammograms. In case a T2-weighted MRI is not available, registration is done between a Fast Low Angle Shot 3D (FL3D) (139) or T1-weighted MRI series and full X-ray mammograms. For the T2-weighted series, the minimum resolution in the x and y direction is 0.5 mm/pixel, while the maximum resolution is 0.88 mm/pixel. The minimum spacing between slices for these cases is $3mm$, while the maximum is $5mm$. The minimum resolution of full X-ray and spot mammograms for the same cases is 0.07 mm/pixel and 0.048 mm/pixel, respectively, while the maximum resolutions are 0.1 mm/pixel and 0.085 mm/pixel, respectively. For the FL3D-weighted series, the minimum resolution in the x and y direction is 0.66 mm/pixel, while the maximum resolution is 0.937 mm/pixel. The minimum spacing between slices for these cases is $0.8mm$, while the maximum is $3mm$. The resolution of full X-ray and spot mammograms for the same cases is the same 0.085 mm/pixel and 0.0488 mm/pixel, respectively. For T1-weighted series, the minimum resolution in the x and y direction is 0.53 mm/pixel, while the maximum resolution is 0.786 mm/pixel. The minimum spacing between slices for these cases is $0.9mm$, while the maximum is $2mm$ (See more details in Appendix B.4). For the image based registration method, see distribution of resolutions in detail in Table 5.3.

5.1.4 Devices

There is also a difference in the available devices for MRI and full X-ray mammograms. For MR images, there are two vendors: Siemens and Philips. For Siemens, there are fourteen different models available in the clinical dataset received from the Medical University of Vienna. The most common model is the Siemens TrioTim model with thirteen patients as shown in Figure 5.4 (right). For Philips, there are two different models available as shown in Figure 5.4 (left). For full X-ray mammograms, there are seven machines: Philips, Sectra Imtec AB, FUJIFILM Corporation, GE Medical Systems, Hologic, Inc., AGFA, and Siemens as shown in Figure 5.5 (left). The most common machine for full X-ray mammograms is Siemens with two different models: Mammomat Inspiration (34 patients) and Mammomat Revelation (1 patient) as shown in Figure 5.5 (right). For spot mammograms, the machine is Siemens whether it is in the upright or prone position.

Table 5.1: Distribution of combination views of the three patients available from the Medical University of Vienna in case of upright position regarding imaging situation for the X-ray guided biopsy

Modality / view	**X-ray**	**Spot mammogram**	*Number of Patients*
right breast	MLO	LM	1
left breast	CC	LM	1
right breast	CC	ML	1

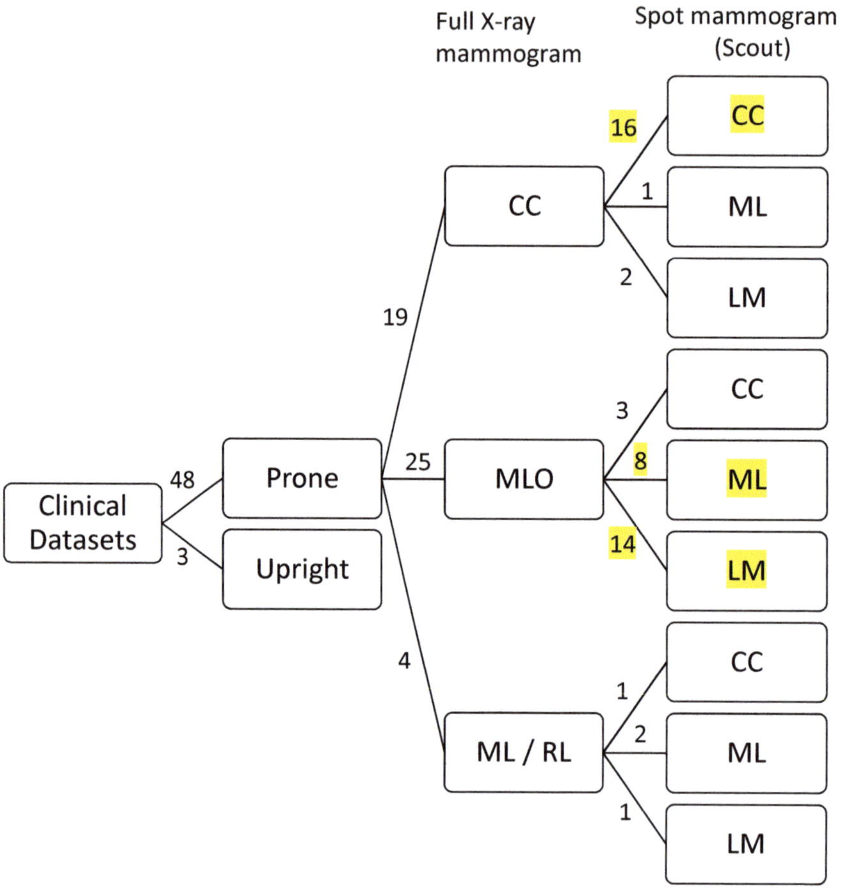

Figure 5.1: Distribution of the combination of views of 51 patients from the Medical University of Vienna specifically for image based registration method. Yellow-marked cases are the most dominant available clinical datasets regarding the combination of views.

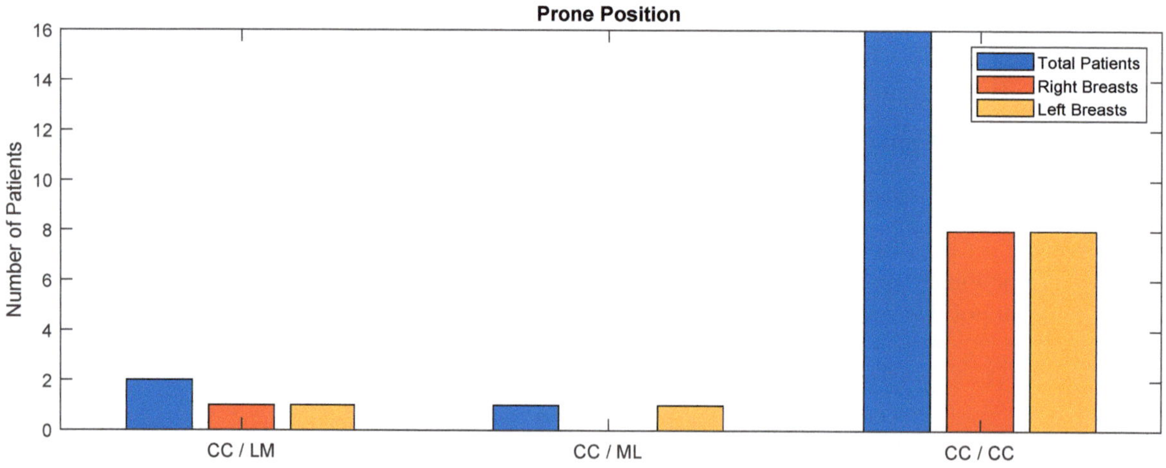

Figure 5.2: Distribution of left and right breasts for the prone position in case of the full X-ray mammogram is in CC view and three views: LM, ML, and CC for the spot mammogram.

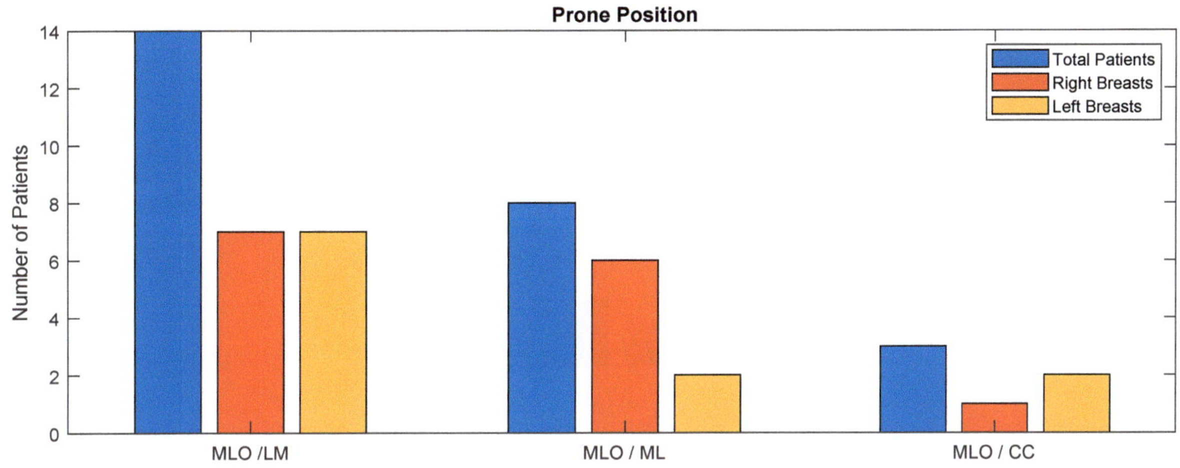

Figure 5.3: Distribution of left and right breasts for the prone position in case of the full X-ray mammogram is in MLO view and three views: LM, ML, and CC for the spot mammogram.

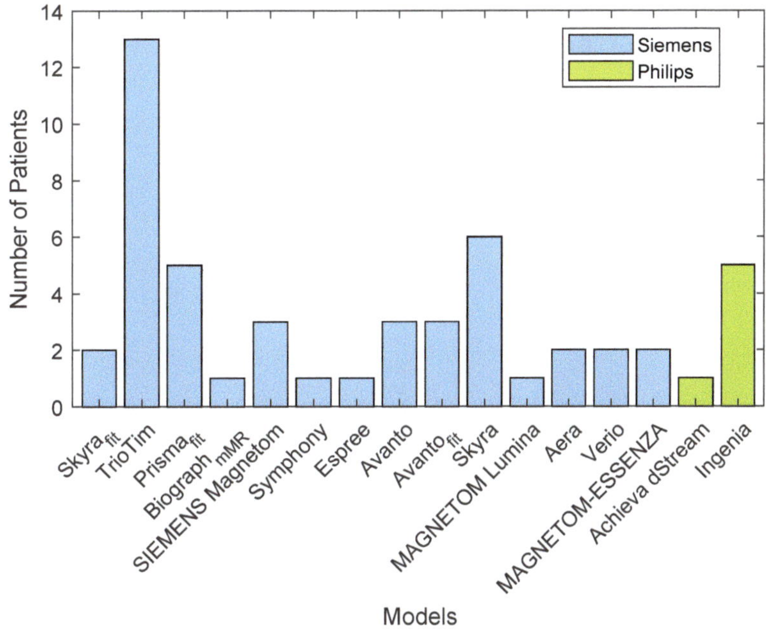

Figure 5.4: Distribution of the models of two machines (Siemens and Philips) available in the clinical datasets for MR images: fourteen models of Siemens and two models of Philips.

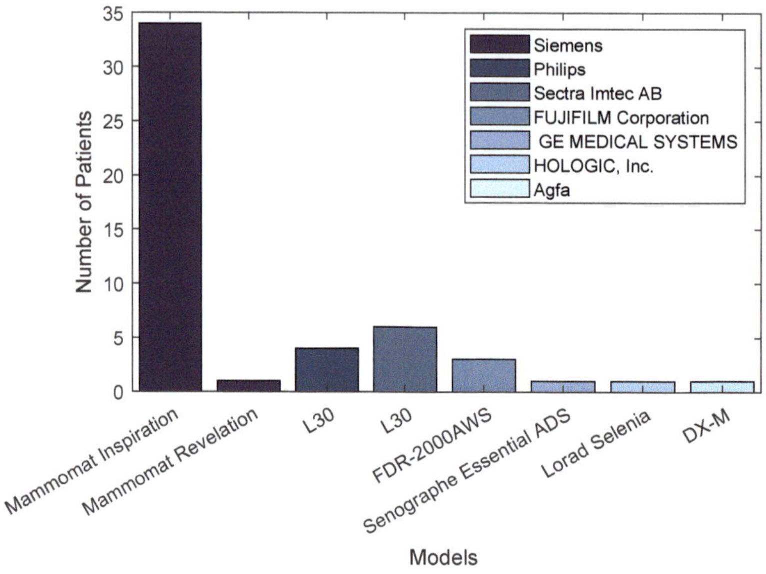

Figure 5.5: Distribution of the machines available in the clinical datasets for full X-ray mammograms. Siemens is the most dominant with two models (35 patients). Six other models from six machines.

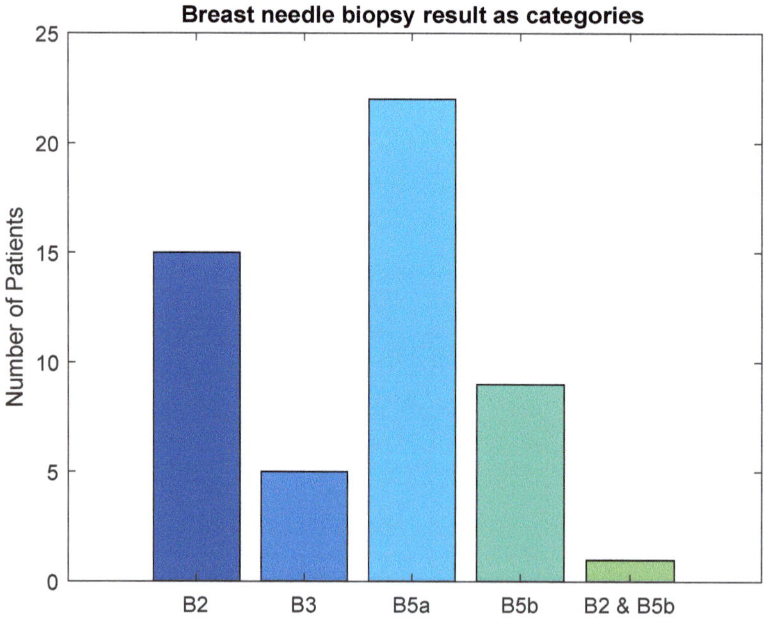

Figure 5.6: Distribution of the 51 patients based on the breast needle biopsy categories according to the Histopathological appearances. The decision for each case is made by the radiologists of the Medical University of Vienna.

Table 5.2: Distribution of combination views of the four patients available from the Medical University of Vienna in the case of prone position (left breasts) regarding imaging situation for the X-ray guided biopsy

Modality / view	**X-ray**	**Spot mammogram**	*Number of Patients*
left breast	ML	ML	1
left breast	ML	LM	1
left breast	ML	CC	1
left breast	RL	ML	1

Table 5.3: Image resolution of X-ray and spot mammogram for image based registration

X-ray / Spot mammogram	Number of Patients
0.1 / 0.0488	1
0.085 / 0.085	2
0.085 / 0.0488	33
0.07 / 0.0488	1
0.05 / 0.0488	13
0.05 / 0.085	1

5.2 Characteristics of Clinical Data Calculated or Informed from Radiologists

The variance regarding the second category will be discussed in this section. Some characteristics are taken from the decision of the expert radiologists that have been used for further analysis such as breast needle biopsies based on histology. Other characteristics such as Percentage Density (PD) were calculated based on the proposed segmentation methods in this thesis. In addition, some characteristics are considered based on both since the lesions have been annotated by radiologists while the size of the annotation and the location of lesions are automatically calculated in this thesis.

5.2.1 Histopathological Diagnosis

Histopathological appearances, as described by radiologists, are based on both pathological and radiological information (10). There are mainly five categories from B1-B5 as described in Table 5.4 (142; 10). These five primary categories are split based on the information about the lesion and its radiological appearance, the localization method, the specimen type, the presence or absence of histology calcification, and the pathologist's judgment (142).

B1 category: is when the pathologist only finds normal breast tissue in biopsy samples and detects no calcium on a specimen (10).

B2 category: This is when the pathologist can safely detect benign changes linked to visible calcification that have been sufficiently sampled. The patient can be confidently released without the need for additional interventions. It is the most frequent outcome of a stereotactic biopsy. Around (66% - 75%) of biopsies, whether employing CB or VAB, is obtained. The breast radiologist's main concern when given a B1 or B2 biopsy result is whether cancer might still be present. Depending on the type of needle used, a false negative result is more or less likely to happen (10).

B3 and B4 categories: these categories mostly include lesions that, upon needle biopsy, may reveal benign histology but are known to exhibit heterogeneity or carry a higher risk of concurrent malignancy. In comparison to B4 (66%), the B3 group has a reduced risk of malignancy on further surgical biopsy (25%). Although surgical excision is required for the majority of B3 lesions, according to Perry et al. (10), a multidisciplinary meeting should be conducted to discuss each case.

B5 category: there are two subtypes available: malignant in situ (B5a) and malignant invasive (B5b). Similar to B3 and B4 lesions, it is possible to underestimate the severity of the disease when a B5a (in situ malignancy) is upgraded to invasive malignancy after surgery. The type of needle that is utilized again affects the underestimation rate (10).

The distribution of the category of needle biopsies of these datasets is shown in Figure 5.6. There are mainly three dominant categories B2, B5a, and B5b which have 15, 22, and 9 patients, respectively.

Table 5.4: Categories of breast needle biopsies regarding histopathological appearances

Category	Description
B1	normal tissue
B2	benign lesion
B3	lesion of uncertain malignant potential
B4	suspicious of malignancy
B5a	malignant in situ
B5b	malignant invasive

5.2.2 Breast Density

There are several techniques for evaluating the density of breasts either manually or automatically. For manual evaluation, it is done using visualization, which requires a lot of time. Automatic methods require accurate commercial tools and to have access to them is too expensive. There are also open source tools such as Laboratory for Individualized Breast Radiodensity Assessment (LIBRA) (92) that work on 2D X-ray mammograms. Unfortunately, it did not provide meaningful results for the available clinical datasets. Based on the literature, there is a survey for measuring breast density using MRI (172). The most common algorithm for measuring the breast density on MRI volume is FCM (172). In addition, Liu et al, (104) suggested calculating PD from reconstructed MRI using Generative Adversarial Networks (GANs) methods. Apart from that, one algorithm was proposed for measuring the breast density based on MR images directly, since segmentation algorithms take a lot of time to generate an automatic segmentation and then auto-correct the inaccurate parts which is done by radiologists (198).

The density of breasts has been analyzed in this thesis based on the proposed segmentation method. Two aspects were calculated: first, the amount of glandular tissue compared to the fatty tissue, and second, the amount of glandular tissue compared to the breast tissue. As shown in Figure 5.7, most of the available datasets (26 patients) lie in the percentage between 0 and 20% regarding the ratio between glandular and fatty tissues (left). In the next three classes from 20-40%, 40-60%, and 60-80%, the number of patients

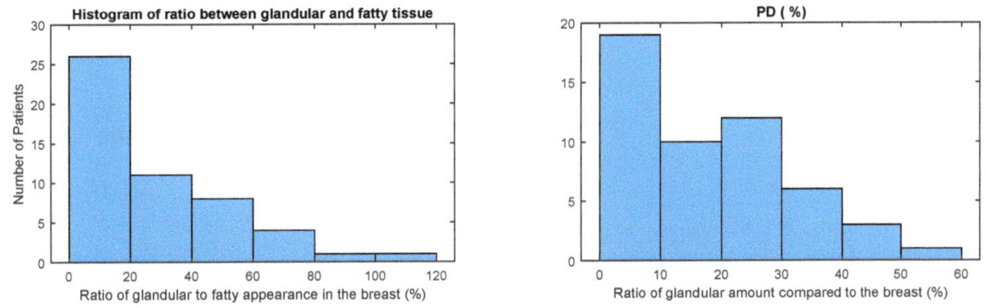

Figure 5.7: Histogram of the clinical datatsets regarding two aspects: ratio of glandular and fatty tissues in percentages (left). The ratio of glandular tissue compared to breast tissue in percentages (Percentage Density-(PD)) (right).

decreases to 11, 8, and 4, respectively. Regarding the amount of glandular tissue compared to the breast tissue, the categories have been split like literature (91) with a step size of 10% between each class. 19 patients lie in the class from 0-10% and 12 patients lie in the class from 20-30%.

5.2.3 Sizes of Lesions

The size of lesions in the three modalities, which have been annotated by radiologists, was calculated. One annotated lesion is considered 3D in MRI while the other two annotated lesions are considered 2D in full X-rays and spot mammograms. Lesion sizes might not correlate due to the errors made by manual annotations. Hence, breast sizes regarding 3D and 2D have been calculated first to validate the correlation between the 3D and 2D images in general, as shown in Figure 5.8. There is a strong correlation between the sizes of breasts in full X-ray (CC or MLO) compared to the sizes of breasts in MRI with an r of 0.91 and 0.82, respectively. The minimum breast size available in 2D (CC and MLO), 3D (MRI) is $5.6x10^3\ mm^2$, $9.8x10^3\ mm^2$, and $1.7x10^5\ mm^3$ while the maximum size is $4.6x10^4\ mm^2$, $3.4x10^4\ mm^2$, and $1.8x10^6\ mm^3$, respectively. A comparison between the size of annotations is done secondly in which a moderate correlation between the area annotation of full X-ray and volume annotation of MRI with an r of 0.64 and 0.56 for CC and MLO views, respectively is found. There is a strong correlation found between the annotation of full X-ray and spot mammograms with an r of 0.83 and 0.76 for CC and MLO views, respectively.

5.2.4 Location of Lesions

The localization of lesions has been analyzed by splitting them into defined regions. For MRI, the breast is split into 12 regions which follow the standard quadrants of the breast (217). Slices of MRI volume, in which the breast shape (field of view (FOV)) is located, have been split first into three regions based on inferior, middle, and superior regions. Then, each slice is split into four regions: upper inner, lower inner, upper outer, and lower outer as shown in Table 5.5. An example of a middle slice in a left breast is shown in Figure 5.11 (left): upper inner (R2), lower inner (R5), upper outer (region 8), and lower outer

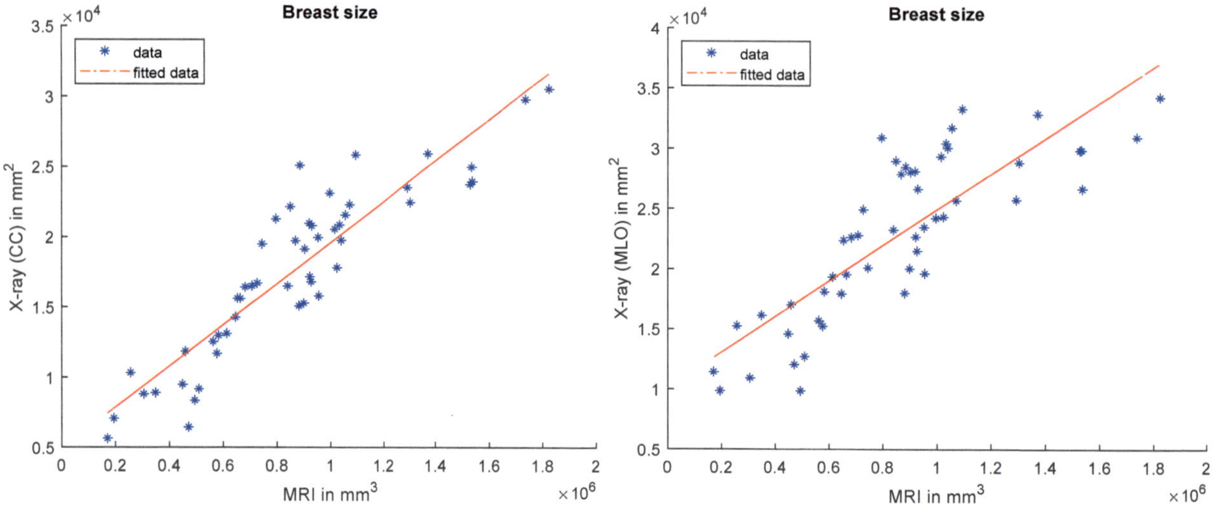

Figure 5.8: Distribution of the breast sizes in full X-ray mammograms (2D): CC and MLO and in MRI (3D). A strong correlation is found between the breast sizes 3D and 2D with an r of 0.91 and 0.82, respectively.

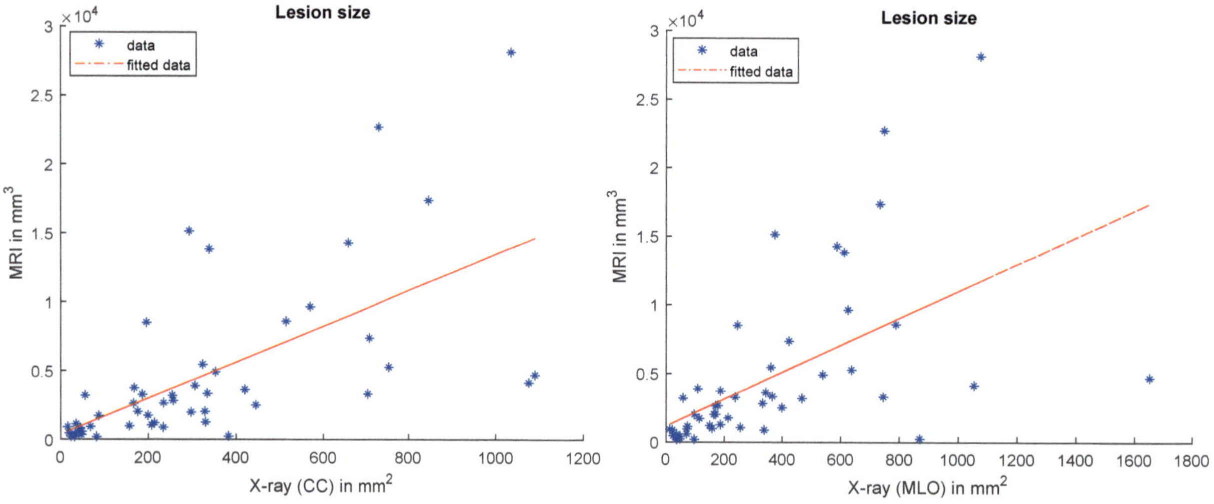

Figure 5.9: Distribution of the lesion sizes annotated by experienced radiologists in full X-ray mammograms (2D): CC and MLO and in MRI (3D). A strong correlation is found between the breast sizes 3D and 2D with an r of 0.64 and 0.56, respectively.

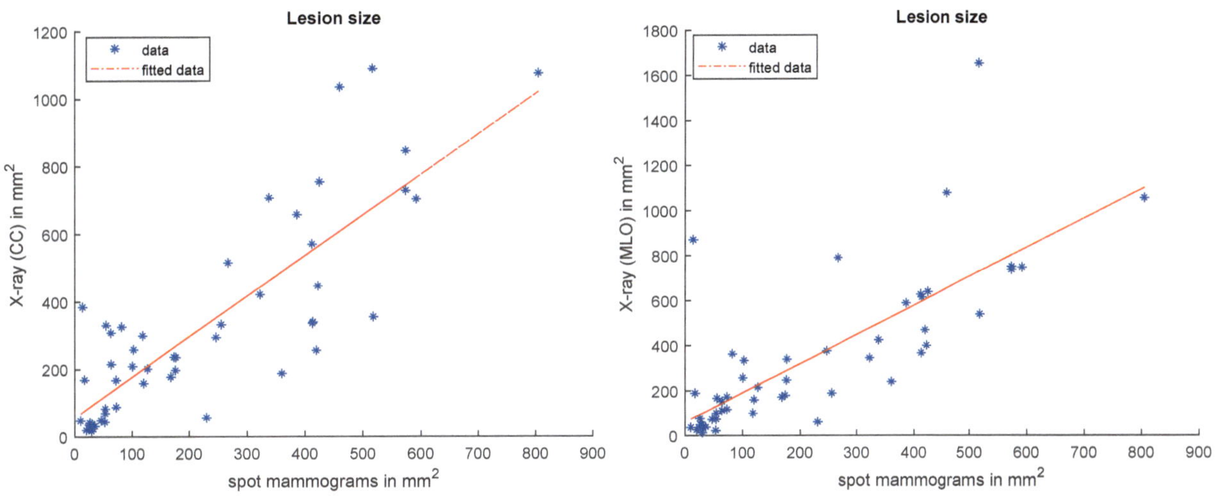

Figure 5.10: Distribution of the lesion sizes annotated by experienced radiologists in full X-ray mammograms (2D): CC and MLO and in spot mammograms (2D). A strong correlation is found between the breast sizes with an r of 0.83 and 0.76, respectively.

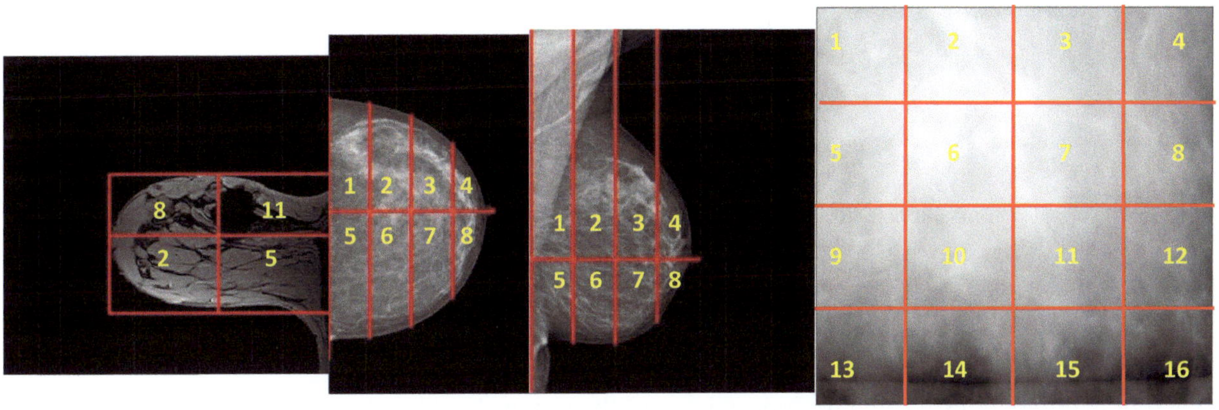

Figure 5.11: Definitions of regions in each modality: MRI: four regions are shown in this example which are R8, R11, R2, and R5 (left). Full X-ray mammogram: all regions available for the two views CC and MLO are shown in this example from R1 till R8 (middle). Spot mammogram: all regions available for spot mammogram are shown in this case from R1 till R16 (right).

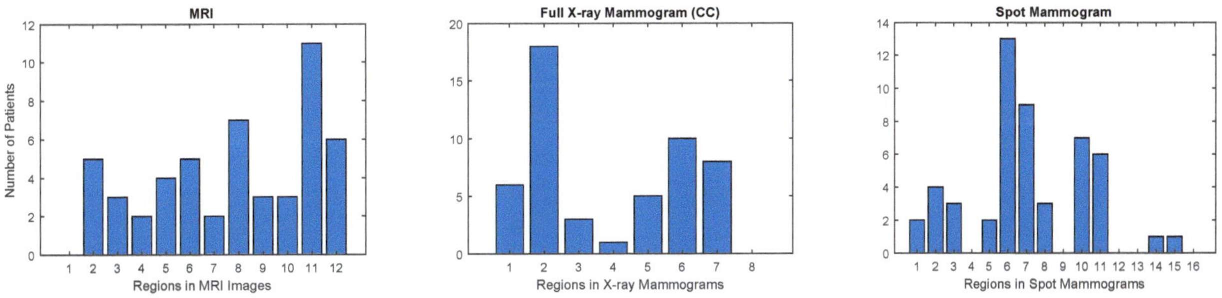

Figure 5.12: Number of patients in each region defined. MRI (12 regions) (left), full X-ray mammograms (8 regions) (middle), and spot mammograms (16 regions) (right).

(R11). For an X-ray mammogram, the breast is split into 8 regions. Two regions towards the muscle in the posterior direction are defined, distinguishing the upper region in the inferior/feet direction (R1) and the lower region in the superior/head direction (R5). Similarly, another two regions towards the nipple (R4 and R8) in the anterior direction are defined. The rest of the regions are in the center part of the breast (R2, R3, R6, and R7) as shown in Figure 5.11 (middle). For the spot mammogram, since it is a square window of 50×50 mm for most of the available cases, 16 regions of small squares are split equally, as shown in Figure 5.11 (left).

Table 5.5: Regions specified for location of lesion in MRI

Regions	Description
R1	upper inner - inferior slices
R2	upper inner - middle slices
R3	upper inner - superior slices
R4	lower inner - inferior slices
R5	lower inner - middle slices
R6	lower inner - superior slices
R7	upper outer - inferior slices
R8	upper outer - middle slices
R9	upper outer - superior slices
R10	lower outer - inferior slices
R11	lower outer - middle slices
R12	lower outer - superior slices

The decision of where each lesion is located is based on these regions. For the cases where lesions exist in more than one region, the decision is made based on the maximum area located. For MRI regions, it is found that 11 patients are located in R11 which is the lower outer-middle slice. For X-ray regions, it was found that 18 patients lie in R2 which is considered the center (first area in the posterior direction). For spot mammograms, ideally for biopsy images, the lesion should be located in the center area which in this case: R6, R7, R10, and R11. It was found that 13, 9, 7, and 6 patients' lesions were located in these regions, respectively.

6 Evaluation Methods

This chapter discusses the generation of ground truth for segmentation and registration methods. Then, quality metrics, which were used for the assessment of the proposed method, are introduced.

6.1 Data Processing

As mentioned in Chapter 5, the data exchange was done via cloud storage in DICOM data format. Having a ground truth is a mandatory step to evaluate the proposed methods. Five proposed methods need to be evaluated: the segmentation algorithm of MR images for three tissues, the biomechanical model based registration solved by FEM between MRI and X-ray mammography, the proposed machine and deep learning algorithms for replacing FEM simulation of the biomechanical model, the image based registration between X-ray mammograms and spot mammograms, and an evaluation of the full proposed processing chain to register MRI and spot mammograms, which is considered the accumulative registration of the two methods.

6.1.1 Generating Ground Truth for Breast MRI Segmentation

For the evaluation of segmentation algorithms, annotation of images of MRI is required. The ground truth was first created by using niftySeg (34) for classifying glandular and fatty tissues. NiftySeg is based on an expectation maximization algorithm. NiftySeg is an independent and open-source tool for pre-segmentation. There were some issues with this open-source tool when used to classify glandular and fatty tissues as shown in Figure 6.1. Also, there was no detection for muscular tissues like the proposed method in this thesis. Therefore, those issues were corrected manually by a medical practitioner using Medical Imaging Interaction Toolkit (MITK) (131) and the breast muscle was segmented manually in each slice.
As was expected, it would take expert radiologists around one to two hours for one dataset to generate ground truth for evaluating the segmentation methods. That is why eight clinical datasets have been annotated for evaluation purposes. Another set of 25 datasets has been evaluated qualitatively, which will be discussed in Section 6.3.1.

6.1.2 Generating Ground Truth for Image Based Registration

For the evaluation of registration algorithms, annotation of landmarks visible in all three modalities is required. In this thesis, it was decided by the medical partners that microcalcifications can fulfill this

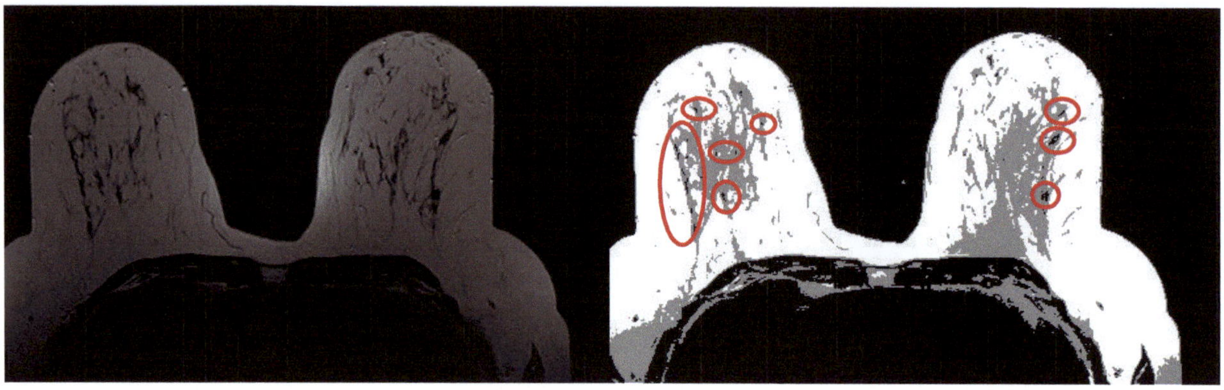

Figure 6.1: Visualization of an example of classification of fatty and glandular tissues using NiftySeg (transverse view). Original image (left). The segmented image (right). Red circles indicate the holes or problems not classified by niftySeg. Additionally, the muscle is not detected.

purpose. There are two options to annotate microcalcifications: deterministic shapes (with sharp edges like squares or circles) or free-hand shapes. Some disadvantages have arisen from having sharp edges, like squares or cubes because microcalcifications can be out of the annotated square, and if bigger squares are marked instead, they are decisive and misleading. The solution for that was annotating a freehand shape based on the exact circumference of a cluster of microcalcifications. To have a free-hand shape, annotations have been performed by an experienced radiologist using Slicer 3D (56) as shown in Figure 6.2.

6.2 Evaluation Metrics

Four main evaluation methods have been used for evaluating the proposed methods. These are True Positive Rate (TPR), DSC, Jaccard Index, and TRE. The equations will be presented first, and then the usage of these evaluation methods for each proposed algorithm will be discussed.

6.2.1 True Positive Rate

The TPR is a metric that is also known as sensitivity or recall between two sets of elements A and B. It calculates the percentage of the actual positives that are correctly detected, as shown in equation 6.1 (125):

$$TPR = \frac{TP}{TP+FN} = \frac{|A \cap B|}{A} \tag{6.1}$$

6.2.2 DICE Coefficient

DSC or F-score measures the similarity between two sets of elements A and B, or pixels: positive and negative. In other words, it is the average of precision and recall, as shown in equation 6.2 and in Figure 6.3 (a). Both of them include true positives (TP), false positives (FP), and false negatives (FN).

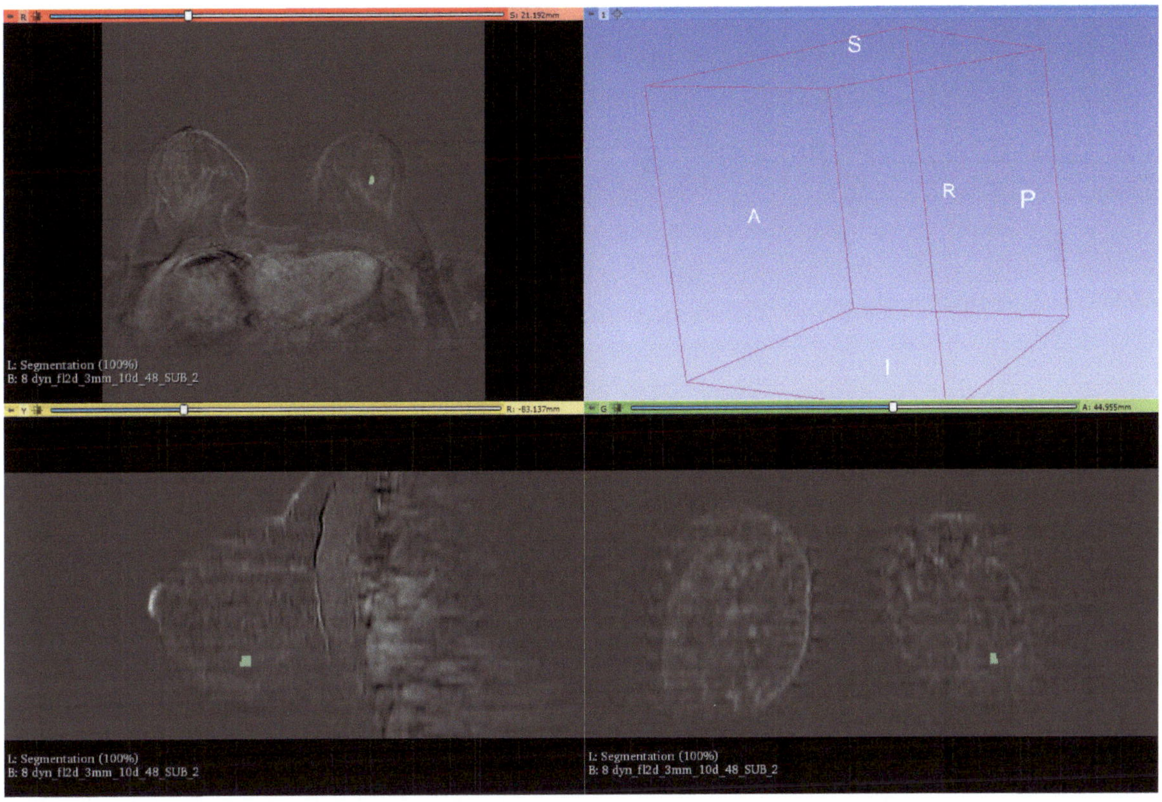

Figure 6.2: Visualization of an example of annotation done by radiologists using Slicer 3D in a subtraction MRI series. The green area is the annotation of radiologists.

The precision (denoted as prec) is the ratio of true positives to all the predicted positives (218). It is calculated by

$$prec = \frac{TP}{TP+FP}$$

The recall or TPR (denoted as rec) measures the percentage of all detected positives that were correctly predicted as explained in 6.2.1 and shown in equation 6.1:

$$DSC = \frac{2TP}{2TP+FP+FN} = \frac{2 \cdot rec \cdot prec}{rec+prec} = \frac{2|A \cap B|}{|A| \cup |B|} \tag{6.2}$$

6.2.3 Jaccard Index

Jaccard Index or Intersection over Union (IoU) metric measures the similarity between two sets A and B. It is a percentage that ranges from 0% to 100% and indicates how comparable two sets of data are, as shown in equation 6.3 and in Figure 6.3 (b) (199):

$$IoU = \frac{TP}{TP+FP+FN} = \frac{|A \cap B|}{|A| \cup |B|} \tag{6.3}$$

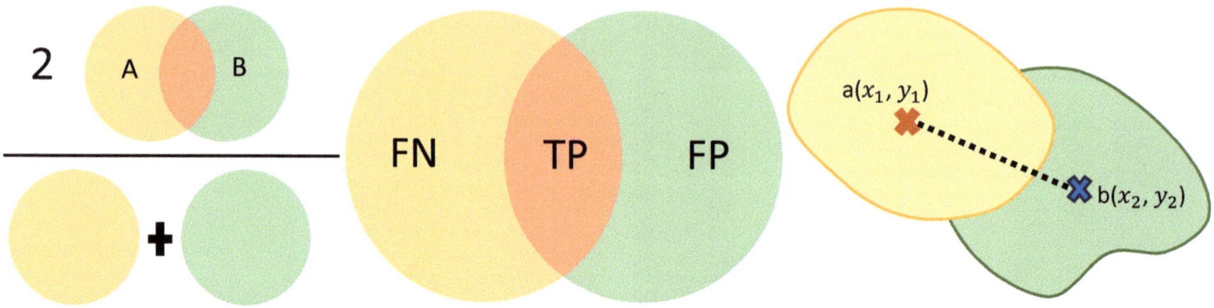

(a) (b) (c)

Figure 6.3: Visualization of the four evaluation methods that have been used in this thesis: the Venn diagram of DSC (a), the Venn diagram of Jaccard Index and TPR (b), and the visualization of Euclidean distance (c). Yellow is the ground truth annotation which is A, green is the predicted annotation which is B, and red is the intersection or overlap between the two annotations.

The difference between DSC and IoU is that DSC emphasizes on true positives rather than only the overlap between sets as declared in the equations.

6.2.4 Target Registration Error

The TRE is the Euclidean distance (d) between two points a and b as shown in Figure 6.3 (c), where a is the actual point and b is the predicted point (58):

$$d = \sqrt{(b_{x_2} - a_{x_1})^2 + (b_{y_2} - a_{y_1})^2} \tag{6.4}$$

6.3 Evaluation Methods

6.3.1 Biomechanical Model Based Registration Method

The biomechanical registration method has been evaluated using TRE which is the Euclidean distance between the center of gravity of the predicted lesion in the registered moving image (synthetic X-ray) generated from the MRI and the center of gravity of the annotated lesion by radiologists in the reference image (original X-ray) as shown in equation 6.5:

$$TRE_{Biomechanical} = ||C_{XR} - C_{MRI}|| \tag{6.5}$$

Moreover, the IoU and DSC have been calculated to evaluate the overlapping between lesion annotations. This was done for further evaluations for clinical use, which will be discussed in Chapter 8.
In addition, TPR has been calculated to avoid the problem of indeterministic shapes. Areas of the annotations in both images might not correlate due to the error in manual annotations.
Also, estimation of DSC is calculated between the similarities of the breast shapes of the two images (synthetic X-ray and original X-ray).

Segmentation

The segmentation methods have been evaluated using DSC and IoU. For this case, the precision is a score that represents how much of e.g., fatty tissue is identified in the segmentation compared to fatty tissue in the ground truth. Similarly, the precision of glandular and muscular tissues was calculated. The recall indicates how much of e.g. fatty tissue in the ground truth was detected by the segmentation. Similarly, the recall of glandular and muscular tissues was calculated.
IoU metric measures the number of pixels common between the segmented image and ground truth image divided by the total number of pixels present across both images for each tissue separately.

Machine and Deep Learning Algorithms for Simulation of Biomechanical Model Deformation

In order to evaluate the performance of the proposed machine and deep learning models, the deformed nodes from the FEM simulation are considered as the ground truth. By that, the RMSE could be calculated for each node between the proposed models as explained in Section 3.2.5 and the ground truth, as shown in equation 6.6. In this equation, ε represents the deviation of a node's position from the ground truth in the x, y, and z directions. Afterward, the mean value of all nodes belonging to one dataset is calculated, where one dataset refers to one breast of one patient. Finally, the mean and median RMSE of all the datasets are calculated in the validation set (see Section 7).

$$RMSE_{node} = \sqrt{\frac{\varepsilon_x^2 + \varepsilon_y^2 + \varepsilon_z^2}{3}} \tag{6.6}$$

6.3.2 Image Based Registration Method

Annotated lesions in the two modalities of full X-ray and spot mammograms were used as a benchmark to assess the accuracy of image based registration algorithms. TRE has been calculated as the Euclidean distance between the center of gravity of the lesion predicted in the registered moving image (spot mammogram) and the center of gravity of the annotated lesion in the reference image (full X-ray mammogram) to evaluate the image based registration method as shown in equation 6.7:

$$TRE_{Imagebased} = ||C_{CR} - C_{XR}|| \tag{6.7}$$

Moreover, the IoU and DSC have been calculated to evaluate the overlapping between lesion annotations. TPR was calculated similarly to the evaluation of the biomechanical model based registration, as explained in Section 6.3.1,

6.3.3 Accumulative Registration of the Two Methods

The same evaluation methods as explained in Section 6.3.2 are used here in calculating the accumulative registration of the two methods. The TRE as shown in equation 6.8, DSC, IoU, and TPR of both registration steps in combination is calculated based on the predicted lesion position originating from the MRI annotation from the first step. This is given as input for the second step (image based registration) to predict the position in the spot mammogram image.

$$TRE_{Total} = ||C_{CR} - C_{CRpredicted}|| \tag{6.8}$$

7 Results

After the two proposed registration methods have been developed, this chapter presents the results obtained from all the proposed methods:

- segmentation of MR images for three tissue types: fatty, glandular, and muscular
- proposed machine and deep learning algorithms for simulations of the biomechanical model
- biomechanical model based registration between MRI and X-ray mammography
- image based registration between full X-ray mammography and spot mammograms
- accumulative registration of the two methods between MRI and spot mammograms

7.1 Segmentation

In this section, the results from the proposed segmentation methods' will be presented qualitatively and quantitatively for T2-weighted MRI. Additionally, three example cases of segmentation for T1-weighted are shown in Figure 7.3. This demonstrates that the method works also for the T1-weighted MRI series.

7.1.1 Qualitative Analysis

The analysis has been evaluated qualitatively for the first approach (segmentation of 3D volume denoted as 3DNNs) by two observers. It has been done using 25 datasets that were available at that time for the T2-weighted MRI series. Each dataset's slices were equally divided into three categories: inferior, middle, and superior slices. Within each category, seven possible issues related to the segmentation of tissues were examined. Three of them are connected to the segmentation of the muscular tissue: distinguishing glandular from muscular tissue, holes in the muscle, and artifacts at the border between glandular and muscular tissue. Two of them are connected to the segmentation of the glandular tissue: noisy classification and the possibility that the breast's borders could be mistakenly identified as glandular tissue due to the partial volume effect. The incorrect overall contour of the breast and missing parts of the breast tissue are the two main problems that were taken into account regarding the segmentation of fatty tissue, as shown in Figure 7.1. For each issue, a score between 1 and 5, in which 1 represents the worst score and 5 represents the best score, was calculated by taking the average of the observations from two people. A maximum score of 35 and a minimum score of 7 were obtained by adding the scores for each of the seven problems. The overall score for each category was then divided into five groups, as indicated in Table 7.1. It has been observed that middle slices have fewer issues than inferior and superior slices. There

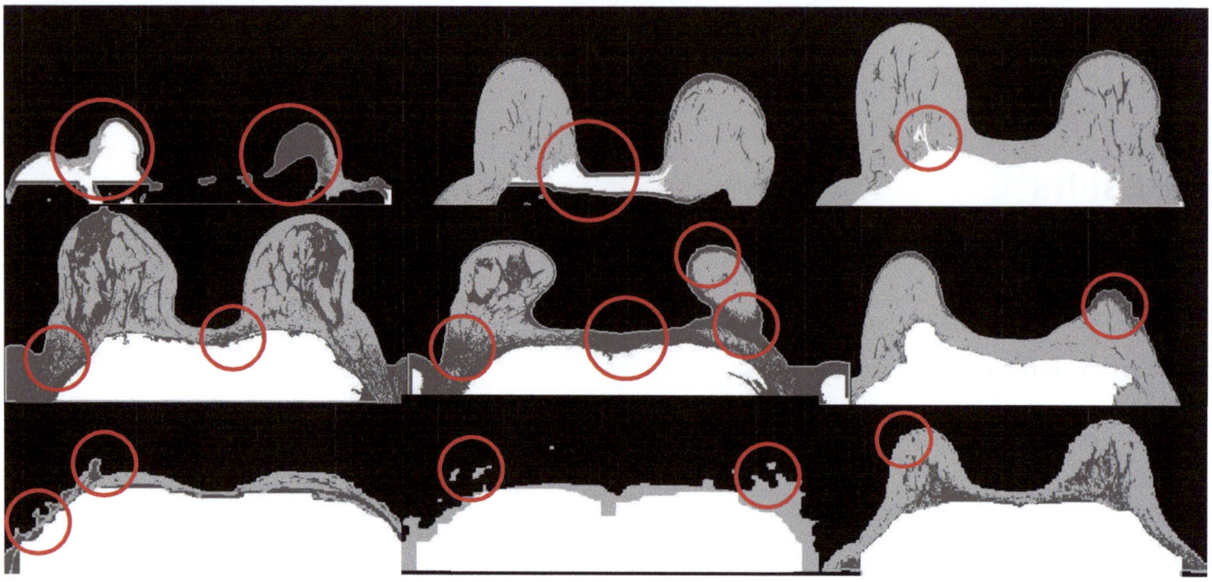

Figure 7.1: Examples of the seven common problems that were qualitatively evaluated; three problems of muscular tissue (first row), two problems of glandular tissue (second row), and two problems of fatty tissue and breast shape (third row).

Table 7.1: Percentage of datasets in the group score of each category of slices

Slices	30-35	25-30	20-25	15-20	7-15
Inferior	56%	32%	-	12%	-
Middle	76%	12%	12%	-	-
Superior	64%	16%	12%	8%	-

are no significant issues in more than 75% of the middle slices (category 30-35). In category 15-20, only 12% and 8% of inferior and superior slices, respectively, have segmentation issues.

Quantitatively, the mean DSC of eight clinical datasets was computed for the three tissues in the inferior, middle, and superior slices as illustrated in Figure 7.2 in order to numerically validate the three categories derived from the qualitative results. It demonstrates that the most reliable category is the middle slices, supporting the assessments of the two observers. The mean DSC achieved is 0.91, 0.64, and 0.93 for fatty, glandular, and muscular tissue, respectively.

7.1.2 Quantitative Analysis

There are three approaches which have been proposed in Section 3.2.1; segmentation of 3D volume (denoted as 3DNNs), segmentation of 2D stacked volume (denoted as 2DNNs), and segmentation of a subset from the volume (denoted as Subset3DNN). The three approaches have been compared to the ground truth of breast MRI, as shown in Section 6.1.1. Eight different datasets of T2-weighted MR images of different resolutions have been evaluated using DSC and Jaccard index, as shown in Section

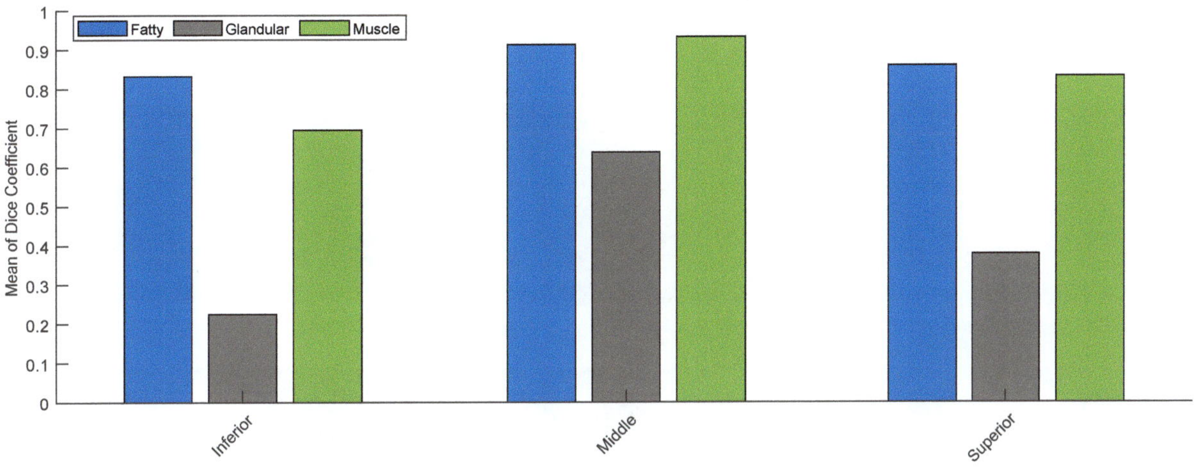

Figure 7.2: Mean DSC of eight clinical datasets for the three tissues: fatty, glandular, and muscular tissues in inferior, middle, and superior slices, respectively.

6.3.1. The proposed approaches ran on an Intel(R) Core i7-9700K CPU with 3.60 GHz in a couple of minutes and on NVidia RTX 2080Ti with a speed-up of 3.

For the first approach (3DNNs), the mean DSC is 0.87 for fatty tissue, 0.45 for glandular tissue, and 0.88 for muscular tissue, as shown in Figure 7.4. The mean of the Jaccard index is 0.78 for fatty tissue, 0.31 for glandular tissue, and 0.79 for muscular tissue, as shown in Figure 7.5. For the second approach (2DNNs), the mean DSC is 0.70 for fatty tissue, 0.18 for glandular tissue, and 0.54 for muscular tissue. The mean of the Jaccard index is 0.56 for fatty tissue, 0.10 for glandular tissue, and 0.40 for muscular tissue. For the third approach (Subset3DNN), the mean DSC for each tissue is 0.82 for fatty tissue, 0.40 for glandular tissue, and 0.85 for muscular tissue. The mean of the Jaccard index is 0.70 for fatty tissue, 0.26 for glandular tissue, and 0.75 for muscular tissue.

It was obvious that mean DSC and Jaccard index in the second approach are low compared to the other two approaches. In the second approach, segmentation was performed slice by slice. Each slice has different labels than the previous slice due to the random initialization of the NN, as shown in Figure 7.6 which makes it hard to have four deterministic labels in all the slices in the postprocessing block. Due to the fact that fatty, glandular, and muscular features could have the same amount of voxels, this makes it hard and leads to a higher probability that some slices could not be assigned to the same class. Therefore, the probability of losing more information is higher in the 2D stacked NN. That is why the values of DSC and Jaccard index in this approach are low compared to the other two approaches.

The mean DSC and Jaccard index of glandular tissue in all the approaches proposed is less than the other two tissues due to the small number of voxels of glandular tissue compared to the total number of voxels. Additionally, segmentation of very fine structures in glandular tissue is challenging while generating the ground truth.

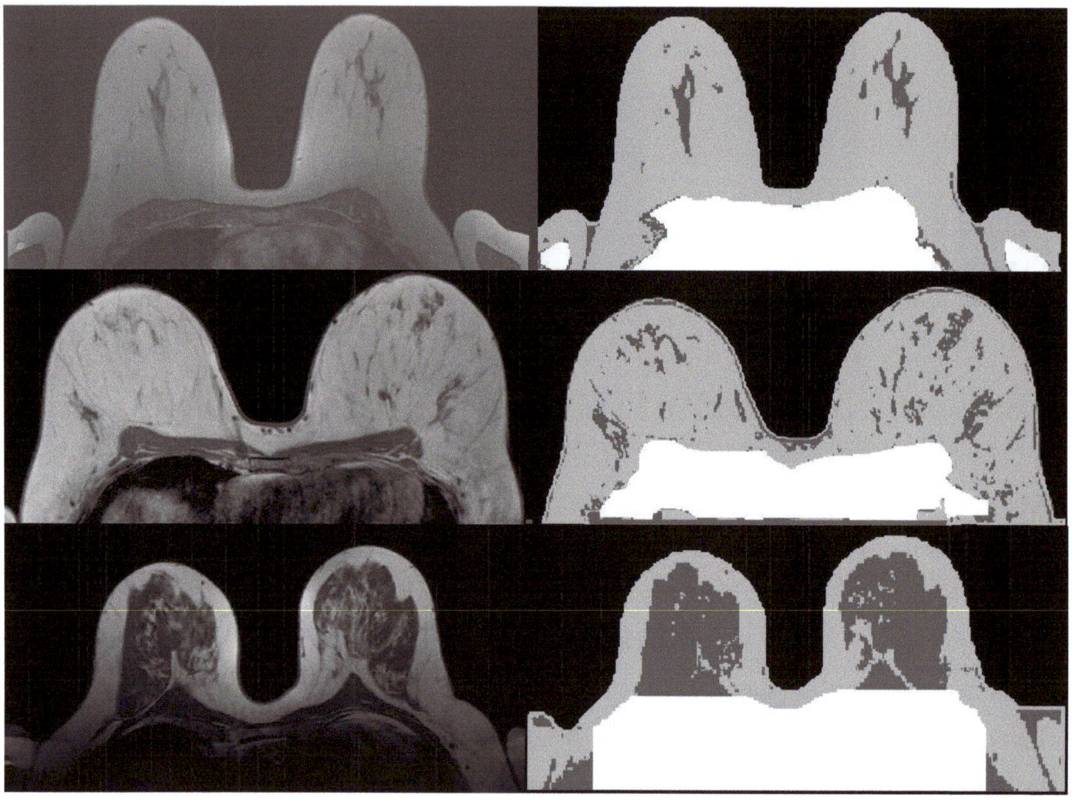

Figure 7.3: Three examples of clinical datasets of T1-weighted MRI series that has been segmented using the proposed methods. Original images (first column). Segmented images (second column).

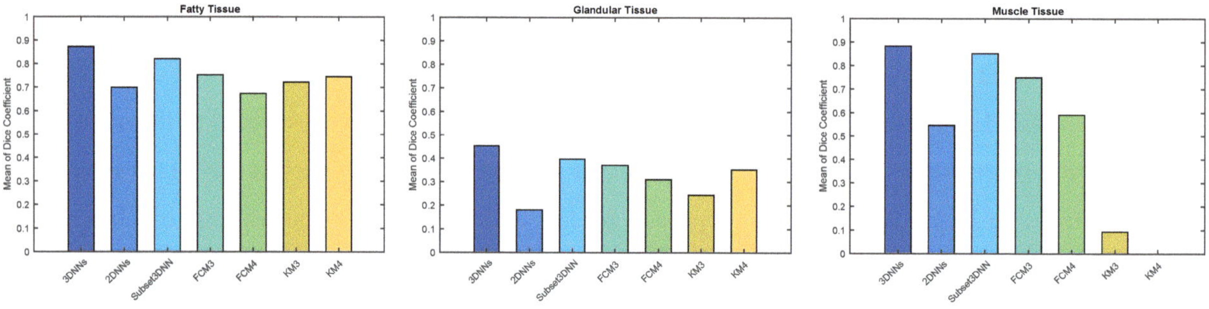

Figure 7.4: Mean DSC of each tissue (fatty, glandular, and muscle) for eight datasets in the proposed algorithm (3DNNs) compared to 2D stacked (2DNNs), a subset from the volume (Subset3DNN), Fuzzy C-means three clusters (FCM3) and four clusters (FCM4) and K-means three clusters (KM3) and four clusters (KM4).

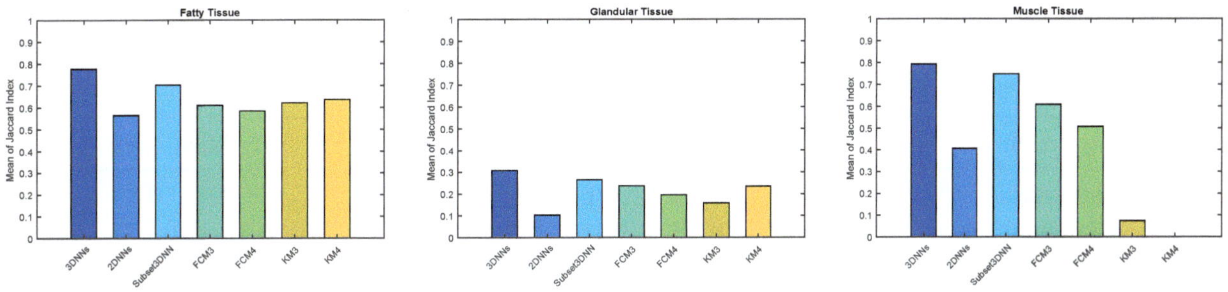

Figure 7.5: Mean Jaccard index of each tissue (fatty, glandular, and muscle) for eight datasets in the proposed algorithm (3DNNs) compared to 2D stacked (2DNNs), a subset from the volume (Subset3DNN), Fuzzy C-means three clusters (FCM3) and four clusters (FCM4) and K-means three clusters (KM3) and four clusters (KM4).

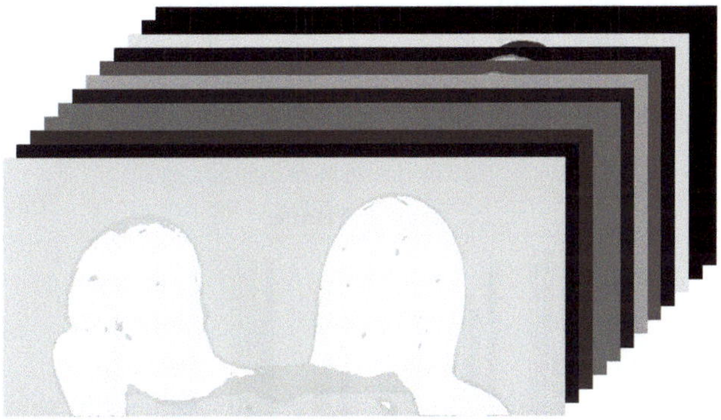

Figure 7.6: An example of one dataset that is segmented by 2D stacked volume. Each slice has different labels than the slice before or after.

It has been realized that the second approach does not detect the muscular tissue compared to the other two proposed approaches as shown in Figures 7.4, 7.5, 7.7, and 7.8. Hence, this approach is considered ineffective.

The third approach can be considered comparable to the first approach in terms of accuracy regarding fatty and muscular tissues. Regarding DSC and Jaccard index of glandular tissues in cases from 3 till 8, there is a huge difference compared to the first approach as shown in Figures 7.7, and 7.8. The DSC of glandular tissue for the first approach in case 3 is 0.36, while the DSC is 0.21 for the third approach. There is a hypothesis that it might depend on the size of the breast which would require different numbers and positions of slices extracted from the volume, i.e. superior, inferior, or middle slices. If the size of the breast is large enough, the middle slice, or in other words, one slice is sufficient for training the approach. But if the size of the breast is too small like for example in Figure 7.9, in the segmentation of the third approach (fourth) compared to the segmentation of the first approach in Figure 7.9 (second), it might be necessary to have more than one slice trained not only the middle slice. This example, which is referred to, is case 3 in Figures 7.7 and 7.8. In the future, the number of slices and different positions in the volume could be increased and changed to improve accuracy instead of using only one slice, or here, the middle slice.

Table 7.2: Accuarcy in terms of DSC of three proposed approaches for segmentation (mean, median) ± standard deviation RMSE (*mm*) in the eight datasets

Approach	3D volume	2D stacked volume	subset from the volume
Fatty tissue	**(0.87, 0.89) ± 0.04**	(0.70,0.79) ± 0.19	(0.82, 0.83) ± 0.07
Glandular tissue	**(0.45, 0.45) ± 0.18**	(0.18,0.14) ± 0.11	(0.40, 0.37) ± 0.2
Muscular tissue	**(0.88, 0.89) ± 0.04**	(0.54, 0.64) ± 0.24	(0.85, 0.87) ± 0.07

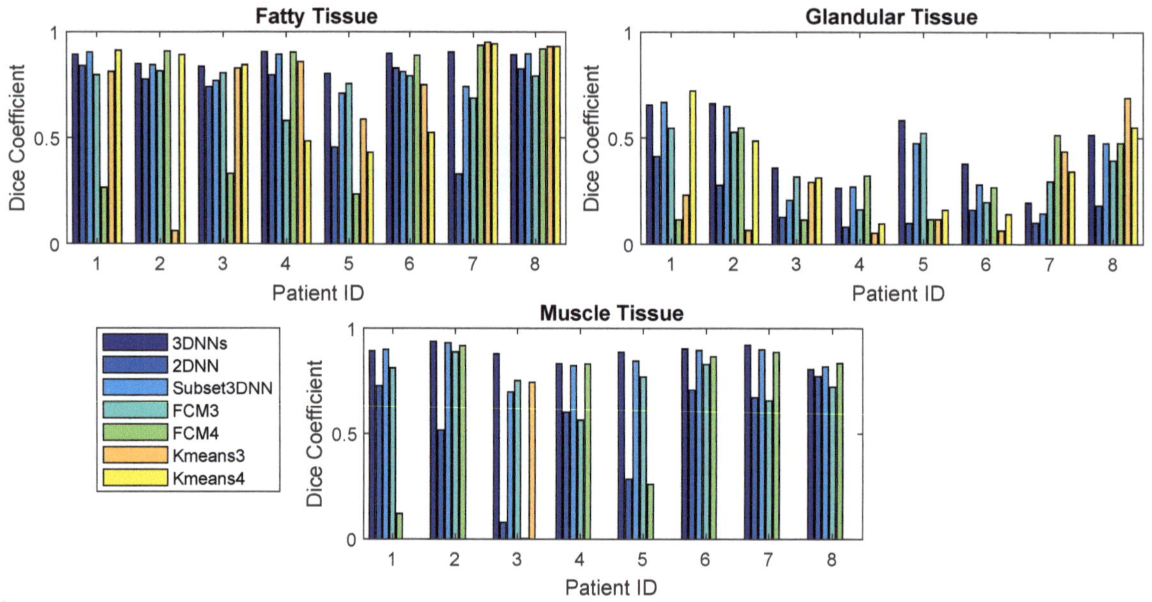

Figure 7.7: Distribution of DSC of each tissue (fatty, glandular, and muscle) across the eight datasets in the proposed algorithm (first approach) (3DNNs) compared to 2D stacked (second approach) (2DNNs), a subset from the volume (third approach) (Subset3DNN), Fuzzy C-means three clusters (FCM3) and four clusters (FCM4) and K-means three clusters (KM3) and four clusters (KM4).

The three approaches have been compared from a computational point of view as shown in Figure 7.10. The computation time varies from one case to another because it depends on the breast size, since in the preprocessing block, all unnecessary background voxels have been removed. The 2D stacked volume not only shows less classification in the tissues of each class but also is slower than the first approach by nearly a factor of 1.3 speedup in CPUs. The third approach has proven to be the fastest algorithm, as it achieves a 45 times speedup compared to the first approach. However, in some cases, the glandular tissue is not classified using the third approach. Therefore, the first approach was used in this thesis since it has the best values regarding mean, median, and standard deviation of DSC in the three proposed approaches, as shown in Table 7.2.

The proposed method was compared to other segmentation approaches by replacing only the three proposed approaches with algorithms used in the literature. This was namely, Fuzzy C-means which was implemented in earlier work for automatic generation of biomechanical models (78) and K-means, shown in Figures 7.4 and 7.5. These methods have been applied once with three clusters and once with four

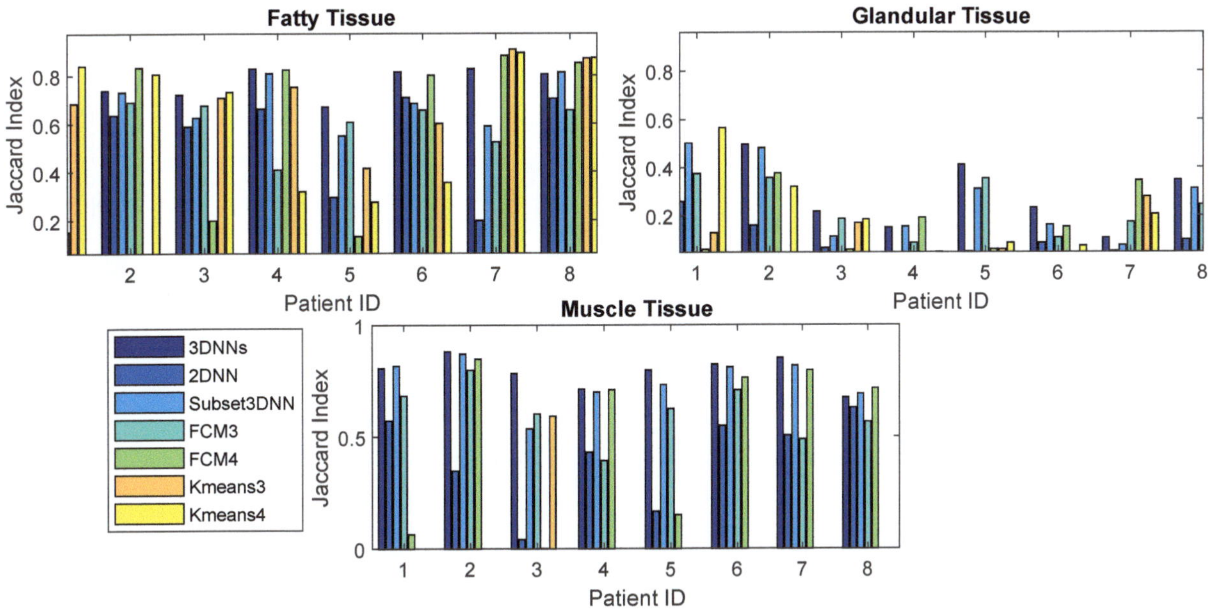

Figure 7.8: Distribution of Jaccard index of each tissue (fatty, glandular, and muscle) across the eight datasets in the proposed algorithm (first approach) (3DNNs) compared to 2D stacked (second approach) (2DNNs), a subset from the volume (third approach) (Subset3DNN), Fuzzy C-means three clusters (FCM3) and four clusters (FCM4) and K-means three clusters (KM3) and four clusters (KM4).

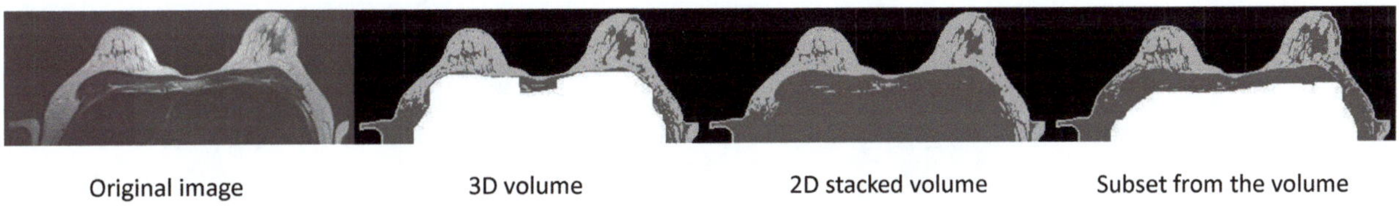

Figure 7.9: An example of segmentation using the three approaches compared to the original image showing that small-size breasts might not work for the third approach of the subset from the volume using one slice. Original MRI from the Medical University of Vienna (first). Segmentation of the three classes using the first approach 3D volume (second). Segmentation of the three classes using the second approach 2D stacked volume (third). Segmentation of the three classes using the third approach subset from the volume (fourth).

clusters. In all of the subsequent comparisons, the result of those algorithms uses the proposed preprocessing and postprocessing. Using the Fuzzy C-means algorithm with three clusters, the mean DSCs obtained for the three tissues is lower than that which was obtained based on 3D NNs. The mean DSC and mean Jaccard index of fatty tissue are 0.75 and 0.61, respectively. The mean DSC and mean Jaccard index of glandular tissue are 0.37 and 0.24, respectively, while the mean DSC and mean Jaccard index of muscular tissue are 0.75 and 0.60, respectively. The results obtained with four clusters are even worse than those obtained with three clusters. K-means with three clusters show less accuracy in all metrics than the proposed method. However, with four clusters, the results for fatty and glandular tissue are comparable to the proposed method. Muscular tissue is however, not detected at all using this algorithm. An exemplary segmentation compared to Fuzzy C-means with three clusters is shown in Figure 7.11.

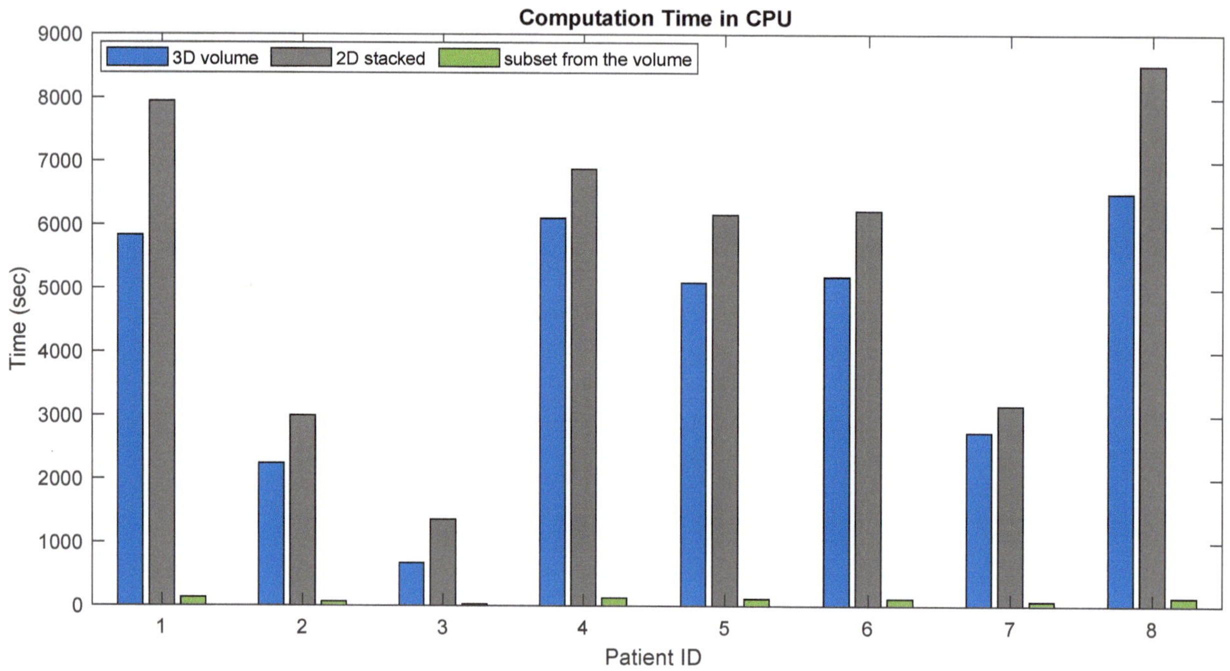

Figure 7.10: Computation time of the three approaches proposed across the eight patients that have ground truth in CPUs. First bar (3D volume), Second bar (2D stacked), and third bar (subset from the volume)

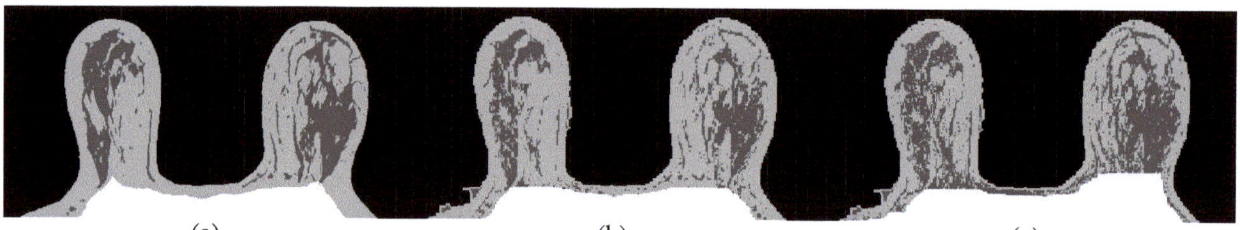

(a) (b) (c)

Figure 7.11: An example of segmentation using the proposed method in comparison to Fuzzy C-means. The ground truth is manually corrected by a medical practitioner (a). The segmentation of the proposed methods (b). The segmentation of Fuzzy C-means with three clusters (c) (162).

7.1.3 Comparison to Literature

While being evaluated on different databases, the proposed method was compared to algorithms from the literature, as shown in Table 7.3 to provide a relation to other approaches. They are all limited to one or two tissues. The first method was used to estimate breast density for two tissues only (67). The Atlas-Basd breast muscle segmentation method was the only method used to detect muscular tissue (66). Two of these methods use supervised learning, which requires a massive amount of labeled data for training (213; 68). In contrast to methods from literature, the proposed method detects three tissues, not just two tissues. The mean DSC achieved for fatty tissue is 0.87, which is better than the one used in Zhang et al. but less than referred to in Gubern-Mérida et al. The mean DSC achieved for glandular tissue is 0.45, which is less than the values in the literature. For muscular tissue, the mean DSC achieved is 0.88, which is better than the one achieved in Gubern-Mérida et al.

Table 7.3: Comparison of the mean DSC for three tissues compared to literature review

Tissue	Gubern-Mérida et al.	Gubern-Mérida et al.	Zhang et al.	Ha et al.	Proposed
Fatty	0.94	-	0.86	-	**0.87**
Glandular	0.80	-	0.83	0.81	**0.45**
Muscular	-	0.72-0.74	-	-	**0.88**

7.2 Machine Learning Algorithms for Simulation of Biomechanical Model Deformation

In this section, the results of biomechanical simulation using machine learning algorithms will be discussed. Since the proposed machine and deep learning algorithms are supervised learning methods, they require a lot of datasets to validate this method and to avoid overfitting problems. Therefore, two groups of datasets were considered: 26 datasets from the Medical University of Vienna with ages ranging from 29 to 78 years, which is a subset of the dataset (Chapter 5) since this subset only has T2-weighted MRI series. Additionally, 232 datasets from the University Hospital of Jena with ages ranging from 23 to 92 years were used. All datasets were collected retrospectively. Patients were treated independently and according to standard procedures that comply with relevant laws and institutional guidelines. The institutional committees approved the collection of data. Datasets were fully anonymized. Each dataset consists of a T2-weighted MRI series collected in clinical practice according to the internal guidelines of the respective hospital. In addition, the model was parameterized using metadata from a corresponding patient's mammogram, such as compression thickness. In total, 258 patients were used.

To increase the size of the database, first, breast meshes from both the left and right breasts of each patient were generated, resulting in a total of 516 datasets. Second, intermediate steps between the initial state and the final state of the deformed breast were retrieved in order to train the model with different deformation states. Any number of intermediate steps could be extracted since the deformation is a function of time in the FEM simulation solved by the dynamic solver. 100 steps were extracted for the estimation of the unloaded state. From the time step when the upper compression plate begins to contact the breast to the time step when the breast is fully compressed according to the metadata of the mammogram, approximately 70 steps were extracted for the compression simulation to provide a good variety of intermediate steps. A total of 70884000 instances were employed for training and validation in the unloaded state step. A total of 48979346 instances were extracted for the compression step.

The three models, ERT, XGBoost, and Att-BLSTM were initially evaluated using one-fold validation in order to optimize the hyperparameters and compare them. The data was divided into 90% for the training set (467 datasets) and 10% for the validation set (49 datasets). The validation set was equally divided into 50% of datasets from the Medical University of Vienna and 50% of datasets from the University Hospital of Jena. The unloaded state estimation and the compression simulation were investigated separately.

Table 7.4: Unloaded state estimation: (mean, median) $\pm$ standard deviation RMSE (mm) of the validation set.

Steps	ERT	XGBoost	Att-BLSTM
All	$(6.8, 5.9) \pm 2.9$	$(6.3, 0.6) \pm 36.9$	$(4.9, 0.6) \pm 26.6$
Last	$(8.1, 6.3) \pm 5.5$	$(7.8, 1.0) \pm 37.3$	$(6.4, 1.1) \pm 26.7$

The evaluation method used was RMSE; see details in Section 6.3.1. In addition to that, the percentage of nodes was categorized from the validation set into classes with an interval of a certain value for the RMSE starting from $0\,mm$ to $22\,mm$ with a step size of $2\,mm$. Since a few percentages of nodes are located in a class with errors larger than $22\,mm$, an additional class for those nodes was added. This categorization provides a representation of the error distribution.

7.2.1 Unloaded State Estimation

For the unloaded state estimation when considering the prediction of all 100 intermediate steps, from Table 7.4 it can be seen that the median RMSE for Att-BLSTM and XGBoost is similar, while Att-BLSTM provides the minimum mean RMSE of $4.9\,mm$. When considering the last step, Att-BLSTM provides the minimum mean RMSE of $6.4\,mm$. As for the median, XGBoost performs slightly better with a median RMSE of $1.0\,mm$ compared to $1.1\,mm$ for Att-BLSTM. The standard deviation for both methods is considerably higher than ERT due to four and six outliers for XGBoost and Att-BLSTM, respectively. The interquartile range for XGBoost is $2\,mm$ and for Att-BLSTM is $1.4\,mm$. As shown in Figure 7.12, the RMSE values for XGBoost and Att-BLSTM are nearly similar based on the categorization of the percentages of nodes into accuracy classes.

Two cases were visualized in the three views of the body plane: the first illustrates how well the proposed models perform visually in comparison to the deformation from the FEM simulations. For this good case as shown in Figure 7.18, 7.19, and 7.20, the minimum error achieved in Att-BLSTM was $0.49\,mm$. The RMSE error achieved was $0.47\,mm$ and $0.49\,mm$ for XGBoost and Att-BLSTM, respectively. Both methods visually perform the same in the three views compared to the FEM simulation (ground truth). However, the ERT model's predicted breast mesh shape with an RMSE of $5.7\,mm$ significantly loses volume, and the displacement of the nodes does not follow the same pattern as the displacement estimated by the FEM simulation. Second, a case with an RMSE of $5.9\,mm$, $1.1\,mm$, and $1.1\,mm$ for ERT, XGBoost, and Att-BLSTM, respectively has been presented for illustration. In this case, due to a non-optimal automatic segmentation of the MRI volume, some elements are incorrectly modeled as glandular tissue on the surface of the breast. Additionally, some elements exceed the quality criterion limits of ABAQUS during the FEM simulation and would need to be re-meshed. Despite these challenges in predicting a non-optimal initial dataset, the behavior of the XGBoost and the Att-BLSTM provides visually convincing results, still similar to the deformation in the FEM simulations, as shown in Figure 7.24, 7.25, and 7.26.

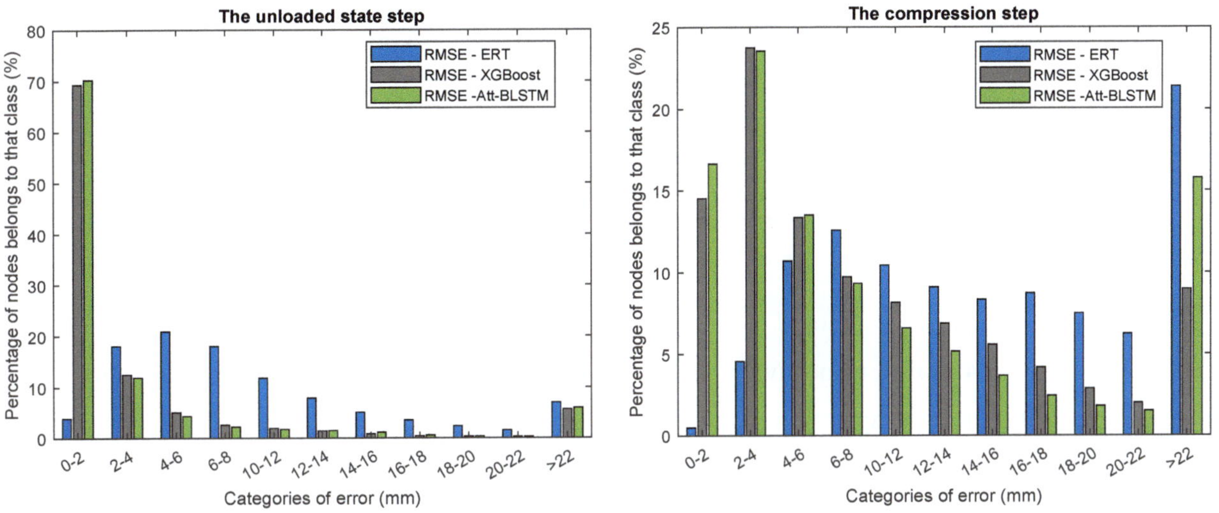

Figure 7.12: Comparison of the percentage of nodes in each category of error for the unloaded state estimation (left) and for the compression step (right). The RMSE values are calculated for the three methods (ERT, XGBoost, Att-BLSTM).

7.2.2 Compression Step

When taking into account the prediction of all 70 intermediate steps for the compression step, it can be seen from Table 7.5 that XGBoost provides the minimum mean and median RMSE of $4.9\,mm$ and $3.5\,mm$, respectively. For the last step, in which the compression is the highest and reaches the compression thickness specified in the mammogram's metadata, the mean and median RMSE are $7.3\,mm$ and $4.3\,mm$, respectively. Compared to these values, Att-BLSTM performs marginally worse with an average RMSE of $6.0\,mm$ overall steps. For the last step, the mean error achieved is $9.1\,mm$. While ERT performs for both, overall steps and the last step, noticeably worse with $10.0\,mm$ and $13.4\,mm$, respectively. According to the error distribution in Figure 7.12 (right), Att-BLSTM predicts slightly less nodes with an RMSE higher than $10\,mm$ compared to the XGBoost method. Both XGBoost and Att-BLSTM have approximately the same number of nodes with an RMSE lower than $10\,mm$. It can also be realized that there is a significant difference in the first class from $0-2\,mm$ and the last class, which summarizes errors greater than $22\,mm$. For the first class, the distribution shows that Att-BLSTM performs slightly better than XGBoost, while the last class indicates that the number of nodes with a large error in Att-BLSTM is considerably higher than in XGBoost.

The same two cases that were presented for the unloaded state estimation in three views that correspond to three body planes are shown in Figures 7.21, 7.22, and 7.23. The first case compares the deformation acquired by the FEM simulation to the performance of the suggested models. It is obvious that the breast mesh predicted by the ERT method has an RMSE of $6.1\,mm$, loses volume, and the deformation of the nodes results in a different breast shape as in the FEM simulations. XGBoost and Att-BLSTM provide similar performance with an RMSE of $1.7\,mm$ and $1.7\,mm$, respectively, in the three views compared to the FEM simulation (ground truth). In the second case, it is shown that neither method's prediction

Table 7.5: Compression step: (mean, median) ± standard deviation RMSE (mm) of the validation set

Steps	ERT	XGBoost	Att-BLSTM
All	(10.0, 8.0) ± 5.4	(4.9, 3.5) ± 4.0	(6.0, 3.6) ± 7.1
Last	(13.4, 10.6) ± 9.9	(7.3, 4.3) ± 7.5	(9.1, 4.7) ± 13.1

yields a visually comparable outcome. In this case, the RMSE for ERT, XGBoost, and Att-BLSTM is $25\,mm$, $20.8\,mm$, and $42.1\,mm$, respectively. This case is one of those with a high compression ratio (67%) since the thickness before compression is around $120\,mm$, as shown in Figure 7.24 and the thickness after compression is approximately $40\,mm$. Furthermore, as previously mentioned, the breast mesh is not ideal since, for example, glandular tissue can be seen on the surface of the breast (dark blue), which is a result of an inadequate automatic segmentation of the MRI. As opposed to the ground truth, Att-BLSTM not only fails to predict the shape of the breast accurately but also causes a significant global displacement of the coordinates. XGBoost also fails to accurately represent the shape of the breast.
In the following subsections, a deeper analysis of the performance of the models with respect to patient characteristics is emphasized.

Correlation with Features of the Breast

In order to determine whether any subgroups are correlated with the RMSE, the results with respect to three characteristics of the breast were investigated: the error as a function of the different tissue types (glandular, fatty, and muscular tissues), the breast volume, and the compression ratio. Based on the validation set of the one-fold validation, the analysis was carried out.

Tissue Type: It was investigated whether there is a correlation between the RMSE and the tissue types that were taken into account in the model, namely the fatty, glandular, and muscular tissues. To accomplish this, the instances were subdivided into nodes labeled with the appropriate tissue type, i.e., the inner nodes of the breast mesh of those elements modeled as fatty, glandular, and muscular tissues. Nodes at the tissue interfaces were not taken into consideration.
A histogram of the RMSE for the three models is plotted for each tissue type as shown in Figure 7.13. Due to the variation in the number of nodes belonging to each class, a log-normal distribution was fitted to the histograms. The μ stands for the RMSE logarithmic values' mean, and the σ for their standard deviation. According to Table 7.6 and Figure 7.13, XGBoost has the smallest μ of 1.76 with a mean RMSE of $9.4\,mm$ for fatty tissue. With a mean RMSE of $8.4\,mm$, XGBoost has the smallest μ of 1.7 for glandular tissue. Although, Att-BLSTM has the smallest μ of 1.41 for muscular tissue, it has the highest σ of 1.07. XGBoost has the minimum mean RMSE of $6.5\,mm$
There is no clear pattern when comparing the various tissue types across all evaluations. Apart from the fact that the RMSE tends to be lower for nodes in muscular tissue than for fatty and glandular tissues.

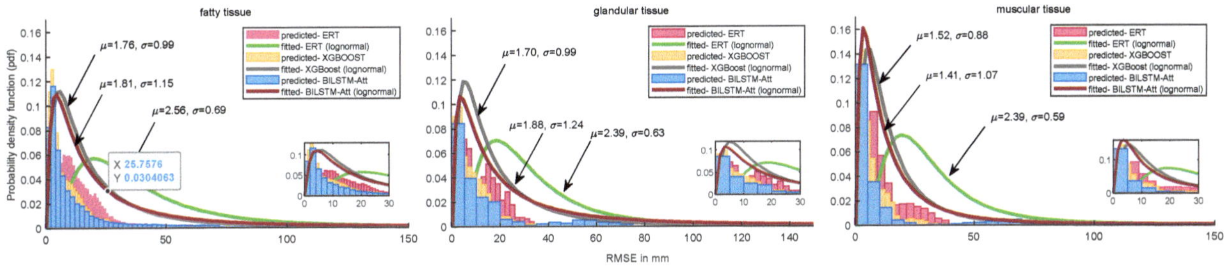

Figure 7.13: Probability density function versus RMSE: the relation between the tissue type and the RMSE of the validation set for the three models: ERT, XGBoost, and Att-BLSTM.

Table 7.6: Average RMSE for tissues in mm

Tissue	ERT	XGBoost	Att-BLSTM
fat	16.6 ± 13.3	**9.4 ± 10.9**	12.1 ± 17.7
glandular	13.1 ± 7.4	**8.4 ± 7.5**	13.3 ± 16.8
muscle	13.1 ± 8.3	**6.5 ± 5.8**	8.4 ± 14.7

This can be because muscular tissue was modeled as a rigid body that can not deform due to the boundary conditions in the biomechanical model. As a result, this behavior might be easier for the machine learning model to learn than the intricate deformation of fatty and glandular tissues.

Breast Volume: In order to examine if larger breast sizes, which deform more during mammographic compression, can be predicted with less accuracy than smaller ones, an investigation was done to examine whether there is a relationship between the breast volume and the RMSE. Figure 7.14 shows the RMSE as a function of breast volume measured in mm^3. There is no discernible trend, and the data points are widely dispersed. A polynomial fitting curve with degree one was used to analyze the correlation. With a maximum r of 0.09, it is seen that the three approaches ERT, XGBoost, and Att-BLSTM have an independent relationship.

Compression Ratio: It was assumed that the error increases with the compression ratio based on the discrepancy between the result obtained by averaging the results of all deformation steps and the final step. For validating this hypothesis, a correlation between the RMSE and the compression ratio of the validation set was analyzed by calculating the compression ratio in intermediate steps until it reaches the desired breast thickness that is recorded in the metadata for each dataset, as shown in Figure 7.15. Then, starting at 1% and going up to 76%, which is the highest compression ratio found in the patient cohort, the mean value for all the datasets at each compression ratio was determined. With a 1% step size, the datasets were binned. The outcomes of this investigation are shown in Figure 7.15 (left). The observed data points were fitted with a second-degree exponential function. A significant correlation between the fitting curve and the data is indicated by the exponential curve fit's r-squared (R^2) values, which are 0.90 for XGBoost, 0.88 for Att-BLSTM, and 0.90 for ERT. This supports the initial hypothesis.

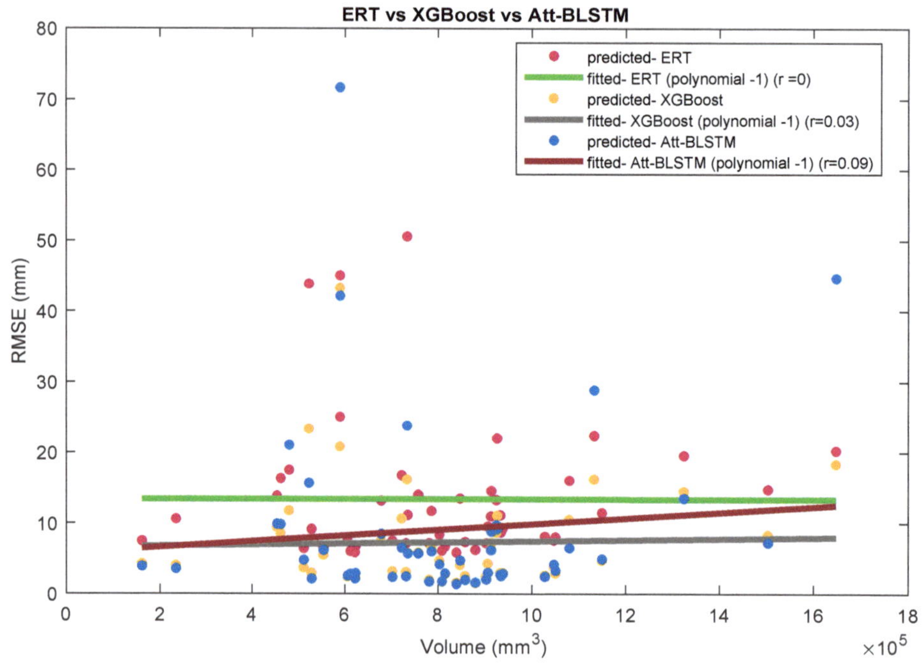

Figure 7.14: Relation between the breast volume and the RMSE of the validation set for the three models: ERT, XGBoost, and Att-BLSTM. Breast volume does not correlate with the RMSE.

For the XGBoost and Att-BLSTM models, which achieve the best overall results, the relationship between the RMSE and the compression ratio for all of the available datasets was also examined with 10-fold cross-validation. Figure 7.15 displays these results (right). The error distribution closely matches the one-fold validation subset evaluation. The relationship between RMSE and the compression ratio develops gradually up to a compression ratio of 53%. It was evaluated whether, in this range, the rise in error with increasing compression ratio may likewise be regarded as linear by performing a linear fit with R^2 of 0.93. Yet, the error substantially increases after 53% compression, which leads us to believe that the entire error distribution has an exponential relationship. The 10-fold validation data in Figure 7.15 (right) were likewise fitted with a two-degree exponential fitting curve, yielding R^2 of 0.83.

The fitted curves in Figure 7.15 show that, for all compression ratios, ERT has the highest error in contrast to the other two approaches. Also, it can be seen that up to a compression ratio of roughly 53%, the performance of Att-BLSTM and XGBoost is almost identical with 5.8 mm. XGBoost outperforms Att-BLSTM at a higher compression ratios with lower RMSE values.

7.2.3 Cross-Validation Analysis

As shown earlier, XGBoost and Att-BLSTM appear to be generally capable of offering a good agreement of predicted breast shapes compared to the FEM ground truth. In terms of the median RMSE for the estimation of the unloaded state, Att-BLSTM and XGBoost nearly get the same results, while XGBoost performs better for the compression step used in the one-fold validation.

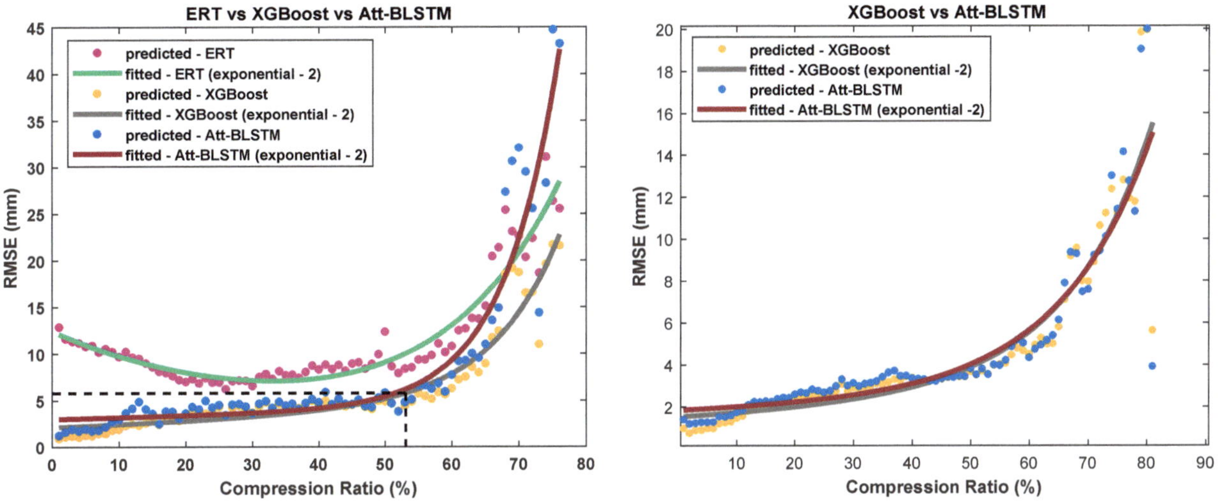

Figure 7.15: Relation between the compression ratio and the RMSE: the validation set (49 datasets) for the three models, ERT, XGBoost, and Att-BLSTM, respectively (left) and 10-fold cross-validation for two models, XGBoost and Att-BLSTM, respectively (right). The compression ratio is extracted for every intermediate step for each dataset until it reaches the desired compression ratio extracted from the metadata. The RMSE is the mean of all patients in each intermediate step corresponding to the compression ratio. The representation of compression ratio and RMSE in the 10-fold validation (right) is the same as the 1-fold validation (left).

Table 7.7: Literature review of machine learning algorithms for breast deformation

Reference	Compression Ratio	Time-FEM	Time-ML	Architecture	#datasets
Martínez-Martínez et al.	20 %	120 min	<0.2 s	DT/ ERT/random forest	Clinical -10
Rupérez et al.	NA	NA	0.43 s	DT/ERT/random forest	Phantom -10
Mendizabal et al.	NA	407.7 ms	3.14 ms	U-Net	Phantom -10
Proposed	20 - 76 %	20 min	5 s	ERT/XGBoost/Att-BLSTM	Clinical -516

In order to expand the database and evaluate the approaches by avoiding over-fitting to only one validation set, 10-fold cross-validation was used for this purpose. It was applied to: XGBoost and Att-BLSTM model for the estimation of the unloaded state and the compression step.

The average RMSE for XGBoost even reduces to $3.4mm$ and $4.7mm$, respectively, as shown in Table 7.8. Additionally, the error distribution for the two simulation steps is greatly improved. Figures 7.16 and 7.17 further illustrate the results. Regarding RMSE and error distribution, Att-BLSTM shows the same behavior.

From a computational perspective, it takes about 4.5 hours to train the model for XGBoost using the 516 breast meshes in a 10-fold cross-validation, whereas predicting a single dataset only takes about 5 seconds. While it takes over 60 hours to train Att-BLSTM using 516 breast meshes in a 10-fold cross-validation, predicting a dataset takes only about 10 seconds. Both models were trained on an RTX 2080TI GPU using CUDA 10.2. This achieves a speedup of factors 240 and 120 for XGBoost and Att-BLSTM, respectively, compared to a runtime of about 20 minutes for the computing duration of a FEM simulation.

Table 7.8: RMSE ((mean, median) ± standard deviation) of 10-fold cross-validation for the unloaded state estimation and the compression step in the last step for XGBoost and Att-BLSTM

Method	Unloaded state	Compression
XGBoost	$(3.4, 0.8) \pm 30.5mm$	$(4.7, 3.4) \pm 4.7mm$
Att-BLSTM	$(3.0, 1.2) \pm 22.3mm$	$(4.9, 3.4) \pm 5.6mm$

7.2.4 Comparison to Literature

For 10 clinical datasets, Martínez-Martínez et al. and Rupérez et al. attained an accuracy of $0.62mm$. For a comparable problem in the compression step, the proposed approach in this thesis achieves mean and median RMSE values for 516 breasts of $4.7mm$ and $3.4mm$, respectively. However, the first key distinction between the study in this thesis and others is the use of a compression ratio of up to 76%, which results in strong non-linear deformations of the breast. The evaluation of (116; 159) was constrained to a 20% compression thickness. With 10-fold cross-validation, the proposed techniques in this thesis achieve mean and median RMSE at a 20% compression ratio of $2.3mm$ and $1.8mm$, respectively. The second difference is the analysis of two simulation steps, not only the compression step but also the estimation of the unloaded state. The third difference is the number of datasets; 516 breasts were used in the evaluation of this study, while the work of (116; 159) was restricted to 10 phantoms and clinical datasets.

Martínez-Martínez et al. and Rupérez et al. reported the time required for prediction in the ML models and very generally, the computation time for one FEM simulation, which may often take 120 minutes (78; 177), (Table 7.7). A speedup of factor approximately 240 is achieved by the proposed models in this thesis when compared to the FEM model simulation. Despite using different breast imaging modalities in comparison to Mendizabal et al., they were 130 times faster than their FEM model simulation.

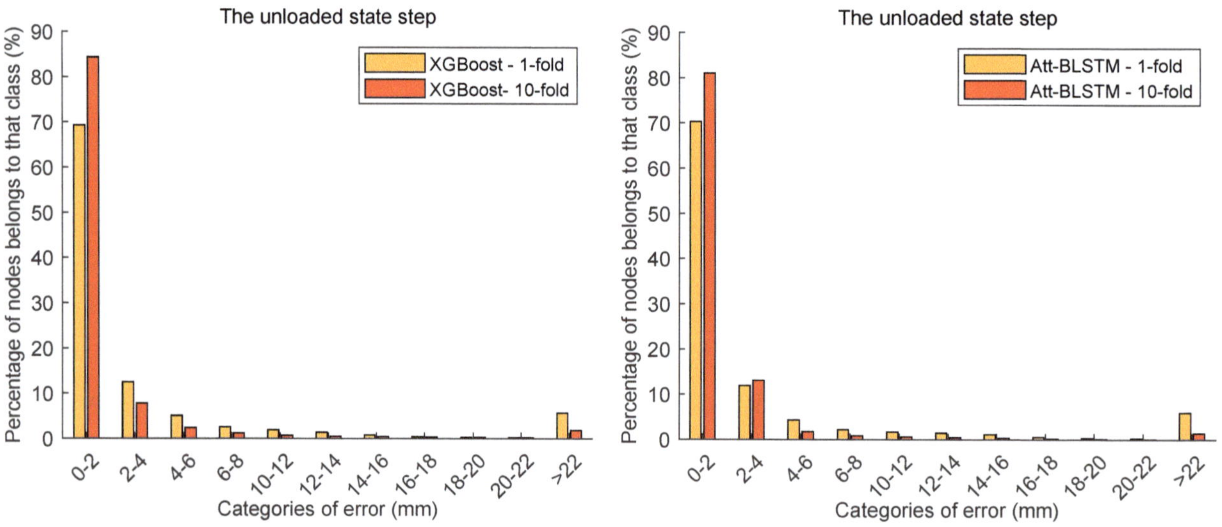

Figure 7.16: Percentages of nodes split into categories for the unloaded state estimation with XGBoost and Att-BLSTM for the 10-fold cross-validation in comparison to the 1-fold cross-validation regarding the RMSE.

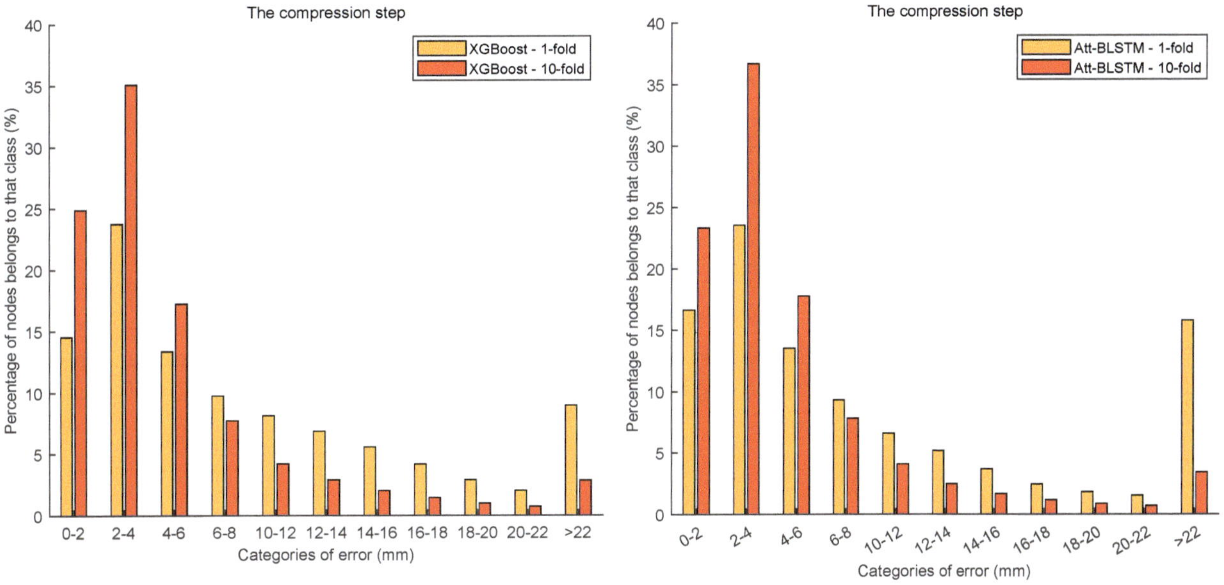

Figure 7.17: Percentages of nodes split into categories for the compression step with XGBoost and Att-BLSTM for the 10-fold cross-validation in comparison to the 1-fold cross-validation regarding the RMSE.

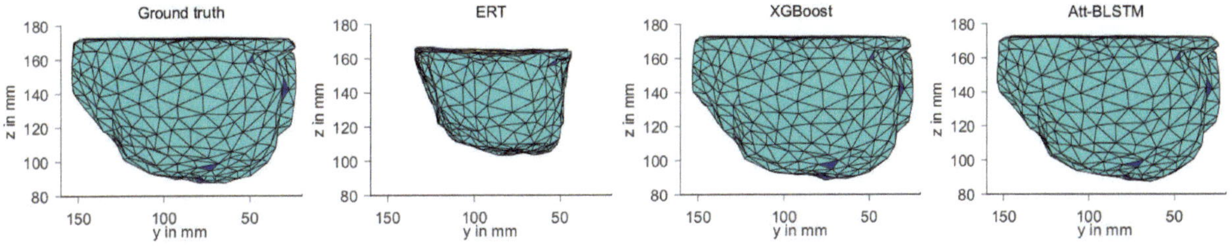

Figure 7.18: Clinical case with a visually good result in the sagittal plane (lateral view): the comparison of the estimation of the unloaded state between the ground truth (FEM simulation) and the prediction by ERT, XGBoost, and Att-BLSTM (164).

7.3 Biomechanical Model Based Registration

In this subchapter, the results of the registration between MRI and X-ray mammography will be presented. TRE, TPR, IoU, and DSC, which are explained in Chapter 6, have been calculated between the predicted annotation and the annotation that has been marked by radiologists. In addition, DSC and IoU will be presented for the shape of the breast between the synthetic X-ray mammography (predicted deformed MRI) created from the biomechanical model based registration method and the original X-ray mammography.

As explained before, a free-hand shape annotation was used. Hence, for calculating the TRE, the center of gravity (COG) of the annotation was calculated to get the error between the predicted annotation and the actual annotation. The annotation of MRI is 3D, while the X-ray annotation is 2D. The annotation was tracked in the 3D volume before and after deformation, as shown in Figures 7.31 and 7.32, and then

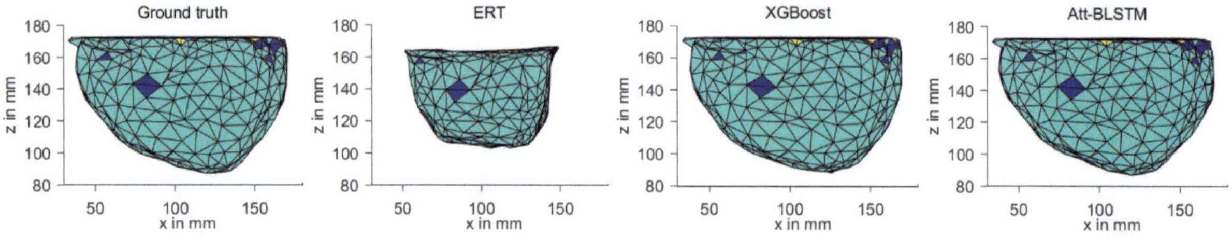

Figure 7.19: Clinical case with a visually good result in the transversal plane (axial view): the comparison of the estimation of the unloaded state between the ground truth (FEM simulation) and the prediction by ERT, XGBoost, and Att-BLSTM (164).

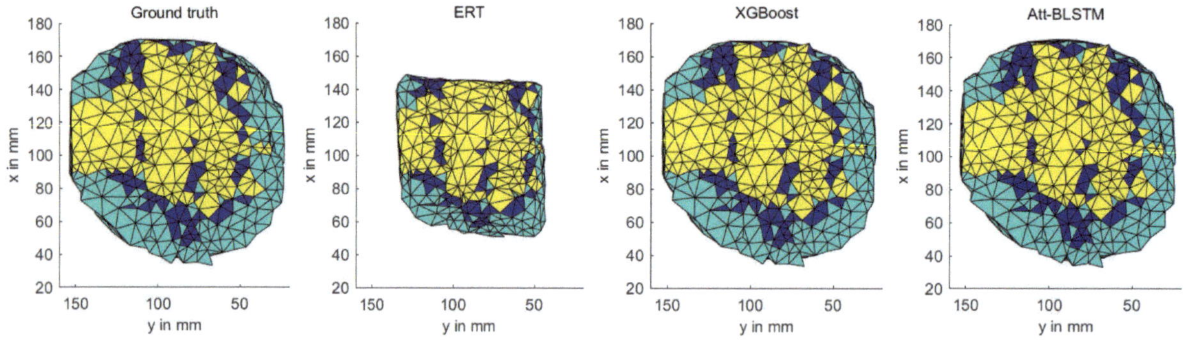

Figure 7.20: Clinical case with a visually good result in the coronal plane (back view): the comparison of the estimation of the unloaded state between the ground truth (FEM simulation) and the prediction by ERT, XGBoost, and Att-BLSTM (164).

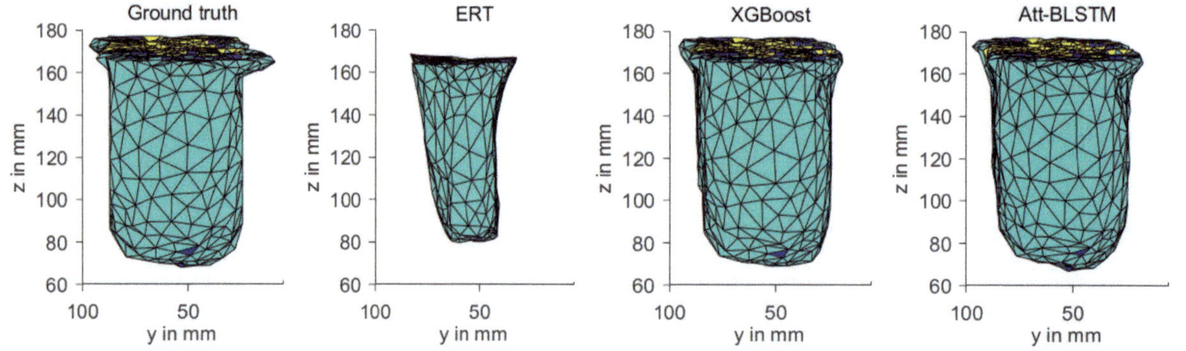

Figure 7.21: Clinical case with a visually good result in the sagittal plane (lateral view): the comparison of the compression step between the ground truth (FEM simulation) and the prediction by ERT, XGBoost, and Att-BLSTM (164).

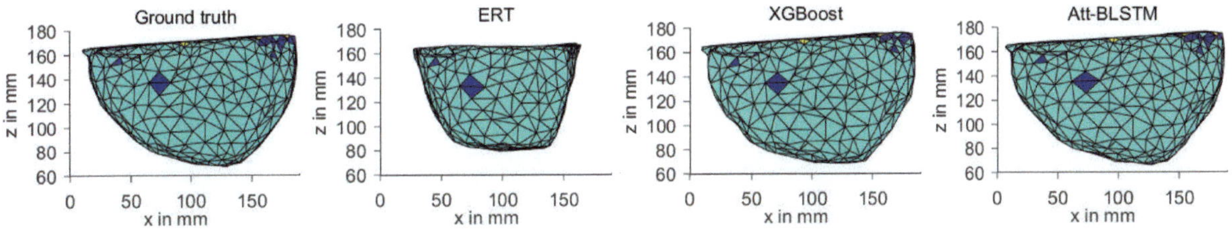

Figure 7.22: Clinical case with a visually good result in the transversal plane (axial view): the comparison of the compression step between the ground truth (FEM simulation) and the prediction by ERT, XGBoost, and Att-BLSTM (164).

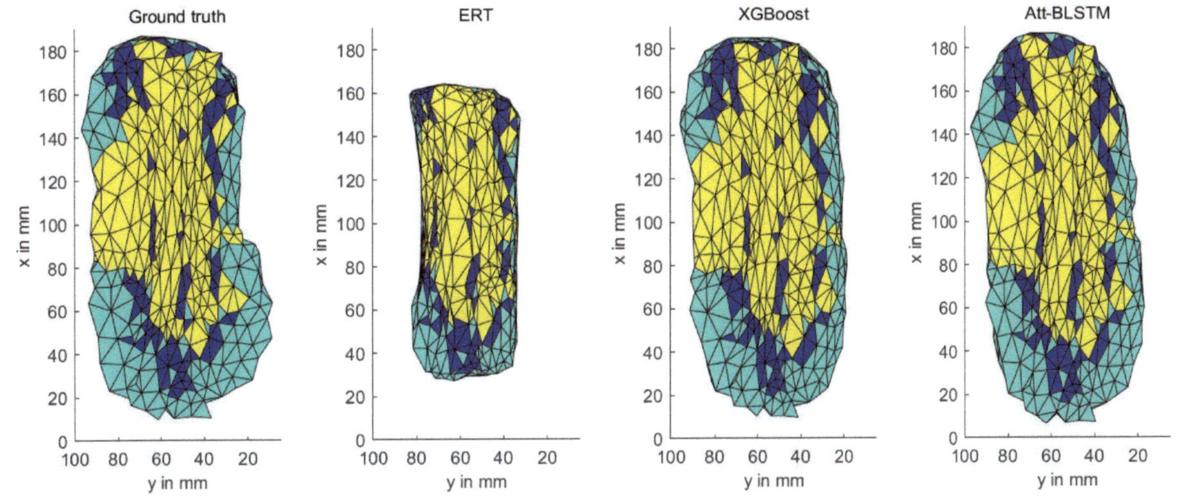

Figure 7.23: Clinical case with a visually good result in the coronal plane (back view): the comparison of the compression step between the ground truth (FEM simulation) and the prediction by ERT, XGBoost, and Att-BLSTM (164).

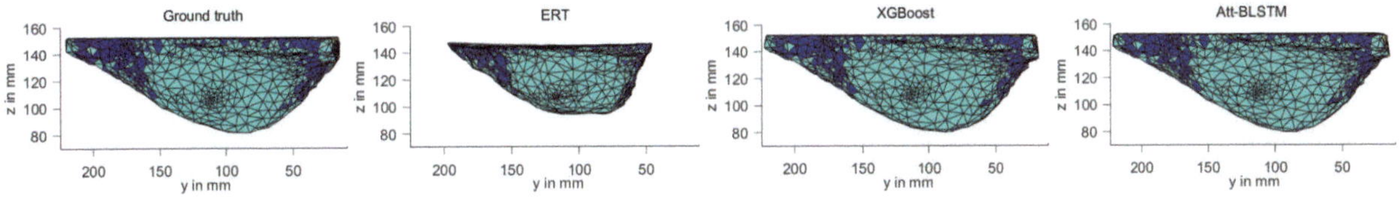

Figure 7.24: Clinical case with visually non-optimal result in the sagittal plane (lateral view): the comparison of the unloaded state estimation between the ground truth (FEM simulation) and the prediction by ERT, XGBoost, and Att-BLSTM (164).

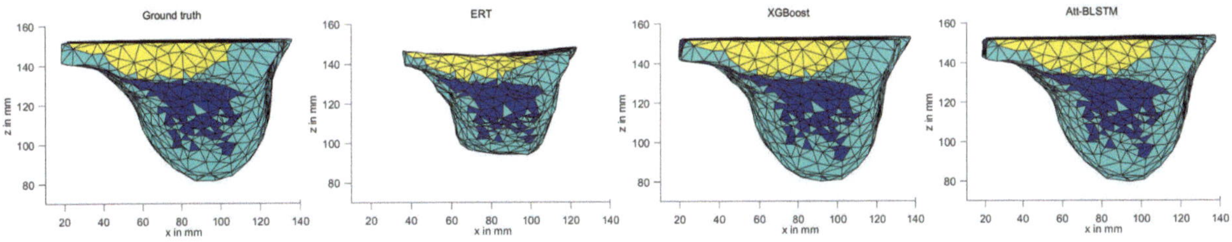

Figure 7.25: Clinical case with visually non-optimal result in the transversal plane (axial view): the comparison of the unloaded state estimation between the ground truth (FEM simulation) and the prediction by ERT, XGBoost, and Att-BLSTM (164).

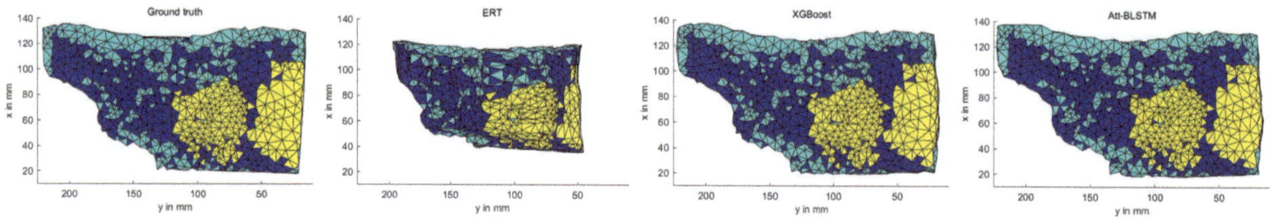

Figure 7.26: Clinical case with visually non-optimal result in the coronal plane (back view): the comparison of the unloaded state estimation between the ground truth (FEM simulation) and the prediction by ERT, XGBoost, and Att-BLSTM (164).

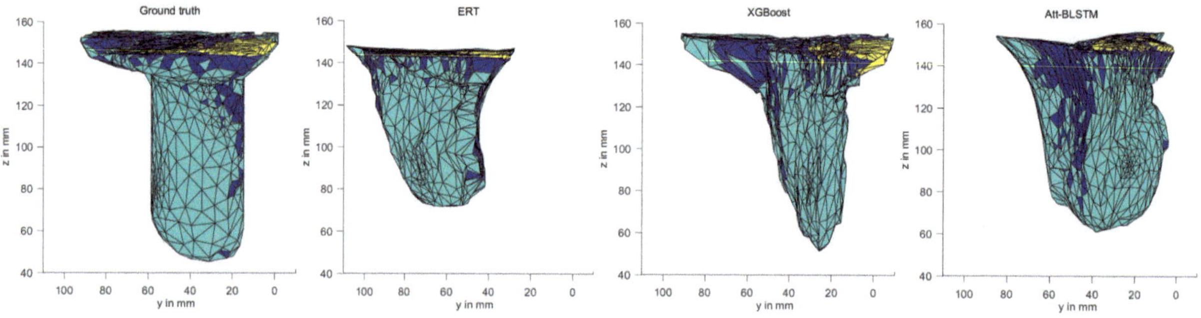

Figure 7.27: Clinical case with visually non-optimal result in the sagittal plane (lateral view): the comparison of the compression step between the ground truth (FEM simulation) and the prediction by ERT, XGBoost, and Att-BLSTM (164).

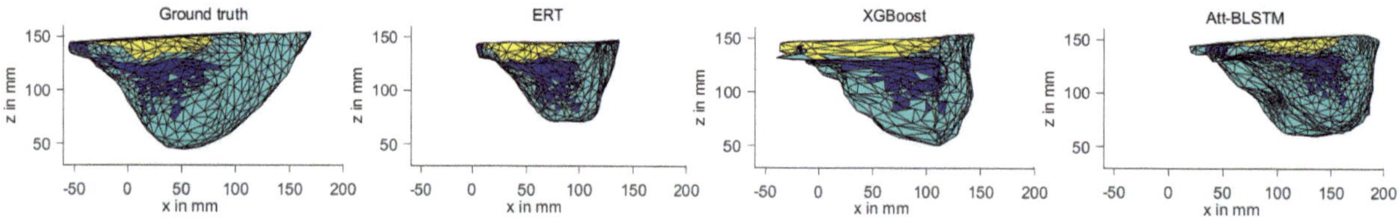

Figure 7.28: Clinical case with visually non-optimal result in the transversal plane (axial view): the comparison of the compression step between the ground truth (FEM simulation) and the prediction by ERT, XGBoost, and Att-BLSTM (164).

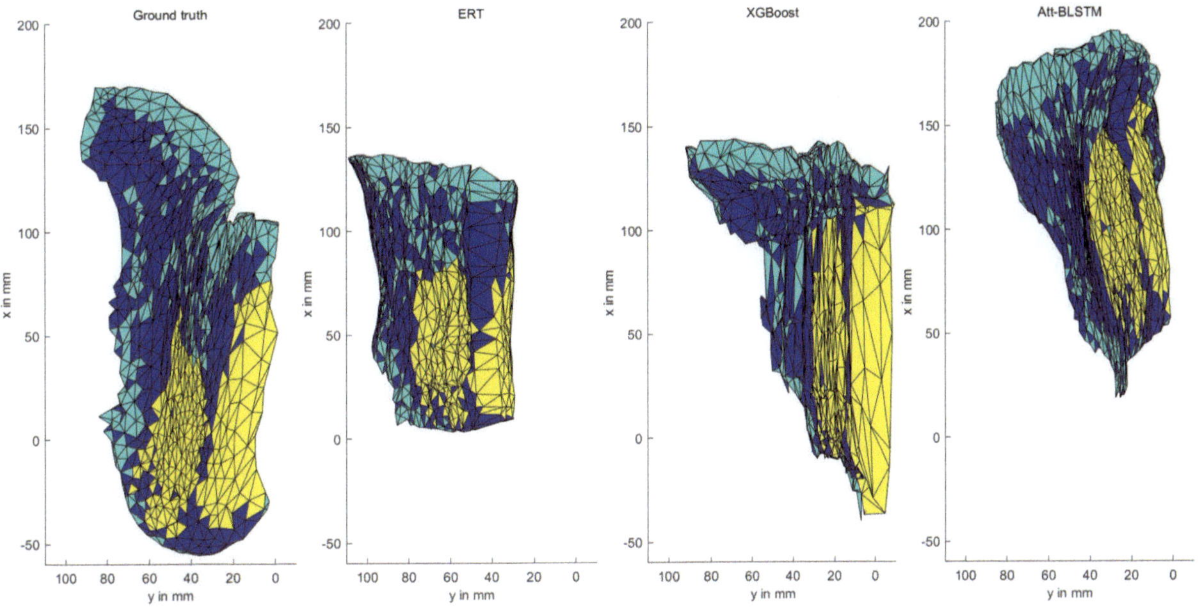

Figure 7.29: Clinical case with visually non-optimal result in the coronal plane (back view): the comparison of the compression step between the ground truth (FEM simulation) and the prediction by ERT, XGBoost, and Att-BLSTM (164).

the 2D projection of this annotation was retrieved. Three examples have been visualized in Figures 7.38, 7.39, and 7.40 for CC cases and similarly for MLO cases. The red lesion represents the lesion that is marked by the radiologist in the full X-ray mammogram and the green lesion represents the predicted lesion based on the annotation in the MR image using biomechanical model based registration. The yellow color represents the overlap of the two lesions.

The proposed biomechanical model based registration method achieves a median error of $31.8\,mm$ for 51 clinical datasets with the standard parameters without optimization, as shown in Table 7.9 and in Figure 7.33 (left). The lower and upper quarters are $16.4\,mm$ and $48.2\,mm$, respectively. While the minimum and the maximum errors are $4.4\,mm$ and $85.92\,mm$, respectively. Such a high error in the case of $85.92\,mm$, as shown in Figure 7.30, was observed for one of the MLO cases in which X-ray mammogram does not look optimum with respect to the two rotation angles of muscle area and rotation around the z-axis for applying compression (first image). Additionally, it might be because of inaccurate segmentation, which causes a hole in the projected deformed MRI (second image). Optimizing the parameters of rotation angles and compression thickness could be used to decrease the error, but due to time constraints, it was not evaluated. For the intersection between the two annotations (the yellow color), three metrics (DSC, IoU, and TPR) have been calculated. It was found that 22 cases out of the 51 cases achieve an intersection, as shown in Figure 7.34. The maximum DSC achieved is 0.623, and the maximum IoU achieved is 0.46 in the same case. TPR has been evaluated to avoid the indeterministic shapes and the non-correlated annotations between MRI and full X-ray mammograms, such as in Figure 7.39. In some

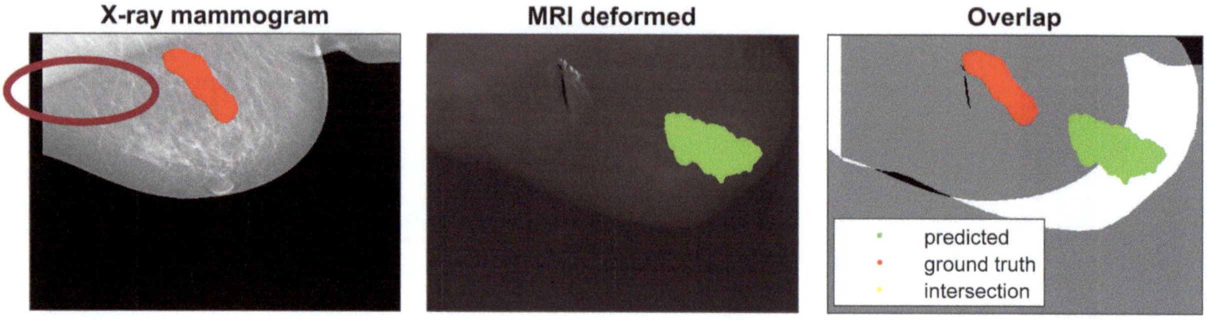

Figure 7.30: An example of a bad registration between MRI and X-ray mammography (Patient ID 31): original X-ray mammography and its problem of a rotation angle of muscle area (see red circle) (first), synthetic X-ray created from the biomechanical model based registration and its problem of an artifact in the image because of inaccurate segmentation (second). Overlap of the two images (third). The predicted lesion that has been annotated by radiologists in the original X-ray (red) and projected tracked annotation of the MRI into the synthetic X-ray (green).

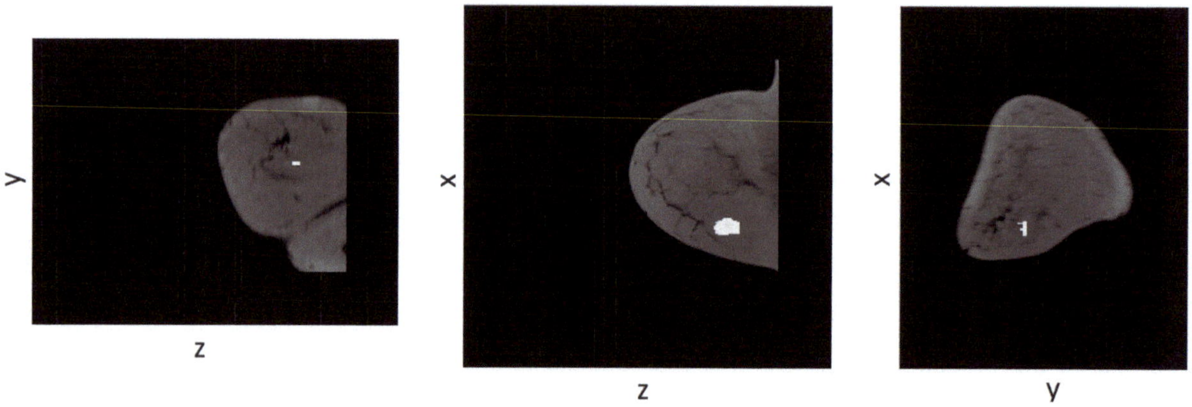

Figure 7.31: An exemplary case shows the annotated lesion by the radiologist in three views of an MR image before applying compression deformation to the breast.

cases, TPR might be much higher than DSC and IoU for such problems, such as in cases 55, 49, 36, 35, 33, 10, and 6, as shown in Figure 7.34. The maximum TPR achieved is 0.97.

In addition for the shape of the breast, the DSC achieves a median of 0.87 as shown in Figure 7.33 (right). The lower and the upper quarters are 0.9 and 0.84, respectively. There is one outlier with a value of 0.7. The minimum and the maximum values are 0.75 and 0.95, respectively. Similarly, for IoU, it achieves a median of 0.77. The lower and the upper quarters are 0.72 and 0.81, respectively. There is one outlier with a value of 0.54. The minimum and the maximum values are 0.6 and 0.9, respectively.

Table 7.9: Accuracy of biomechanical model based registration between MRI and X-ray mammography

Accuracy in *mm*	mean	median	std	min	max	75%	25%
TRE	33.12	31.76	18.95	4.4	85.92	48.2	16.4

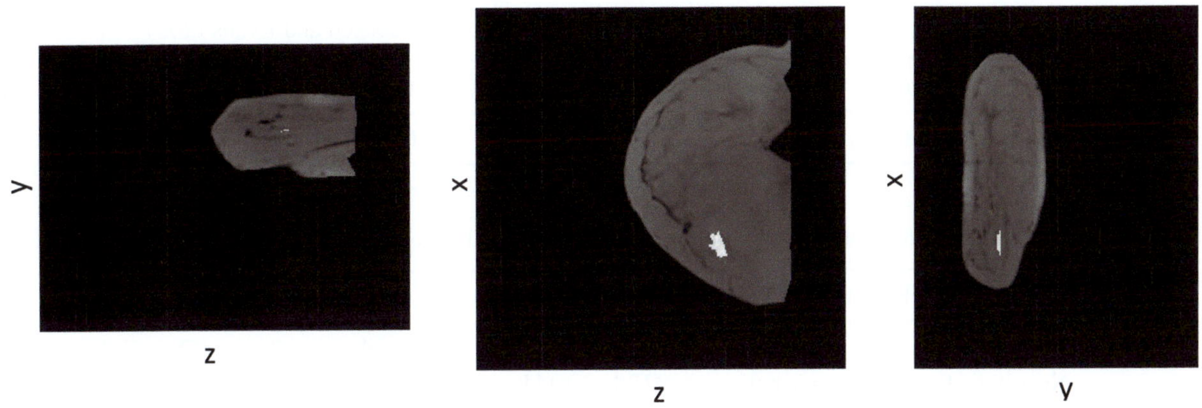

Figure 7.32: The exemplary case shows the tracked annotated lesion by the radiologist in three views of an MR image after applying the compression deformation to the breast.

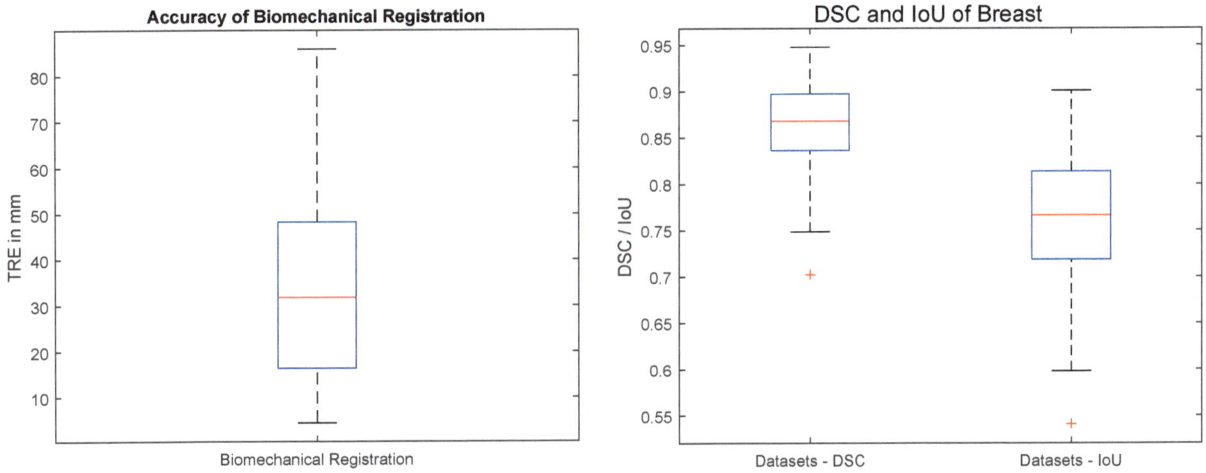

Figure 7.33: Analysis of TRE for the biomechanical model based registration method of 51 cases received from the Medical University of Vienna (left). Analysis of the DSC and IoU for the shape of the breast between the produced synthetic X-ray mammography (predicted deformed MRI) and the original X-ray mammography (right).

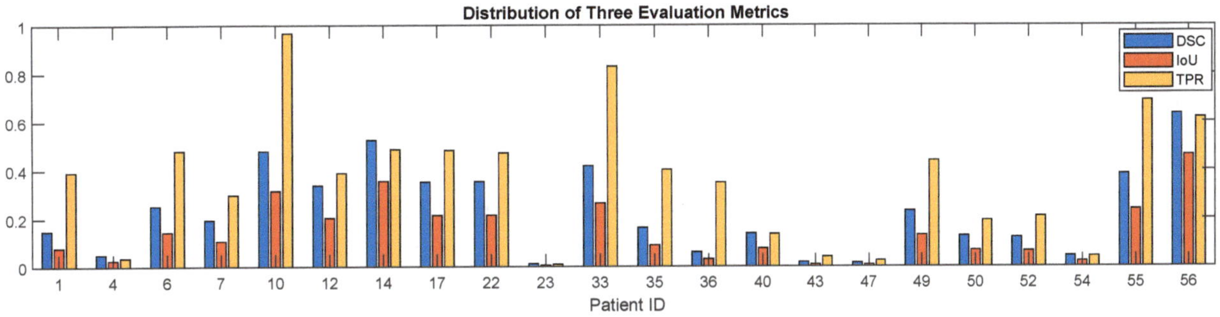

Figure 7.34: Distribution of the three evaluation metrics that have been calculated for the intersection between the actual and the predicted lesion using biomechanical model based registration method (the yellow color in Figures 7.38, 7.39, and 7.40). 22 out of 51 cases achieved an intersection between the two annotations.

Table 7.10: The correlation strength into categories based on Wuensch and Evans (209)

Category	very weak	weak	moderate	strong	very strong
Range	0 - 0.19	0.20 - 0.39	0.40 - 0.59	0.60 - 0.79	0.80- 1.0

Table 7.11: The number of elements in the initial mesh exceeded the limit of the minimum angle (less than <10°)

Elements per Patient ID	11	12	21	22	28	29
Number of distorted elements	2	2	2	4	2	2
Number of elements per patient	9124	2941	7500	2279	8579	6886

7.3.1 Mesh Quality and Volume Loss by FEM

The biomechanical model based registration has been analyzed using seven characteristics that correspond to the variations in datasets as explained in Chapter 5 to investigate if subgroups are correlating with the accuracy of the method. The strength of the correlation is split into categories based on Wuensch and Evans as described in Table 7.10.

Additional features have been investigated such as the quality of initial meshes before being solved by FEMs, the total loss of volume after the deformation of the breast, and the loss of volume for each tissue of three segmented classes of fatty, glandular, and muscular tissues.

Quality of meshes

As explained in Section 3.2.4, there are different quality metrics for evaluating the quality of the initial mesh, and the most effective one is the ratio radius of the tetrahedron mesh. The total number of elements for all the available data of 51 patients is 656960 elements. Figure 7.35 (left) shows the distribution of the percentage of distorted elements in each initial mesh for each clinical case using the best parameters found in the methods. It was observed that 20 cases have 1% distorted elements and 22 cases have 2 % distorted elements, which is considered 82.3% from the available data. Also, other metrics such as Joe Li, minimum angle, and maximum angle have been calculated for the initial mesh. Interestingly, it is found that no patient is exceeding the limits of Joe Liu's and maximum angle. Only six patients out of the 51 clinical datasets have elements that exceed the desired limit of the minimum angle (see Table 7.11).

Additionally, it was investigated if the quality of the mesh using the ratio radius of the tetrahedron mesh has an impact on the accuracy of the biomechanical model based registration. It was observed that there is a very weak correlation between them with an r of -0.04 and R^2 of 0.001, as shown in Figure 7.36 (left).

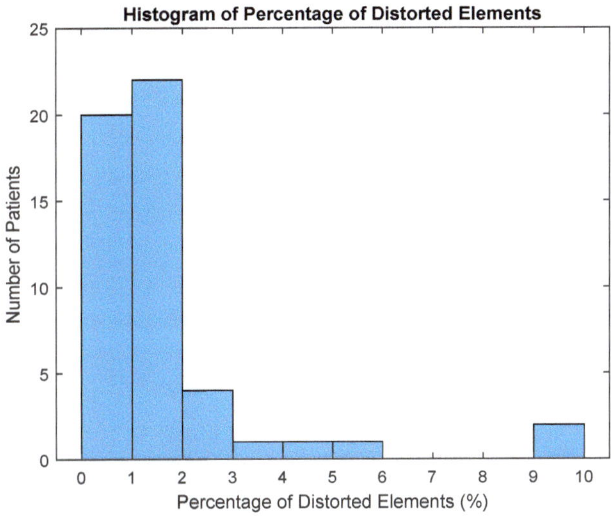

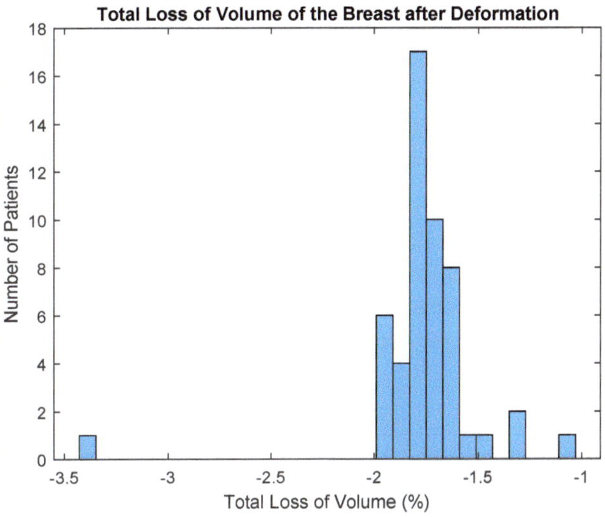

Figure 7.35: Histogram of percentage of distorted elements for the initial mesh in each patient for 51 clinical datasets received from the Medical University of Vienna (left). Histogram of the total loss volume of MRI breast after being solved by FEMs (right).

Loss of Volume

Another aspect that has been evaluated regarding the deformation of the biomechanical model of the MRI breast solved by FEM is the loss of volume which occurred since the material is considered nearly incompressible. Figure 7.35 (right) shows the histogram of total loss of volume between the initial mesh before deformation and the deformed mesh. It was observed that most of the clinical datasets (47 patients) lie in the percentage of total loss of volume between 1.5% and 2%.

Additionally, it has been investigated if the total loss of volume has an impact on the accuracy of the biomechanical model based registration. It was observed there is a very weak correlation between them with an r of -0.16 and R^2 of 0.02, as shown in Figure 7.36 (right).

Moreover, the loss of volume for three segmented tissues fatty, glandular, and muscular tissues has been analyzed to find a correlation between these losses and the accuracy of the biomechanical model based registration method. For fatty tissue, it is considered a very weak correlation in the negative direction with an r of -0.12 and R^2 of 0.01, as shown in Figure 7.37 (a). For glandular tissue, it is considered a weak correlation in the negative direction with an r of -0.20 and R^2 of 0.04 as shown in Figure 7.37 (b). For muscular tissue, it is considered a very weak correlation with an r of -0.08 and R^2 of 0.006.

7.3.2 Correlation with Features of the Breast

Patient Age

The age of the patients in the clinical datasets has been investigated. A polynomial fitting curve (first degree) has been used to evaluate how the datasets are distributed compared to the TRE of the biome-

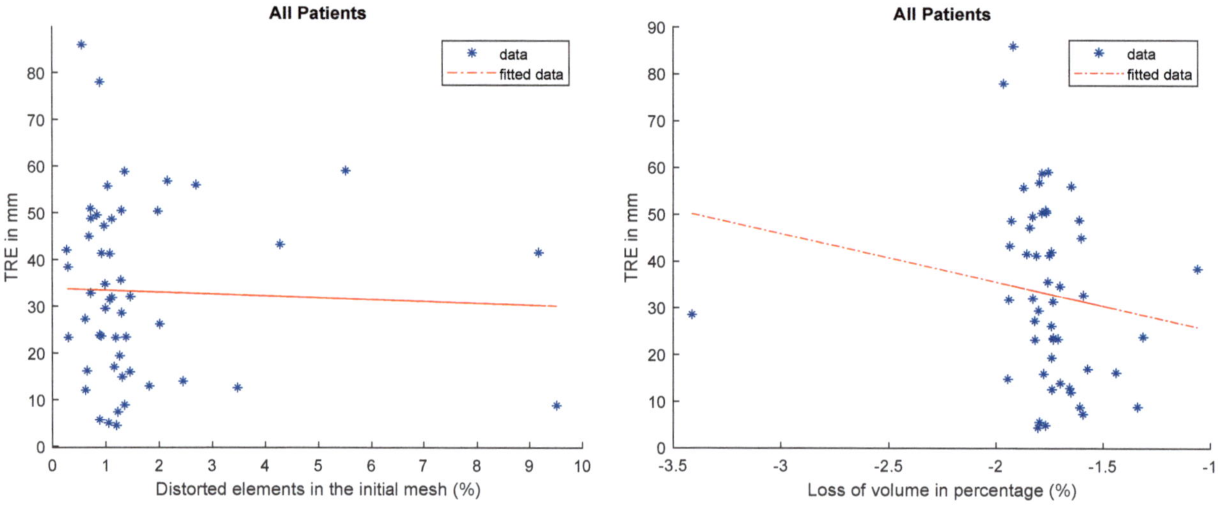

Figure 7.36: A very weak correlation is achieved between the accuracy of the biomechanical model based registration method and the percentage of distorted elements in the initial mesh across all the patients with an r of -0.04 (left). A very weak correlation is found between the total loss volume of MRI breasts and the TRE with an r of -0.16 (right).

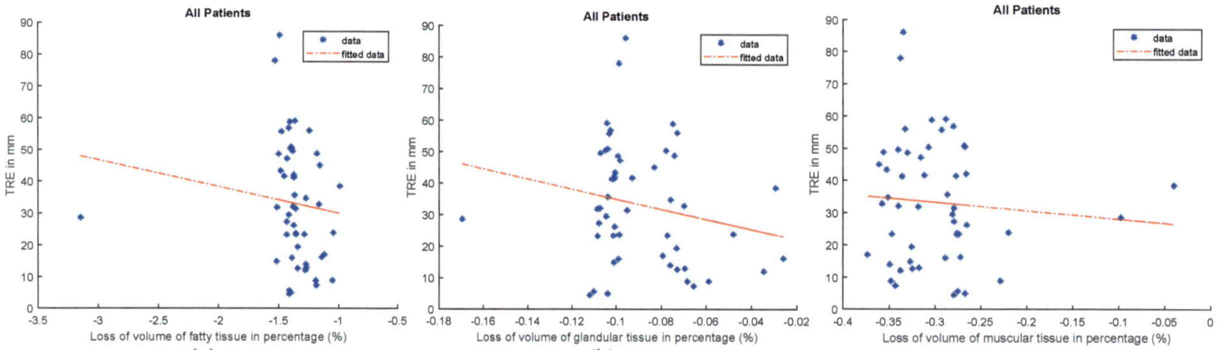

(a) (b) (c)

Figure 7.37: Relationship between the loss volume for the three tissues and the accuracy of the biomechanical model based registration method. A very weak correlation is found for loss of volume in fatty tissue with an r of -0.12 (a). A weak correlation is found for loss volume in glandular tissue with an r of -0.20 (b). A very weak correlation is found for loss in muscular tissue with an r of -0.08 (c).

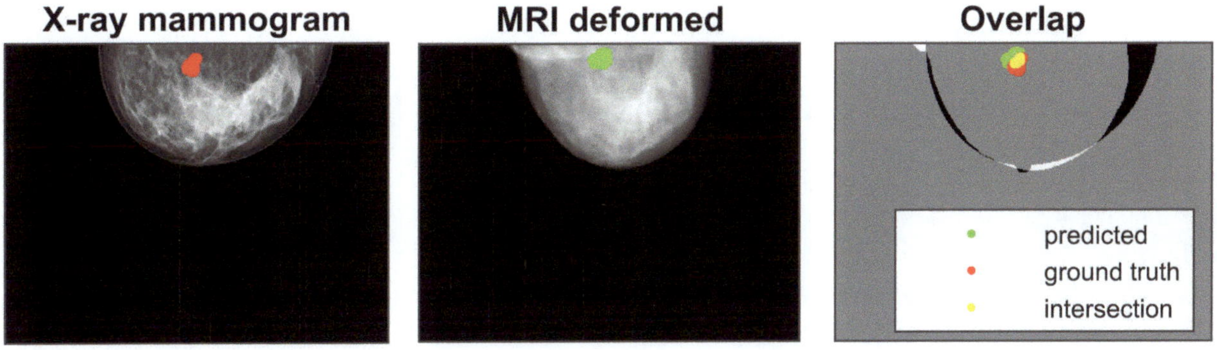

Figure 7.38: An example of registration between MRI and X-ray mammography (Patient ID 40): original X-ray mammography (first), synthetic X-ray created from the biomechanical model based registration (second), and overlap of the two images (third). Ground truth annotation marked by radiologists in original X-ray (red), projected tracked annotation of the MRI into the synthetic X-ray (green), and overlap between the two lesions (yellow).

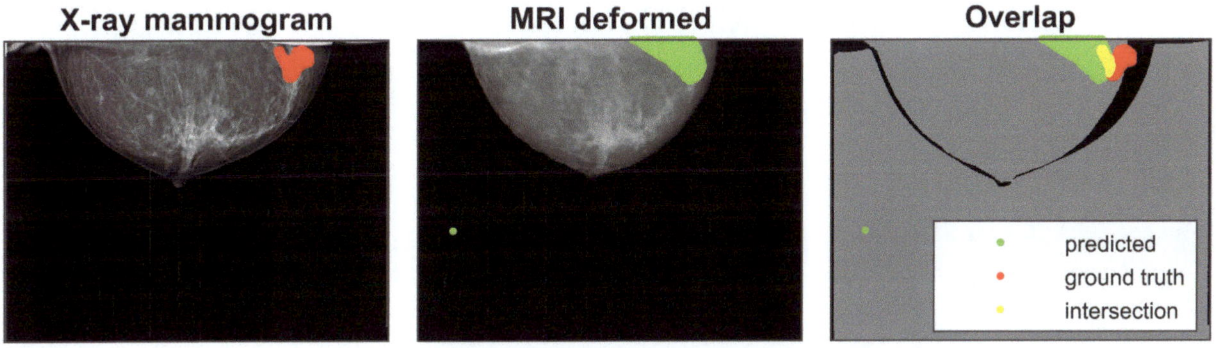

Figure 7.39: An example of registration between MRI and X-ray mammography for a lesion at the back of the breast (Patient ID 52): original X-ray mammography (first), synthetic X-ray created from the biomechanical model based registration (second), and overlap of the two images (third). Ground truth annotation marked by radiologists in original X-ray (red), projected tracked annotation of the MRI into the synthetic X-ray (green) and overlap between the two lesions (yellow).

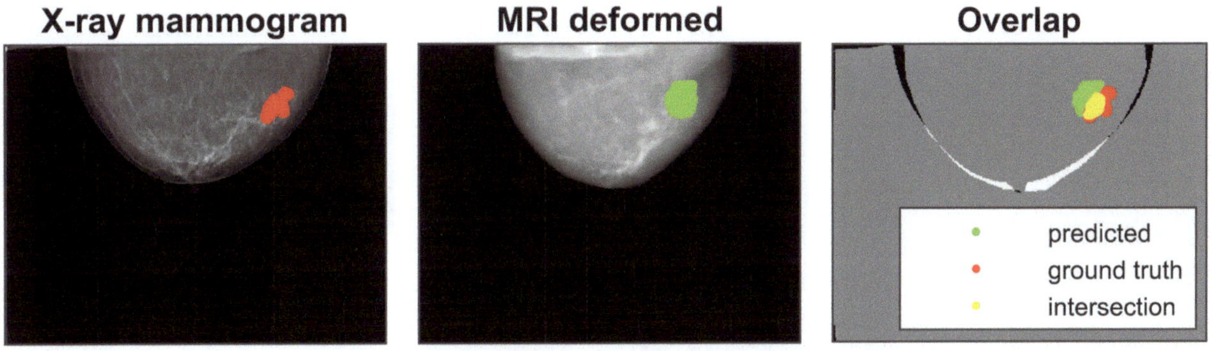

Figure 7.40: An example of registration between MRI and X-ray mammography for an ideal result (Patient ID 12): original X-ray mammography (first), synthetic X-ray created from the biomechanical model based registration (second), and overlap of the two images (third). Ground truth annotation marked by radiologists in original X-ray (red), projected tracked annotation of the MRI into the synthetic X-ray (green), and overlap between the two lesions (yellow).

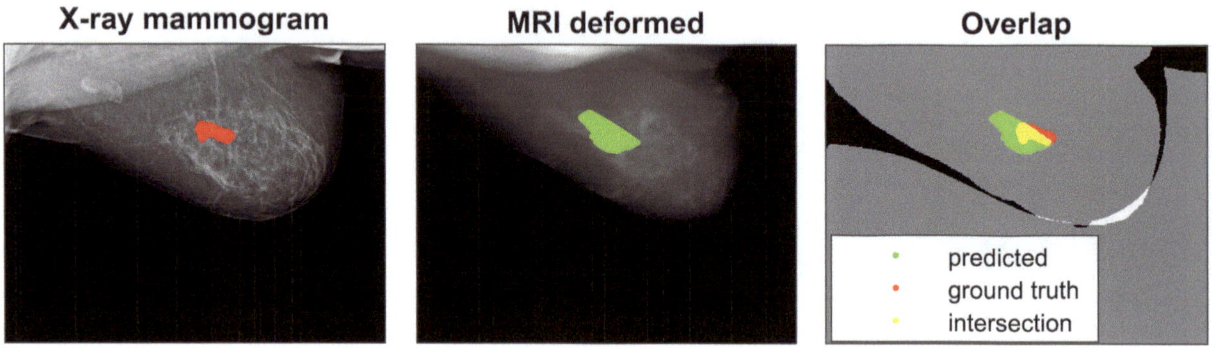

Figure 7.41: An example of registration between MRI and X-ray mammography (Patient ID 7): original X-ray mammography (first), synthetic X-ray created from the biomechanical model based registration (second), overlap of the two images (third). Ground truth annotation marked by radiologists in original X-ray (red), projected tracked annotation of the MRI into the synthetic X-ray (green), and overlap between the two lesions (yellow).

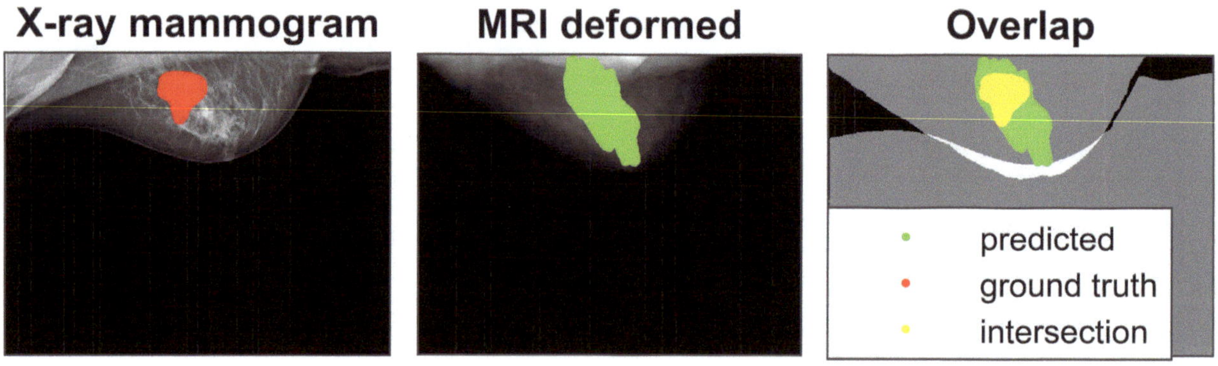

Figure 7.42: An example of registration between MRI and X-ray mammography (Patient ID 10): original X-ray mammography (first), synthetic X-ray created from the biomechanical moel based registration (second), overlap of the two images (third). Ground truth annotation marked by radiologists in original X-ray (red), projected tracked annotation of the MRI into the synthetic X-ray (green), and overlap between the two lesions (yellow).

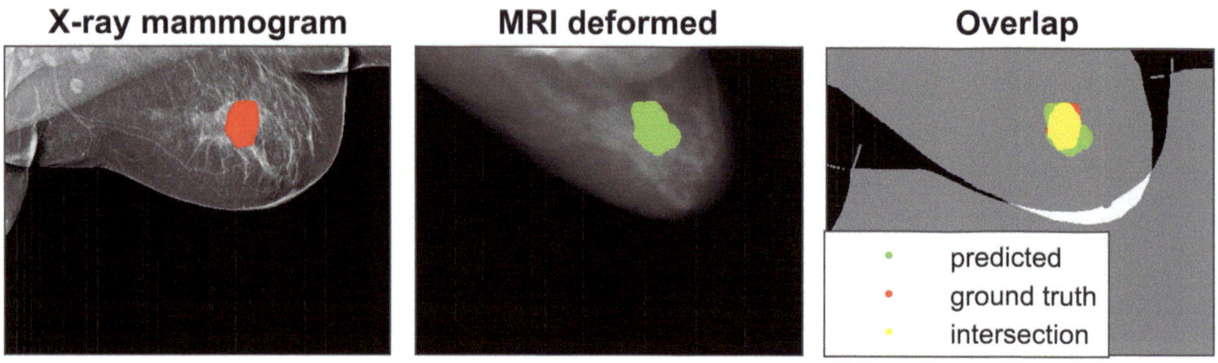

Figure 7.43: An example of registration between MRI and X-ray mammography (Patient ID 56): original X-ray mammography (first), synthetic X-ray created from the biomechanical model based registration (second), overlap of the two images (third). Ground truth annotation marked by radiologists in original X-ray (red), projected tracked annotation of the MRI into the synthetic X-ray (green), and overlap between the two lesions (yellow).

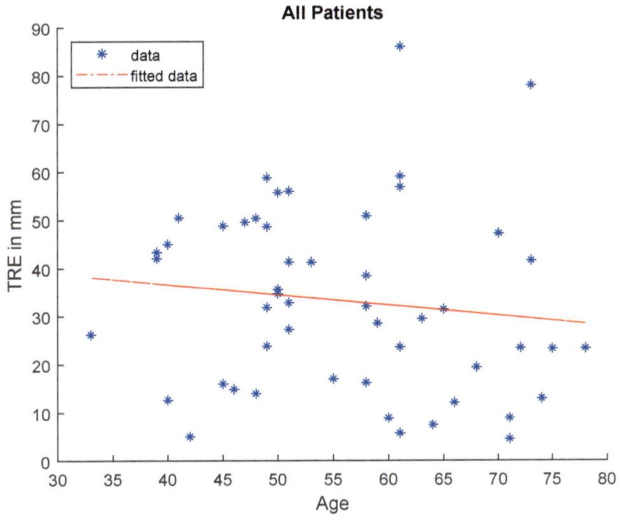

Figure 7.44: A very weak correlation between the accuracy of the biomechanical model based registration method and the age of the patients with an r of -0.13.

chanical model based registration method as shown in Figure 7.44. It was observed that age has a very weak correlation in the negative direction in relation to the TRE with an r of -0.13 and R^2 of 0.02.

Views

The second characteristic, which was investigated, is the views of full X-ray mammography. Boxplots have been used for evaluating the correlation.

It was observed that the median TRE of CC (21 datasets) is less by a factor of 2 than the median TRE of MLO (26 datasets) as shown in Table 7.12. For the TRE of ML and RL cases as shown in Table 7.13, the number of datasets (four datasets) is likely too small for a conclusion. Hence, Figure 7.47 shows only the TRE (left), DSC and IoU of the breast shapes (right) of CC and MLO cases which shows that the biomechanical model based registration method works better for CC cases. In addition, the number of datasets with an overlap of lesion annotations for CC cases is thirteen cases out of 21 cases as shown in Figure 7.45 while the number of datasets with an overlap of lesion annotations for MLO cases is seven cases out of 26 cases as shown in Figure 7.46.

For CC cases, the lower and the upper quarters are 11.65 mm and 29.4 mm, respectively. The minimum and the maximum errors are 4.4 mm and 56.7 mm, respectively. For the intersection between the two annotations as shown in Figure 7.45, the maximum DSC achieved is 0.52 and the maximum IoU achieved is 0.36 in the same case. The maximum TPR achieved is 0.83.

For MLO cases, the lower and the upper quarters are 23.8 mm and 49.5 mm, respectively. The minimum and the maximum errors are 5.6 mm and 85.9 mm, respectively. For the intersection between the two annotations as shown in Figure 7.46, the maximum DSC achieved is 0.63 and the maximum IoU achieved is 0.46 in the same case. The maximum TPR achieved is 0.97.

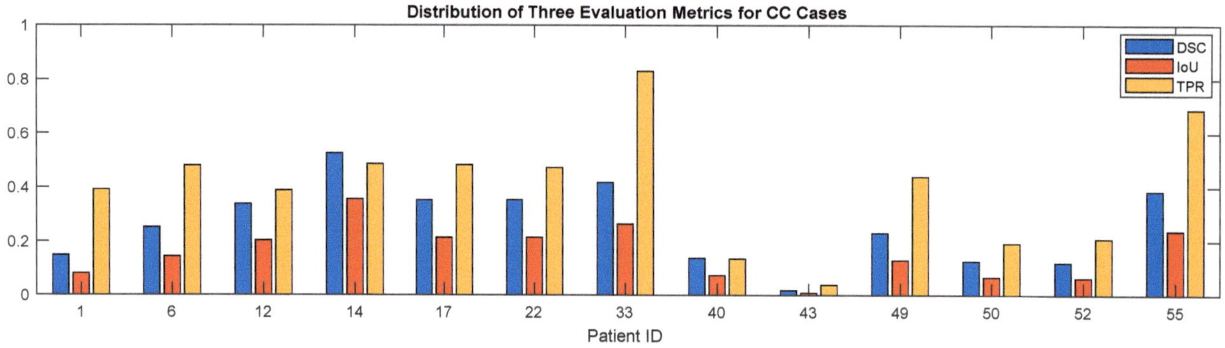

Figure 7.45: Distribution of the three evaluation metrics for the CC cases that have been calculated for the intersection between the actual and the predicted lesion using biomechanical model based registration method (the yellow color in Figures 7.38, 7.39, and 7.40). Thirteen cases out of 21 cases achieve an intersection.

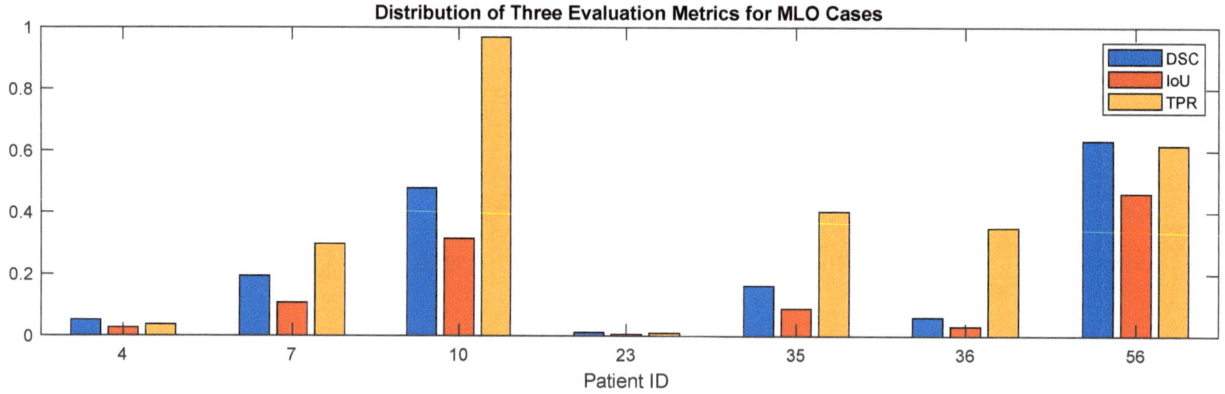

Figure 7.46: Distribution of the three evaluation metrics for the MLO cases that have been calculated for the intersection between the actual and the predicted lesion using biomechanical model based registration method. Seven cases out of 26 cases achieve an intersection.

In conclusion, the accuracy of the biomechanical model based registration method correlates with the views of full X-ray mammography regarding CC and MLO cases.

Table 7.12: TRE of biomechanical model based registration regarding views

Accuracy in *mm*	mean	median	std	cases
TRE - CC	23.27	23.29	15.6	21
TRE - MLO	39.95	41.42	19.2	26
TRE - ML/RL	40.38	40.83	11.56	3 /1

Histopathological Diagnosis

The third characteristic is in terms of the histopathological diagnosis of the breast needle biopsies as explained in Section 5.2.1. Based on the available data, there are five classes: B2 (15 cases), B3 (5 cases), B5a (21 cases), B5b (9 cases), and B2 and B5b at the same time (1 case). The highest number of cases belongs to both B2 and B5a classes, in which it has been observed that median TRE is 31.8 *mm*

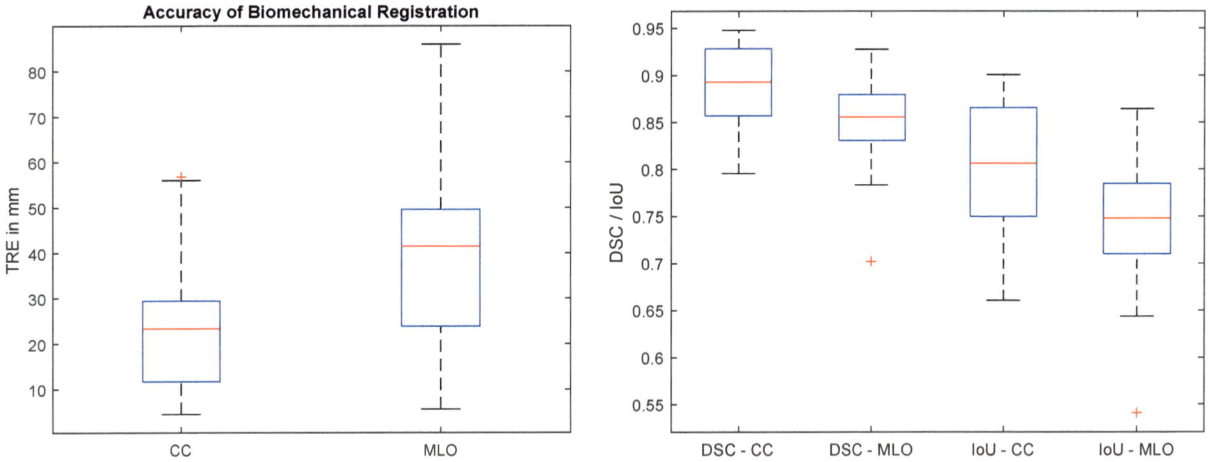

Figure 7.47: Analysis of TRE for the biomechanical model based registration method regarding CC and MLO cases (left). Analysis of the DSC and IoU for the shape of the breast between the produced synthetic X-ray mammography (predicted deformed MRI and the original X-ray mammography for the same cases (right).

Table 7.13: Evaluation of biomechanical model based registration regarding ML and RL cases

Evaluation Metrics	DSC-shape	IoU-shape	TRE	DSC	IoU	TPR
Patient ID 24 (ML)	0.83	0.70	50.3 *mm*	0	0	0
Patient ID 41 (ML)	0.84	0.73	50.41 *mm*	0	0	0
Patient ID 47 (ML)	0.89	0.81	31.35 *mm*	0.01	0.008	0.02
Patient ID 54 (RL)	0.75	0.60	29.4 *mm*	0.04	0.02	0.04

and 32.7 *mm*, respectively as shown in Figure 7.48. In addition, the minimum, maximum, lower quarter, and upper quarter of both classes, as shown in Table 7.14, are quite comparable. It could be concluded that there is no correlation with respect to the histopathological appearance.

Table 7.14: Detailed analysis of boxplots of class B2 and B5a versus the accuracy of the biomechanical model based registration

	min	max	lower quarter	upper quarter
B2	4.43 *mm*	77.9 *mm*	16.8 *mm*	54.4 *mm*
B5a	5.6 *mm*	85.9 *mm*	15.1 *mm*	48.9 *mm*

Breast Density

The breast density of clinical datasets is detailed in Section 5.2.2. Two aspects were taken into consideration: the glandular tissue related to the total breast volume and the ratio between glandular and fatty tissues in the breast. A polynomial fitting curve (first degree) has been used to evaluate how the datasets are distributed compared to the TRE of the biomechanical model based registration method as shown in

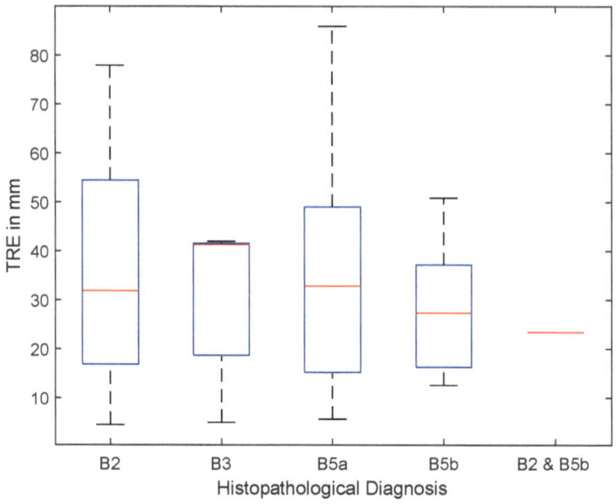

Figure 7.48: Analysis of TRE for the biomechanical model based registration method regarding the breast needle biopsies into categories with no correlation: B2 (15 cases), B3 (5 cases), B5a (21 cases), B5b (9 cases), and B2 and B5b (1 case).

Figure 7.49. It was observed that both aspects have a very weak correlation with the TRE with an r of 0.06 and 0.05, respectively, and R^2 of 0.004 and 0.003, respectively.

Sizes of Breasts and Lesions

The size of breasts and lesions is the fifth characteristic from Section 5.2.3 that has been investigated. A polynomial fitting curve (first degree) has been used to evaluate how the datasets are distributed compared to the TRE of the biomechanical model based registration method, as shown in Figures 7.50 and 7.51. Since it was observed in Section 5.2.3 that sizes of breasts in MRI (3D) and full X-ray mammography (2D) are correlating, it is expected that sizes of breasts will have the same behavior in relation to the TRE of the biomechanical model based registration method. There is a very weak correlation in the negative direction with an r of -0.06, -0.08, and -0.15 and with R^2 of 0.004, 0.006, and 0.02 for MRI, full X-ray mammography CC and MLO, respectively, as shown in Figure 7.50.

To analyze the impact of the sizes of lesions, a polynomial fitting curve (first degree) has been used to evaluate how the lesion size correlates to the TRE of the biomechanical model based registration method. Also, it was observed in Section 5.2.3 that sizes of lesions in MRI and full X-ray mammography are correlating. Hence, it is expected that sizes of lesions in MRI or full X-ray mammography will have the same behavior in relation to the TRE of the biomechanical model based registration method. There is a very weak correlation with an r of -0.09, -0.09, and 0.07 and with R^2 of 0.008, 0.009, and 0.004 for MRI, full X-ray mammography CC and MLO, respectively, as shown in Figure 7.51.

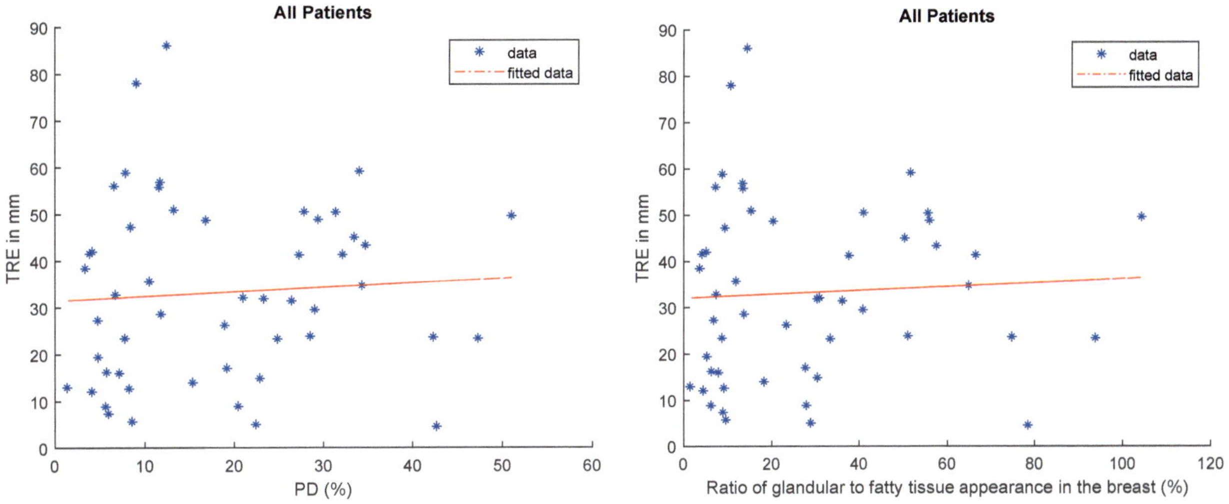

Figure 7.49: A very weak correlation between the accuracy of the biomechanical model based registration method and the distribution of glandular tissue in the breast with an r of 0.06 (left). A very weak correlation between the accuracy of the biomechanical model based registration method and the distribution of the ratio between glandular and fatty tissues in the breast with an r of 0.05 (right).

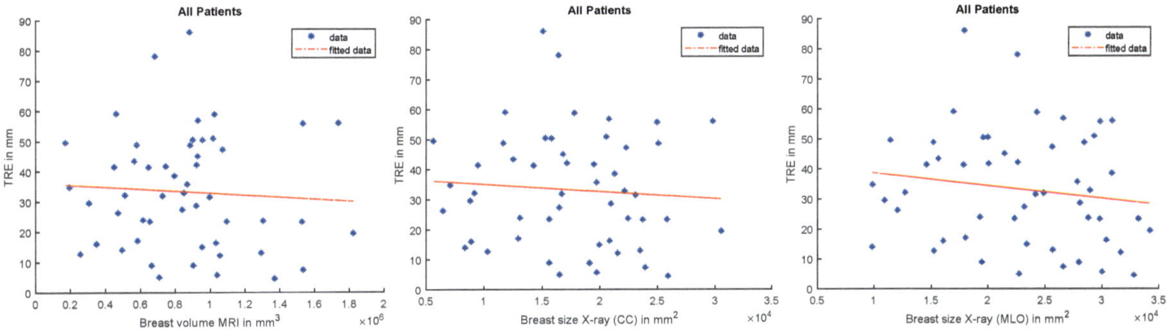

Figure 7.50: Relation between the accuracy of the biomechanical model based registration method and distribution of the breast sizes in MRI (3D) (first) and full X-ray mammograms (2D): CC (middle) and MLO (last). A very weak correlation is found between the breast sizes 3D and 2D and the TRE with an r of -0.06, -0.08, and -0.15, respectively.

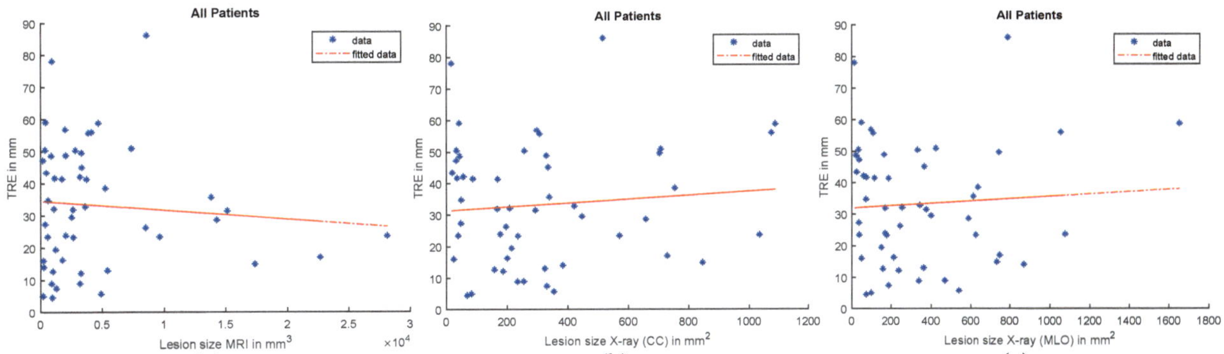

(a) (b) (c)

Figure 7.51: Relation between the accuracy of the biomechanical model based registration method and distribution of the lesions sizes in MRI (3D) (first) and full X-ray mammograms (2D): CC (middle) and MLO (last). A very weak correlation is found between the sizes of lesions 3D and 2D and the TRE with an r of -0.09, 0.09, and 0.07, respectively.

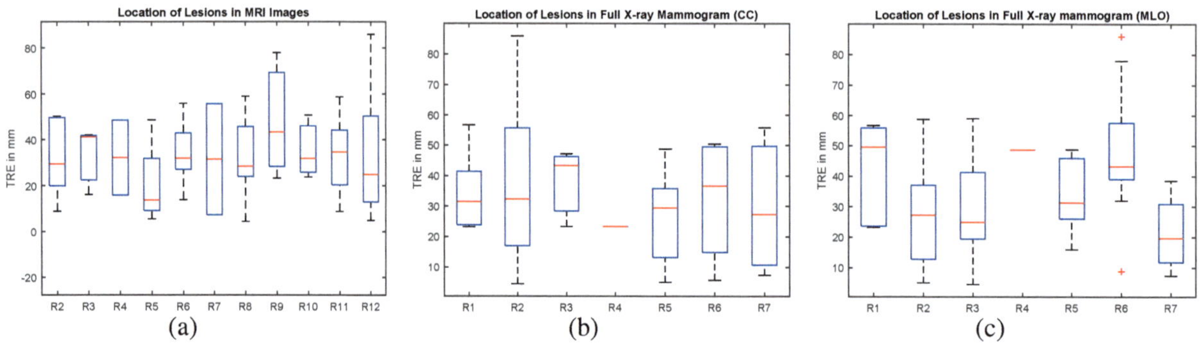

Figure 7.52: Relation between the location of lesions in MRI and full X-ray mammography in both views and the accuracy of the biomechanical model based registration method. R5 of four cases in MR images is the least median error, which corresponds to lower inner middle slices (first). R7 in both views of full X-ray mammograms (CC 8 cases (middle)) (MLO 4 cases (last)) is the least median error, which corresponds to the area of the center of the breast.

Location of Lesions

The location of lesions in MRI and full X-ray mammography is the sixth characteristic that has been evaluated. Boxplots have been used for finding a correlation between locations of lesions and the TRE of the biomechanical model based registration. MR images were split into twelve regions as explained in Section 5.2.4. X-ray mammography has been split into eight regions. For MRI, only two cases have annotated lesions in region R4. R3, R9, and R10 have only three cases, so it is hard to get a conclusive observation since the case numbers located in these regions are likely too small. R5 though, has also a small number of cases (four cases only). Nevertheless, it is expected that median TRE would be small because the deformation in this region, that belongs to the lower inner middle slices, is relatively small compared to other areas. The median TRE achieved is $13.7\,mm$. The highest number of cases belongs to R11 (eleven cases), in which the median TRE is $34.7\,mm$, as shown in Figure 7.52 (first) and it corresponds to lower outer middle slices. In addition, R11 in MRI corresponds to R2 in X-ray mammography which achieves a median of $32.3\,mm$ in the case of CC and a median of $27.2\,mm$ in the case of MLO.

For full X-ray mammography, similarly, some regions cannot be considered as conclusive such as R4 for both views (one case only) and R3 (three cases) in case of CC view. It is clear that lesions at the back of the breast near the muscle in the posterior direction (R1, R2, R5, and R6 in both views) are harder to register than the lesions at the center of the breast (R7): which achieves the least median of $27.3\,mm$ for eight cases in CC view and the median of $19.7\,mm$ for four cases in MLO view.

7.4 Image Based Registration

In this section, the results of aligning between full X-ray mammograms and spot mammograms will be presented. Similarly, TRE, TPR, IoU, and DSC which are explained in Chapter 6, have been calculated between the predicted annotation and the annotation marked by radiologists. In addition, image quality assessment for image contrast and the algorithm for removing the muscle area from the full X-ray mammograms have been evaluated.

As explained before, a free-hand shape annotation was used. Hence, for calculating the TRE, the COG annotation was used to calculate the Euclidean distance between the predicted annotation and the actual annotation. In this case, both annotations are 2D. The same three cases for CC mammograms have been visualized for image based registration in Figures 7.76, 7.75, and 7.77. Similarly, the same three cases for MLO have been visualized for image based registration in Figures 7.78, 7.79, and 7.80.

In these visualizations, the red lesion represents the lesion that is marked by the radiologist, while the green lesion represents the predicted lesion based on the annotation in the spot mammogram using the proposed image based registration. The yellow color represents the overlap of the two lesions.

7.4.1 Image Quality Assessment for the Contrast of Spot Mammograms

As explained in Section 4.2.1, there are four properties in the GLCM matrix; contrast, correlation, energy, and homogeneity utilized for evaluating the image quality of the spot mammogram. Those four properties have been calculated for the six methods proposed for evaluating the quality improvements of spot mammograms using the 51 available clinical datasets. Then afterwards, the mean, median, and standard deviation of each parameter were calculated. Across the six proposed methods (denoted from 1 to 6) in this thesis, it was observed that correlation and homogeneity do not have an impact on the quality of the image because there is no significant difference in ranges achieved (see Figure 7.53). However, contrast and energy vary between the six methods. In addition, methods 1 and 6 could be excluded since the values of contrast are low and the values of energy are high compared to the other methods, as shown in Figure 7.53. The number of outliers in method 1 is too high compared to the other methods. The interval between 25% and 75% is high in method 6 compared to the other methods. The standard deviation of those two methods (1 and 6) is high compared to others. Additionally, it has been observed visually in Figure 4.6 that methods 1 and 6 provide images that are very dark and bright, respectively. From the literature (130), the values of contrast in breast X-ray mammography range from 0.01 to 0.05 for tumor tissues, while for normal tissues, they range from 0.02 to 0.07. The values of energy in breast X-ray mammography range from 0.2 to 0.7 for tumor tissues, while those for normal tissues range from 0.2 to 0.4.

The evaluation of the quality of the images is done for spot mammograms in which there might be tumor tissue. The ranges of contrast and energy should be like X-ray mammography for tumor tissues. For getting a reference value to calculate the Euclidean distance, the midpoint of this range was assumed to be this value. Afterward, the Euclidean distance was calculated between the reference value and the actual value of each parameter and summed up for each dataset. The median, mean, and standard deviation across all the datasets were then calculated, as shown in Figure 7.54 for the six methods.

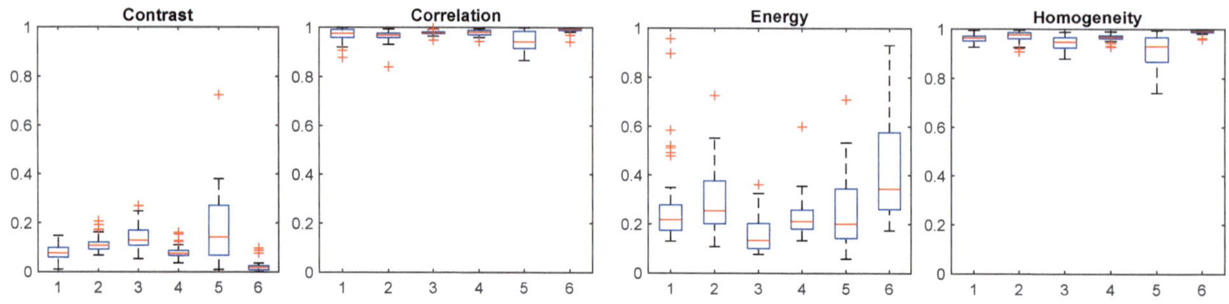

Figure 7.53: Boxplots of the four properties extracted from GLCM matrix for the six methods (denoted from 1 to 6) that have been tested for enhancing image contrast of spot mammograms.

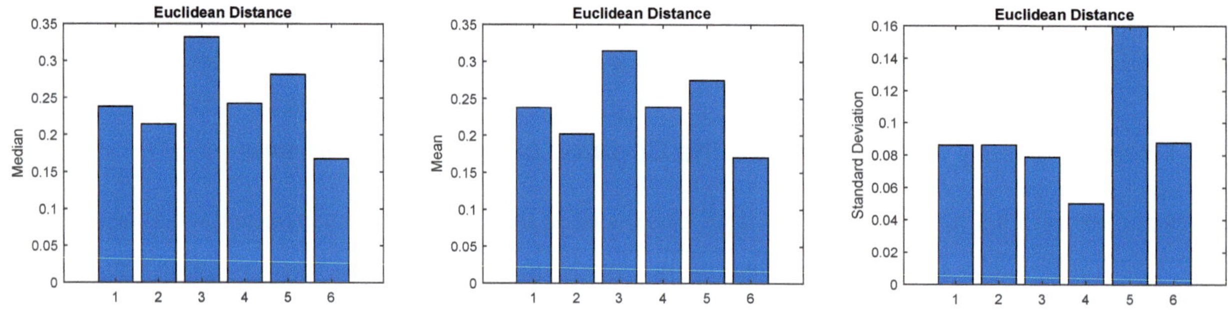

Figure 7.54: The Euclidean distance between the ideal value (reference value taken from literature) and the actual values for contrast and energy parameters. The median of the Euclidean distance for all the patients (first). The mean of the Euclidean distance for all the patients (middle). The standard deviation of the Euclidean distance for all the patients (last).

Additionally, it was observed that method 5 could be excluded as well since the standard deviation is high compared to other methods.

Additionally, the distribution of contrast and energy values for three methods (2, 3, and 4) for the 51 clinical datasets has been visually represented in Figure 7.55 and in Figure 7.56. It has been realized that contrast values are always less in method 4 than in the other two methods, and energy values are always higher in method 2 than in the other two methods. Those three methods, 2, 3, and 4, provide approximately similar results. Hence, an analysis of TRE for the different image similarity metrics that have been explained in Section 4.1, has been investigated for the three methods of image contrast enhancement for the spot mammogram, as shown in Figures 7.57, 7.58, and 7.59. Through this analysis, it was observed that method 2 has the least outliers in all the metrics and the least median error in most of the metrics as shown in Figure 7.57. Therefore, method 2 was selected for the purpose of enhancing the image quality of spot mammograms in this thesis.

7.4.2 Evaluation of Cropping the Muscle Area from MLO Cases

To assess the accuracy of removing the muscle from the full X-ray mammogram, manual annotation was carried out for six cases from the available data. These annotations were marked utilizing a feature that

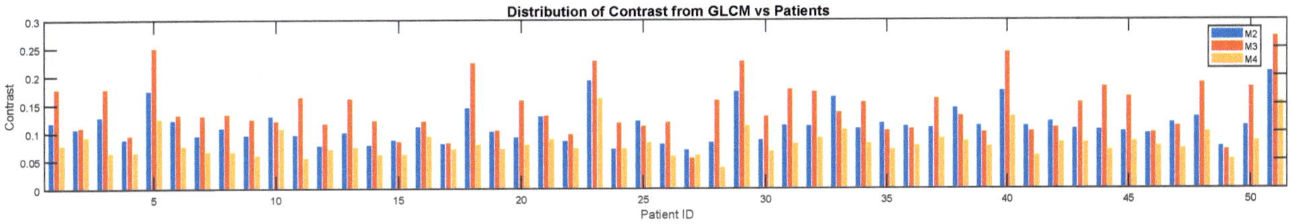

Figure 7.55: Distribution of the contrast property extracted from GLCM matrix for all the available datasets for the three most robust image contrast methods: method 2 (denoted as M2), method 3 (denoted as M3), and method 4 (denoted as M4) that have been tested from enhancing image contrast of spot mammograms.

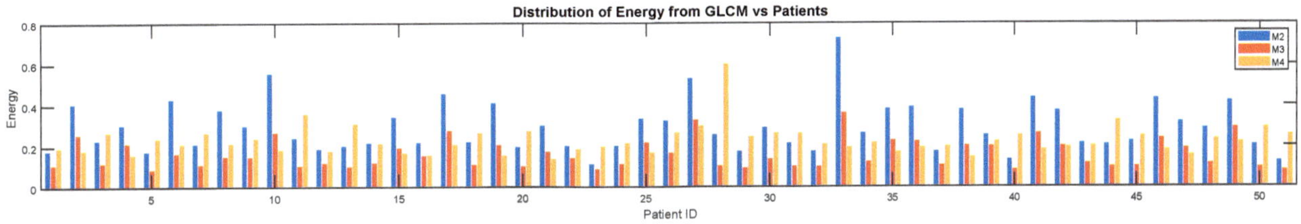

Figure 7.56: Distribution of the energy property extracted from GLCM matrix for all the available datasets for the three most robust image contrast methods: method 2 (denoted as M2), method 3 (denoted as M3), and method 4 (denoted as M4) that have been tested from enhancing image contrast of spot mammograms.

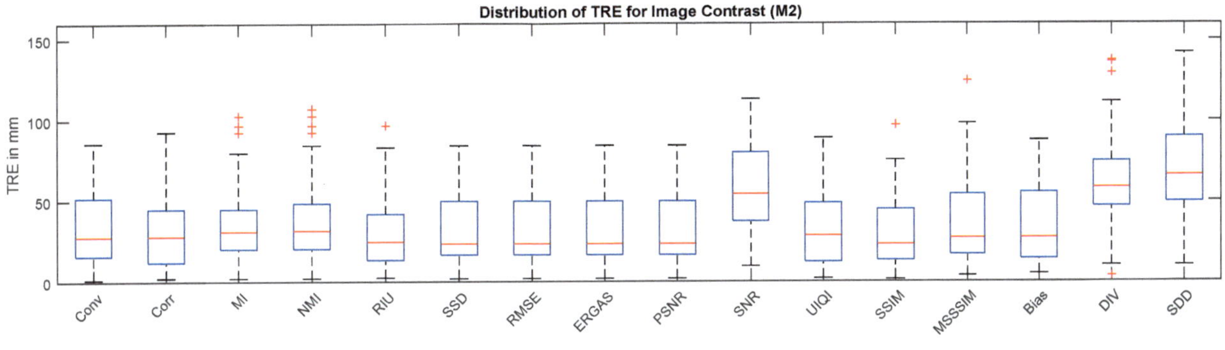

Figure 7.57: Analysis of different image similarity metrics with 51 patients using method 2 of image contrast. Distribution of the TRE for all tested image similarity measures.

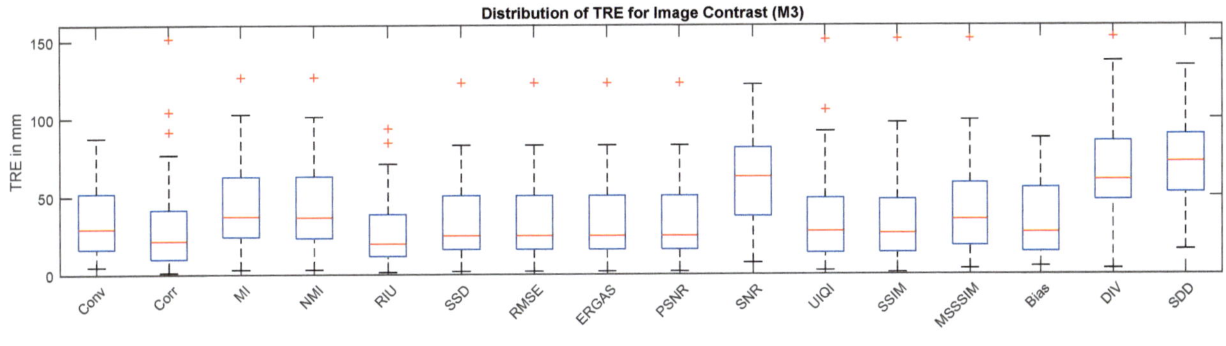

Figure 7.58: Analysis of different image similarity metrics with 51 patients using method 3 of image contrast. Distribution of the TRE for all tested image similarity measures.

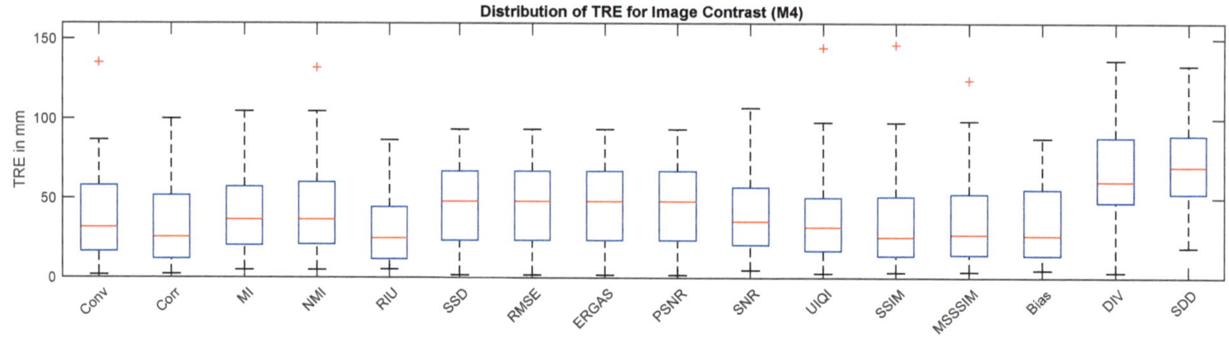

Figure 7.59: Analysis of different image similarity metrics with 51 patients using method 4 of image contrast. Distribution of the TRE for all tested image similarity measures.

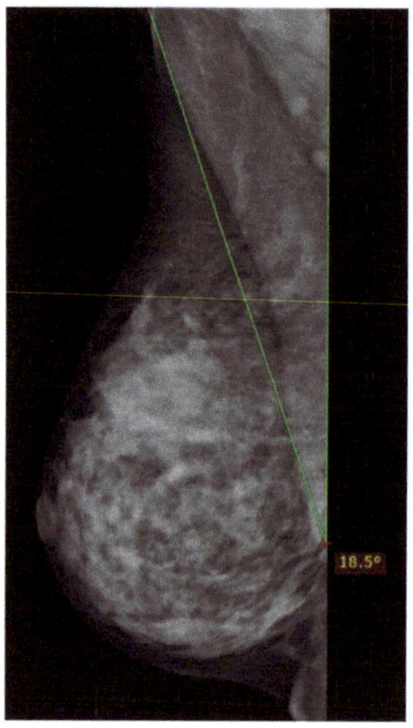

Figure 7.60: An example of the manual annotation of the rotation angle of the muscle area used as a ground truth in RadiAnt DICOM Viewer.

calculates the rotation angle in RadiAnt DICOM Viewer, as shown in Figure 7.60. Since the ground truth was generated manually, a subset from the available datasets was randomly chosen. These values have been compared to the automatic calculation of the proposed methods in Section 4.2.1. Also, the absolute error between the annotated value (manually) and the predicted value (automatically by the proposed methods) has been calculated (third column). The lowest and highest absolute errors achieved are 0° and 2.7°, respectively.

7.4.3 Optimization

After discussing the evaluation results of the preprocessing step, in this section evaluating and combining image similarity metrics will be discussed.

Table 7.15: Rotation angle of the muscle area in the full X-ray mammography

cases	**rot. angle manually**	**rot. angle automatically**	**Absolute Error**
1	13.8°	13.8°	0°
2	18.5°	18.1°	0.4°
3	17.5°	17.2°	0.3°
4	20.2°	19.2°	1°
5	14.5°	12.9°	1.6°
6	15.4°	18.1°	2.7°

The entire X-ray mammogram

First, the image similarity metrics have been calculated between the full size of the X-ray mammogram (the entire X-ray mammogram) and the spot mammograms sliding the spot mammogram over the full X-ray mammogram step by step for all the 51 clinical datasets, as shown in Figure 7.61. Furthermore, three evaluation metrics have been calculated for each image similarity separately, as shown in Figure 7.64. It has been realized that 35 cases could have an overlap between the two annotations if there is an ideal image similarity metric applied. Considering the tested image similarity metrics, RMSE can align the images such that 15 cases show an overlap of annotations, as shown in Figure 7.67 (a). RMSE has also the least median TRE of $22.2mm$, as shown in Figure 7.61. Figure 7.64 shows that RMSE, MSE, ERGAS, and PSNR are mathematically equivalent since all of them are based on MSE, as explained in Section 4.1.

There are two disadvantages of sliding the spot mammogram over the full size of an X-ray mammogram. First, computation time is very high in some cases. It may take over 100 hours to calculate all the image similarity metrics desired (see Figure 7.71 (first)). The computation time varies from patient to patient because of the different sizes of breasts since all the image similarity metrics are calculated after the preprocessing step in which the image is cropped till the nipple position. Second, It slides over the full size X-ray which is not focused on an exact ROI. That might affect the accuracy by finding a local minimum or a local maximum value instead of a global value based on the optimum value of the applied image similarity metric and as a consequence, the overlap between annotations is not achieved as desired. That is why it was proposed to select a ROI first, instead of applying the image similarity analysis of the full size of the X-ray.

Select an ROI

From the results of applying image similarity metrics over the full size of full X-ray mammogram, convolution, RMSE, and UIQI could be selected as constraints to determine ROI since they don't have outliers and nearly have the same lower and upper quarter. Additionally, other metrics such as Corr, SSIM, RIU, MI, and NMI have been examined at the beginning to select the ROI for all 51 patients. Based on the available data, the convolution operation stands out as the most robust image similarity for

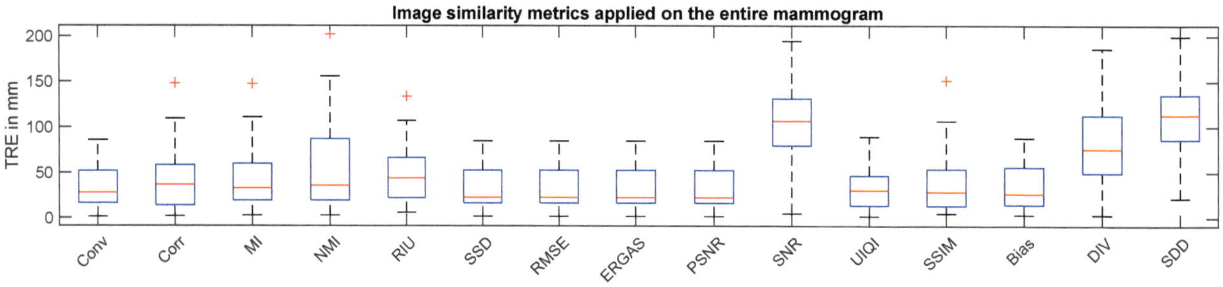

Figure 7.61: Analysis of different image similarity metrics for 51 patients in CC/MLO/ML and CC/ML/LM views for full X-ray mammograms and spot mammograms, respectively in two imaging situation positions: prone and upright. Distribution of the TRE for all tested image similarity measures applied for the full size of X-ray mammograms.

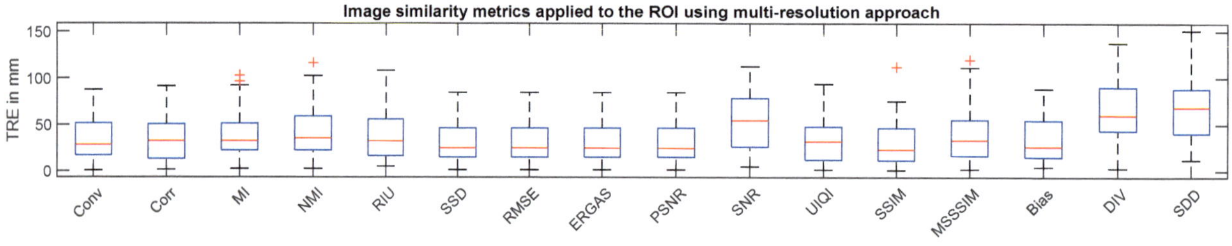

Figure 7.62: Image similarity metrics of 51 patients using image based registration based on dynamic step size with convolution constraint (three iterations). All of the datasets are rescaled for processing with a factor of 0.1 after determining the ROI using the dynamic stepping.

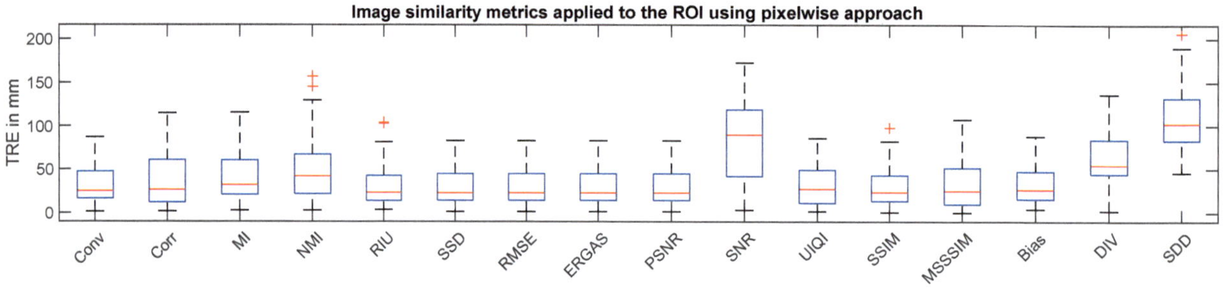

Figure 7.63: Image similarity metrics of 51 patients in CC/MLO/ML views and left/right breast for full X-ray mammograms and CC/ML/LM views and left/right breast for spot mammograms in two imaging situation position: prone and upright using image based registration based on a pixelwise approach. All of the datasets are rescaled for processing with a factor of 0.1 after getting 50% of the maximum values from the convolution operation.

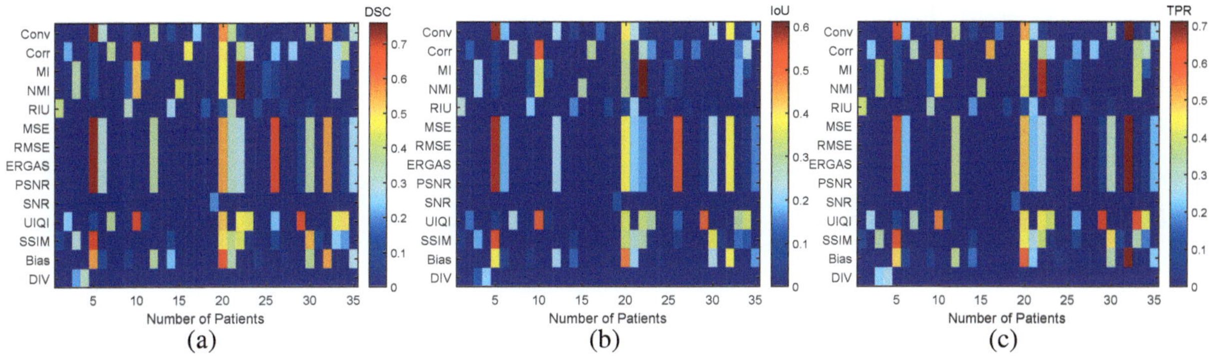

Figure 7.64: Analysis of different image similarity metrics for 51 patients in CC/MLO/ML and CC/ML/LM views for full X-ray mammograms and spot mammograms, respectively. Distribution of the three evaluation metrics: DSC, IoU, and TPR for all tested image similarity metrics applied for the full size of X-ray mammogram. 35 cases can be achieved if an ideal image similarity metric is applied.

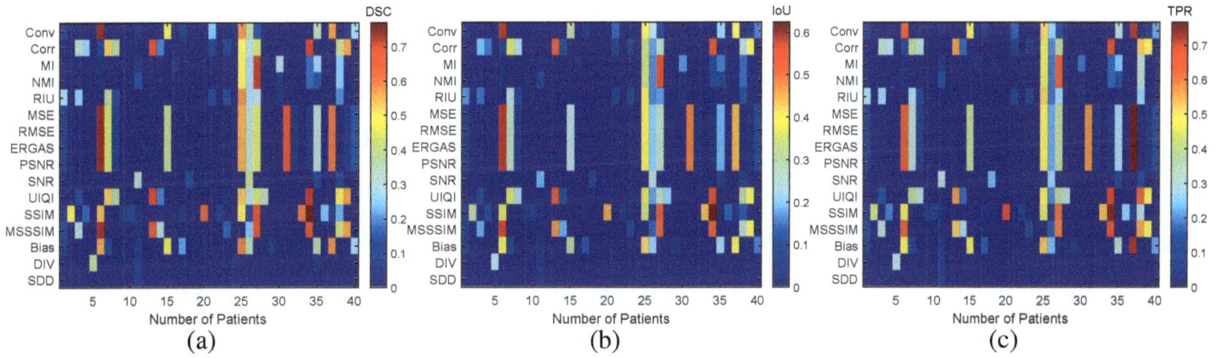

Figure 7.65: Analysis of different image similarity metrics for 51 patients in CC/MLO/ML and CC/ML/LM views for full X-ray mammograms and spot mammograms, respectively. Distribution of the three evaluation metrics: DSC, IoU, and TPR for all tested image similarity metrics applied for the ROI using a multi-resolution approach. 40 cases can be achieved if an ideal image similarity metric is applied.

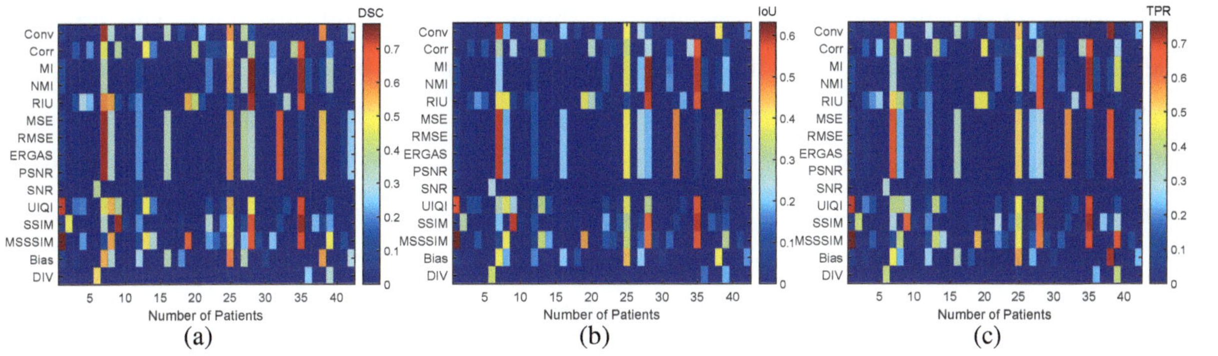

Figure 7.66: Analysis of different image similarity metrics for 51 patients in CC/MLO/ML and CC/ML/LM views for full X-ray mammograms and spot mammograms, respectively. Distribution of the three evaluation metrics: DSC, IoU, and TPR for all tested image similarity metrics applied for the ROI using pixelwise approach. 42 cases can be achieved if an ideal image similarity metric is applied.

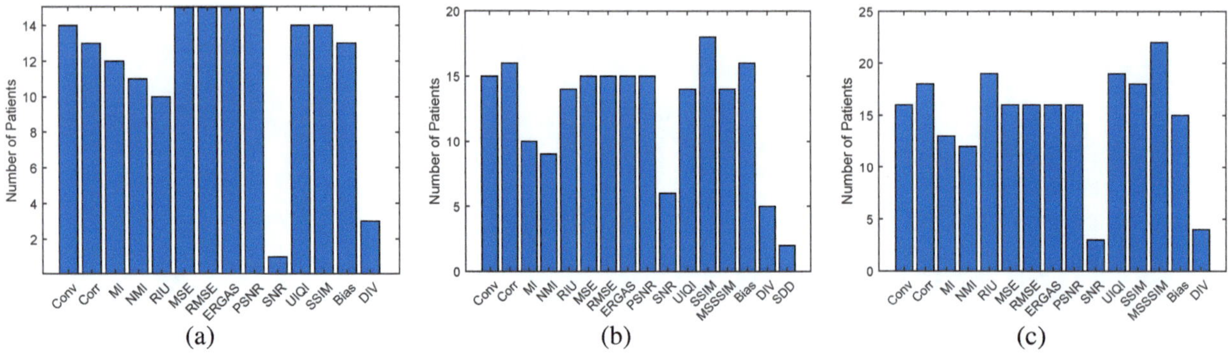

Figure 7.67: Number of datasets with an overlap of lesion annotations between the predicted and the ground truth annotation in each experiment: using the entire X-ray mammogram (a), by getting an ROI using multi-resolution approach (b), and by getting an ROI using a pixelwise approach (c).

the particular problem, making it the preferred choice for selecting the ROI. Furthermore, it is one of the metrics that does not have outliers. The median TRE for 51 patients is $27.6mm$. The lower and the upper quarter are $16.5mm$ and $51.5mm$, respectively, as shown in Figure 7.61.

Although there are some cases using the convolution operation that have high errors, it does not have an effect on selecting the correct ROI. The decision of selecting correct ROI is based on, where the spot mammogram potentially exists. The number of datasets with an overlap of lesion annotations is 14 cases using the entire X-ray mammogram, as shown in Figure 7.67 (a). It has been evaluated visually that using the convolution as a metric to determine the ROI, yields the best results. Only four patients present bad ROI in which the ROI deviates from the location where the spot mammogram should be located. To avoid the problem of having a bad ROI, there are two options. First, optimizing the stepping sizes of X and Y coordinates in the full X-ray mammogram and the number of iterations could be enhanced to select the best ROI.

Second, instead of using a multi-resolution approach, applying convolution operation over the full X-ray mammogram pixel by pixel could be an option to determine an ROI by taking the area that is greater than the mean value of the convolution operation, as shown in Figure 7.69 (first row). This proposal would help in not having bad ROI, but it will increase computation time by a factor of 83x, which is not desirable, as shown in Figure 7.68. Furthermore, the accuracy of using the second option to determine the ROI did not improve that much, as shown in Figures 7.63, 7.66, 7.67 (c). The number of datasets increased by two (42 cases in total) only compared to the proposed mutli-resolution approach (40 cases in total). 42 and 40 cases have an overlap from one of the image similarity metrics, as shown in Figures 7.65 and 7.66, respectively. For further optimization of combining the image similarity as will be discussed in the next section, it achieves similar accuracy but costs more computation time. That is why the multi-resolution approach was selected for getting the ROI.

After calculating all the image similarity metrics between the ROI and spot mammograms, the two disadvantages of full size mammograms are improved. First, it has been realized that computation time decreased significantly, which will be discussed in the next paragraph. Second, as the image similarity

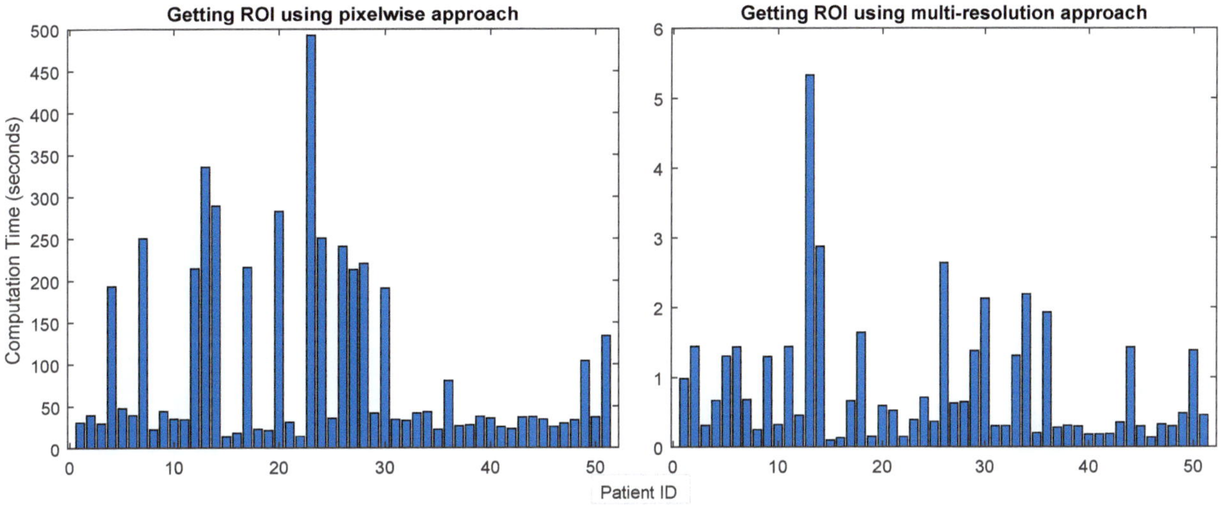

Figure 7.68: Comparison between computation time of getting ROI using two approaches. The first approach uses a pixelwise approach (left) and the second approach uses multi-resolution approach (right).

calculation focuses on a region of interest, the accuracy increased, as shown in Figure 7.62. Most of the accuracy of the lower quarter of the image similarity metrics (RMSE, SSIM, UIQI, Corr, RIU) decreased. As an example, the TRE using RMSE as similarity metric decreased from 16.2 mm to 14.6 mm, and SSIM decreased from 13 mm to 11.2 mm, RIU decreased from 22 mm to 15.8 mm. Additionally, the number of datasets with an overlap of lesion annotations increased from 35 to 40 cases if there is an ideal image similarity metric to be applied and the maximum for each image similarity metric separately is 18 cases such as SSIM image similarity in Figure 7.67 (b).

Rescaling Factor

To speed up the method, the second step that has been proposed is applying a rescaling factor to the images without affecting the accuracy. An experiment has been analyzed for selecting rescaling factors from 0.1 to 1 with a stepsize of 0.1 on a subset from the available datasets (17 patients), which was available at that time, for a subset from the image similarity metrics. This experiment has been tested using the entire X-ray mammography. The median, mean, and standard deviation of TRE achieved has been calculated at each step size of the rescaling factor. From this analysis, as shown in Figure 7.70, it has been realized, that whatever the rescaling factor is selected, it does not have a huge impact on the accuracy. For example, the median TRE achieved using the convolution operation of a rescaling factor of 0.1 till 0.8 is the same with a value 52.4 mm. Since the computation time is high, a 0.1 rescaling factor is selected for decreasing time. By selecting this rescaling factor, the computation time decreased from 2 hours to ten minutes.

It has been realized that computation time decreased significantly, as shown in Figures 7.71 (b) and (c), for example, in the cases that need 100 hours to compute, by applying multi-resolution approach in addition to the rescaling factor, 15 minutes is only required. Also for the cases that needed 3.6 hours,

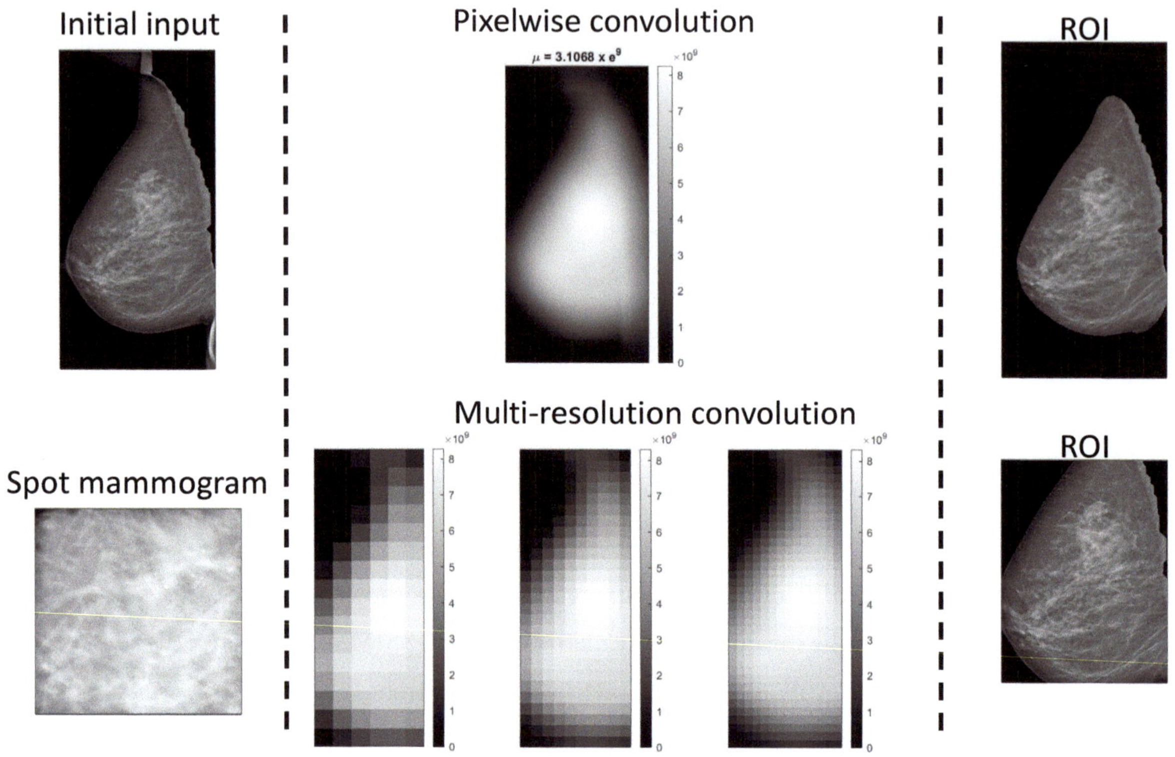

Figure 7.69: An example of applying the convolution operation in order to get the ROI: pixelwise approach (first row) versus multi-resolution approach using three iterations (second row). The third column is the ROI retrieved from both operations, respectively.

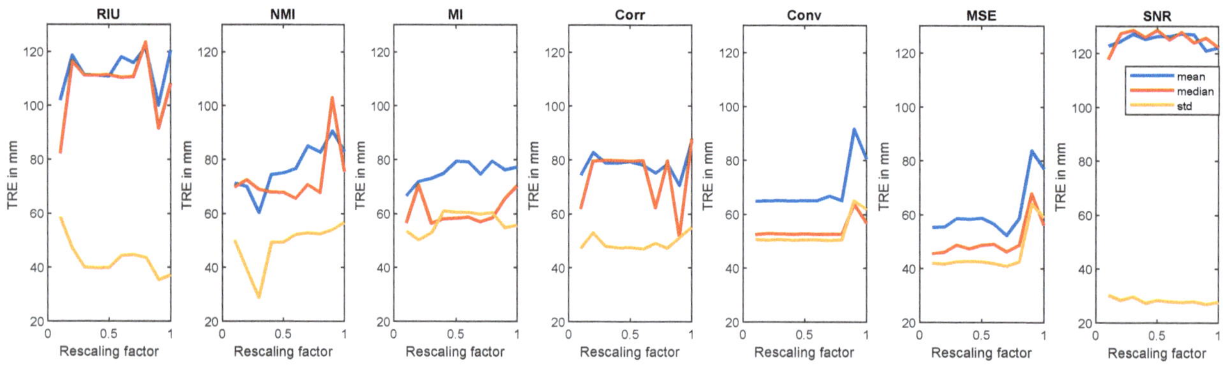

Figure 7.70: An experiment for testing the most robust rescaling factor using a subset of patients. A subset of image similarity metrics (RIU, NMI, MI, Corr, Conv, MSE, and SNR) is examined from a scale factor of 0.1 to 1 with a step size of 0.1. The mean, median, and standard deviation of the subset of the patients are shown in these image similarity metrics. There is no big difference between selecting the rescaling factor.

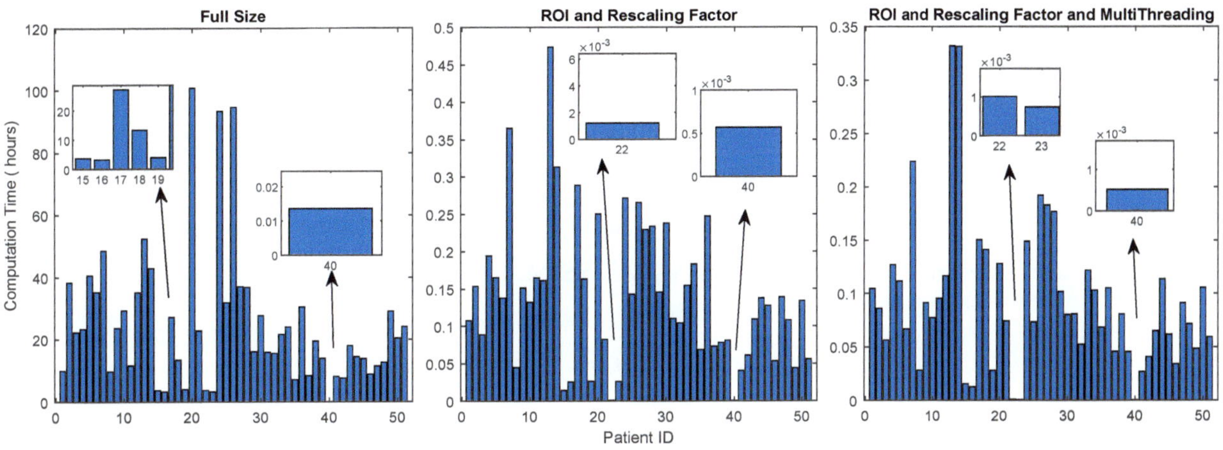

Figure 7.71: The comparison of computation time of applying all the image similarity metrics using the full size of X-ray mammograms or ROI. The computation time of applying all image similarity metrics using the full size of X-ray mammogram (a). The computation time of applying all image similarity metrics using ROI and rescaling factor instead (b). It decreased from a couple of hours of a maximum of 100 hours to a maximum of 15 minutes (b). The computation time like (b) in addition to applying a multi-threading looping (c).

after this proposal, it needs only 4.2 seconds. The difference between Figure 7.71 (b) and 7.71 (c) is applying additional multithreading for the loops, which decreased the computation time more without affecting the accuracy.

7.4.4 Combination of the Most Robust Image Similarity Metrics

Each image similarity metric separately achieves a maximum of 18 datasets with an overlap of lesion annotations like SSIM. It has also been realized that some image similarity metrics work well for some clinical data and other image similarity metrics work well for other clinical data. For robustifying the method, there are five proposed combinations of the most robust image similarity metrics. The most robust image similarity metrics were selected based on similar median error achieved, the least number of outliers, and the lower quarter achieved in addition to empirical experiments for the available data set. For each of the following proposed combinations, the most robust image similarity was selected based on the four mentioned parameters.

The first proposed combination added up the most robust image similarity metrics together as shown in equation 7.1 (denoted as Img_{Uni1}). For the image similarity that has the minimum value as the optimum instead of the maximum value, such as RIU, RMSE, and BIAS, the image similarity metrics were inverted and then added. Then, the position of the optimum value was extracted and considered to be the best match. In this case, the optimum value was considered to be the maximum. Based on the available clinical datasets, the most robust image similarity metrics are Conv, Corr, SSIM, MS-SSIM, RIU, and MSE.

$$Img_{Uni1} = Conv + Corr + SSIM + MSSSIM + RIU + MSE \tag{7.1}$$

The second proposed combination subtracted the most robust image similarity metrics from each other as shown in equation 7.2 (denoted as Img_{Sub}). Similarly, for the image similarity that has the minimum value as the optimum value, they were inverted and then subtracted. Then, similarly, the position of the optimum value was extracted and considered to be the best agreement. In this case, the optimum value was considered to be the minimum. Similarly, for subtraction, the most robust image similarity metrics which have been selected are Conv, Corr, SSIM, MS-SSIM, RIU, and MSE.

$$Img_{Sub} = Conv - Corr - SSIM - MSSSIM - RIU - MSE \tag{7.2}$$

The third and fourth proposed combination determined the weight of each pixel in an image (denoted as img) based on the gradient magnitude at that pixel and then returned the weight array (denoted as W) as shown in equation 7.3. σ was used as the standard deviation for the derivative of Gaussian to compute the image gradient.

To apply to the particular problem, the image (img) was assumed to be the most robust selected image similarity metric and two σs have been examined. First, σ was assumed to be the standard deviation (std) of the applied image similarity metric for example, convolution is selected, as shown in equation 7.5. Second, σ was assumed to be the mean (μ) of the applied image similarity metrics as shown in equation 7.4. So, in total two image weights (W) have been calculated for each robust image similarity. All the weighted images with $\sigma = \mu$ as in equation 7.4 were added up together and all the weight images with $\sigma = std$ as shown in equation 7.5 were added up together. It was empirically tested that the best combination of image similarity metrics based on the available datasets is Conv, Corr, SSIM, MS-SSIM, RIU, MSE, UIQI, NMI, and Bias. Then afterward, the third proposed combination subtracted the first proposed method (denoted as Img_{Uni1}) from the total of all the image similarity metrics of the weighted images (denoted as $Img_{MeanTot}$) as shown in equation 7.6 (denoted as Img_{Uni2}). The fourth proposed combination added the first proposed method (denoted as Img_{Uni1}) to the total of all image similarity metrics of the weighted images (denoted as $Img_{MeanTot}$) and subtracted the total of all image similarity metrics of the weighted images (denoted as Img_{StdTot}) from them as shown in equation 7.7 (denoted as Img_{Uni3}). Then, for both these two proposed combinations, the position of the optimum value was extracted and considered to be the best match. In this case, the optimum value was considered to be the maximum.

$$W = gradientweight(img, \sigma) \tag{7.3}$$

$$Img_{MeanConv} = gradientweight(Conv, mean2(Conv)), \quad Img_{MeanTot} = Img_{MeanConv} + Img_{MeanCorr} + .. \tag{7.4}$$

$$Img_{StdConv} = gradientweight(Conv, std2(Conv)), \quad Img_{StdTot} = Img_{StdConv} + Img_{StdCorr} + .. \tag{7.5}$$

Table 7.16: The second criteria for selecting a good cluster

std	minDis	Class
0	0	take the second class
0	1	take subset from the positions of second class
1	0	probability to happen too low (take the second class)
1	1	take that class

$$Img_{Uni2} = Img_{Uni1} - Img_{MeanTot} \tag{7.6}$$

$$Img_{Uni3} = Img_{Uni1} + Img_{MeanTot} - Img_{StdTot} \tag{7.7}$$

The fifth proposed method extracted the optimal values of the calculated image similarity metrics and registered the moving image to the reference image using the optima of the image similarity metrics as a reference point. They were clustered using K-means into two classes: one is considered a good cluster and the other is considered a bad cluster. There are several ways to decide the criteria of a good cluster. One criterion could be getting the majority vote (denoted as combMat1). This work has been published in (161; 163). The second criterion could be based on standard deviation and the minimum distance between the positions of the optima value in the same cluster as shown in Table 7.16 (denoted as combMat2). The third criterion could be based on six characteristics of the positions to select the optimum value as illustrated and explained in the upcoming points in Table 7.17 (denoted as combMat3). For those two tables, one indicates that the values of the second class in any of the following listed parameters are less than those of the first class, and zero indicates vice versa. For example, the standard deviation between points (denoted as std in Table 7.16) and the minimum distance (denoted as minDis) are zeroes, which means that the "std" and "minDis" of the first class is higher. So based on these criteria, the second class is chosen.

- Standard deviation between points in the same cluster (denoted as std)
- Distance between points in the same cluster
  - Minimum distance in the same cluster (denoted as minDis)
  - Maximum distance in the same cluster (denoted as maxDis)
  - Variance between the distance of points in the same cluster (denoted as variance)

Based on the decision of a good cluster, those detected positions with the best agreement according to the respective metric were averaged. The combined metric is denoted as combMat1, combMat2, and combMat3. Based on the available clinical datasets, the most robust image similarity metrics are Conv, Corr, SSIM, MS-SSIM, RIU, MSE, and Img_{Uni1}.

Table 7.17: The third criteria for selecting a good cluster

std	variance	minDis	maxDis	Class
0	0	0	0	take the second class
0	0	0	1	remove outlier and check again
0	0	1	0	remove outlier and check again
0	0	1	1	take the second class
0	1	0	0	remove outlier and check again
0	1	0	1	remove outlier and check again
0	1	1	0	remove outlier and take that class
0	1	1	1	remove outlier and take that class
1	0	0	0	remove outlier and check again
1	0	0	1	remove outlier and check again
1	0	1	0	remove outlier and take that class
1	0	1	1	remove outlier and take that class
1	1	0	0	take subset from that class
1	1	0	1	take subset from that class
1	1	1	0	take subset from that class
1	1	1	1	take all the points in that class

After evaluation, for those five proposed combinations as shown in Figure 7.72, it was found that the combined image similarity metrics using K-means clustering for the third criteria results in the least median error of 20.2*mm* and the maximum number of cases with an overlap of lesion annotations is 26. For all the other combinations except the subtraction, the number of cases with an overlap of lesion annotations achieved is 20 while for subtraction, the number achieved is 19.

The proposed image based registration method achieves a median error of 20.2*mm* for 51 clinical datasets as shown in Table 7.18 and in Figure 7.73. The lower and the upper quarters are 11.6*mm* and 39.1*mm*, respectively. The minimum and the maximum errors are 1.5*mm* and 87.5*mm*, respectively. For such a high error for the case of 87.5*mm*, it might need to optimize the combined metrics or select a good ROI, or even use another enhancement of the image contrast from the proposed ones since this case is from a Philips machine, of which four cases only were included in the dataset as explained in Section 5.1.4. Due to time constraints and the small number of available cases taken from this machine, this could be followed up in the future. It would be interesting to optimize the three mentioned blocks. For the intersection between the two annotations (the yellow color), three metrics (DSC, IoU, and TPR) have been calculated. It was found that 26 cases out of the 51 cases achieve an intersection as shown in Figure 7.74. The maximum DSC achieved is 0.73 and the maximum IoU achieved is 0.58 in the same case. Since there is a correlation between the annotation in the full X-ray mammogram and spot mammogram and both of the annotations are in 2D, the maximum TPR achieved is 0.63 in the same case.

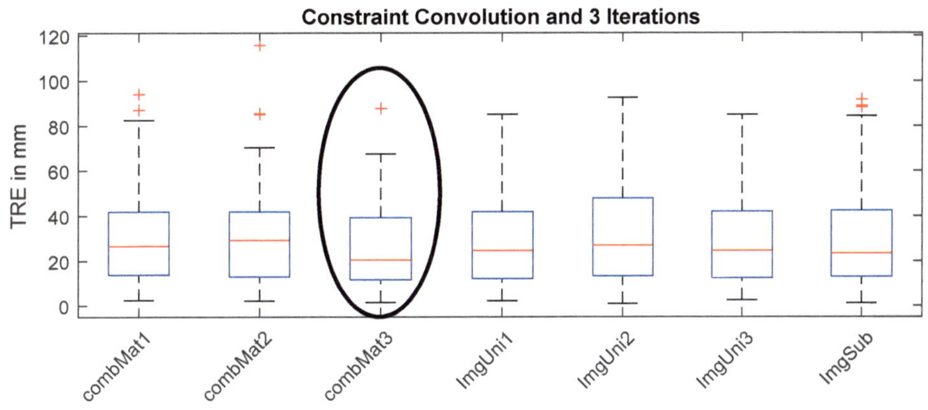

Figure 7.72: Analysis of TRE of the five proposed combined metrics for the image based registration method of 51 cases received from the Medical University of Vienna.

Table 7.18: Accuracy of image based registration between full X-ray mammography and spot mammograms

Accuracy in *mm*	mean	median	std	min	max	75%	25%
TRE	25.6	20.2	18.95	1.5	87.5	39.1	11.6

7.4.5 Correlation with Features of the Breast

The method of the image based registration has been analyzed using seven characteristics that correspond to the variations in datasets as explained in Chapter 5 to investigate if any subgroups correlate with the accuracy of the method.

Patient Age

A polynomial fitting curve (first degree) has been used to evaluate how the datasets are distributed compared to the TRE of the image based registration method as shown in Figure 7.81. It was observed that age has a very weak correlation in the negative direction with the TRE with an r of -0.06 and R^2 of 0.003.

Views

The second characteristic, which was investigated, is views of full X-ray mammography.
The views include the variance in the projection angle which may differ in full X-ray and spot mammograms and also in the side of the breast, whether left or right breasts. As explained before in Section 5.1.2, there are twelve combinations for the views for the prone position for spot mammograms and the same combinations for upright position for spot mammograms. Based on the available datasets, three combinations for the prone position are dominant. For CC (full X-ray mammograms) and CC (spot mammograms), there are sixteen patients. For MLO (full X-ray mammograms) and ML (spot mammograms), there are eight patients. For MLO (full X-ray mammograms) and LM (spot mammograms),

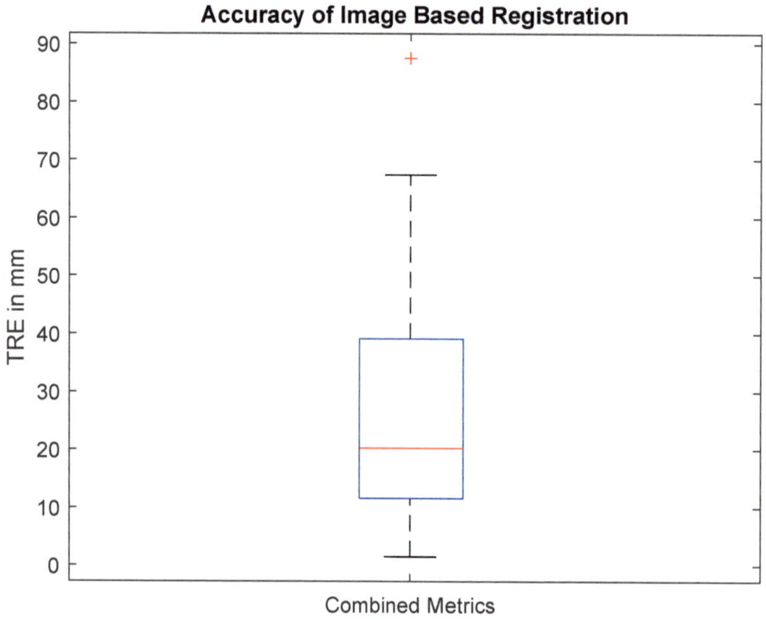

Figure 7.73: Analysis of TRE of the combined metrics for the image based registration method of 51 cases received from the Medical University of Vienna.

there are fourteen patients. There are six patients in the other classes of views for the prone position (see Table 7.19). For the views of ML and RL of full X-ray mammography, four patients are included in these views as shown in Table 7.20. Only three patients are available in the upright position (see Table 7.21). The inconsistency in the views between full X-ray and spot X-ray might increase the error of registration. That is because of the variance in the imaging situation between the two modalities. The variance in projection angle in each imaging situation causes to image the breast tissue from different angles which can't be handled using affine or projective transformation due to the lack of missing information that could be extracted from metadata to apply these transformations. There is likely no conclusive correlation

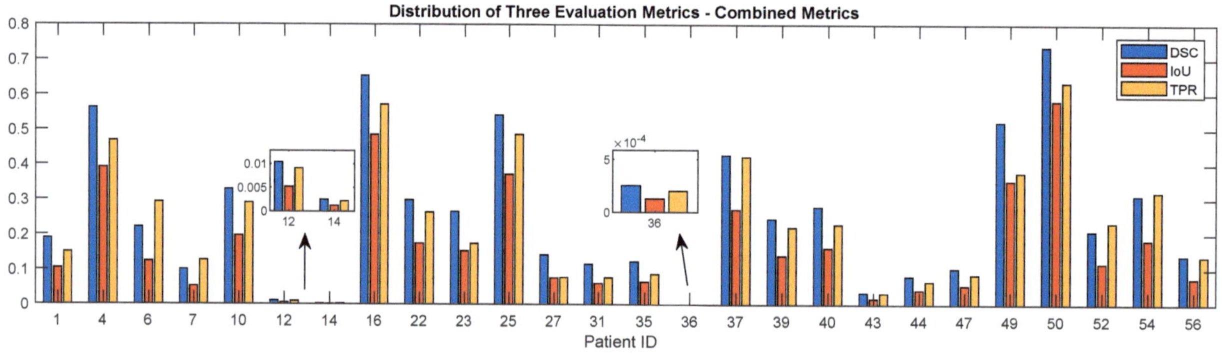

Figure 7.74: Distribution of the three evaluation metrics that have been calculated for the intersection between the actual and the predicted lesion using image based registration. 26 out of 51 cases achieve an intersection between the two annotations.

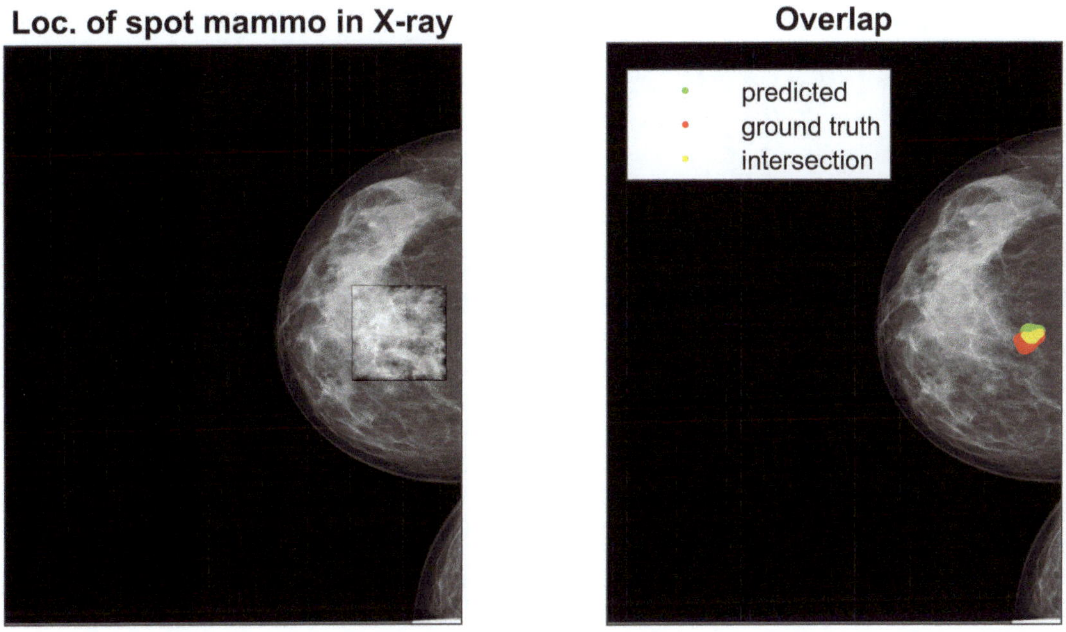

Figure 7.75: An example of registration between X-ray mammography and spot mammogram for the same showed case in the biomechanical model based registration (Patient ID 40): location of spot mammogram in full X-ray mammogram (left). Overlap between the lesion that has been annotated by radiologists in the original X-ray (red), prediction of the position of the lesion using image based registration (green), and overlap between the two lesions (yellow) (right).

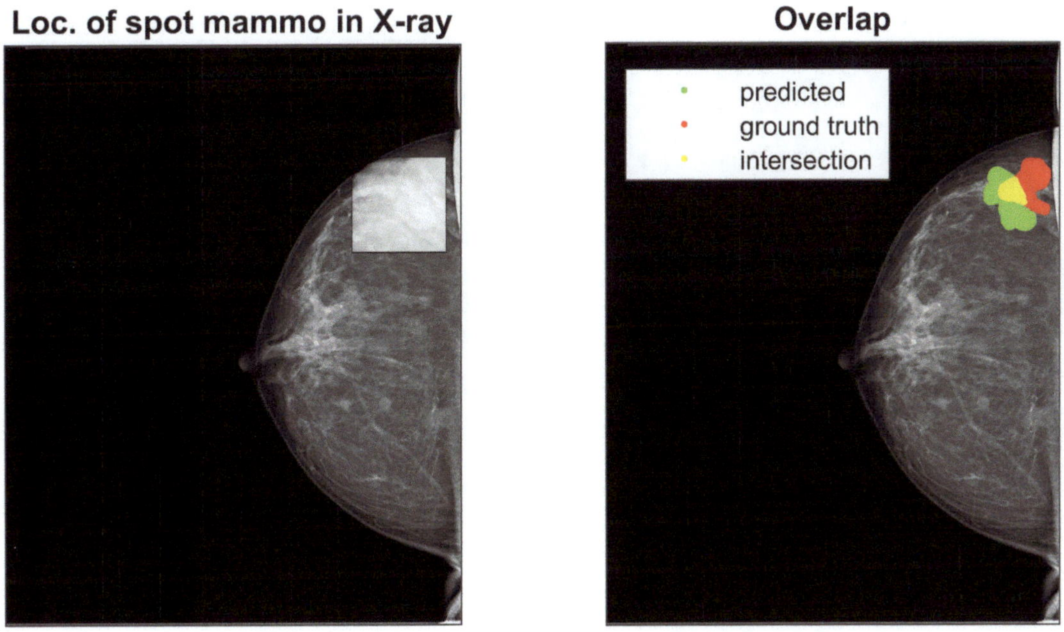

Figure 7.76: An example of registration between X-ray mammography and spot mammogram for the same showed case in the biomechanical model based registration (Patient ID 52): location of spot mammogram in full X-ray mammogram (left). Overlap between the lesion that has been annotated by radiologists in the original X-ray (red), prediction of the position of the lesion using image based registration (green), and overlap between the two lesions (yellow) (right).

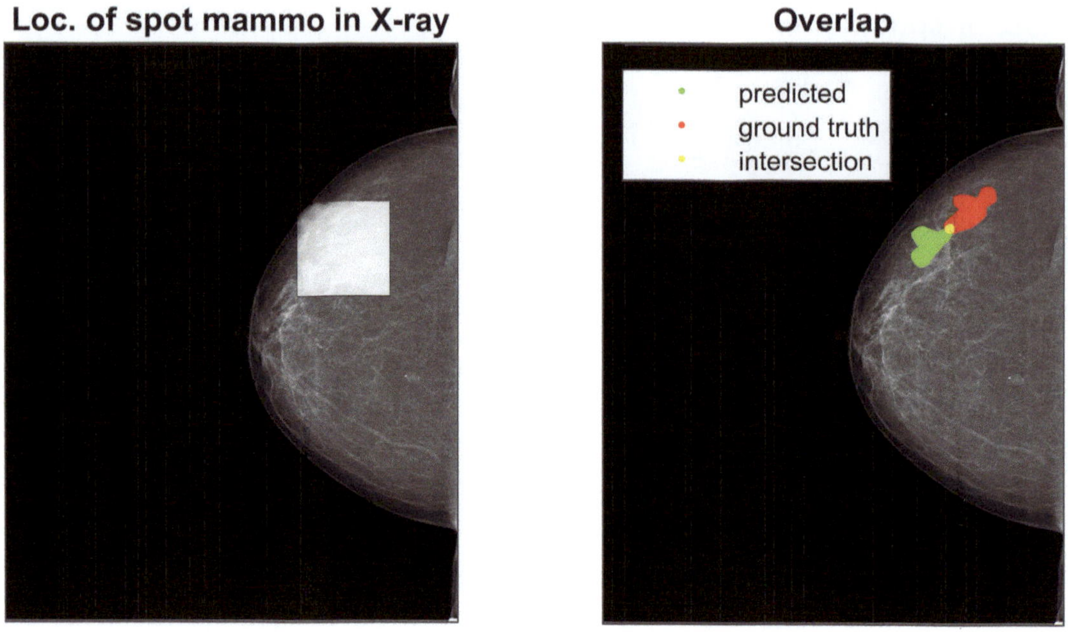

Figure 7.77: An example of registration between X-ray mammography and spot mammogram for the same showed case in the biomechanical model based registration (Patient ID 12): location of spot mammogram in full X-ray mammogram (left). Overlap between the lesion that has been annotated by radiologists in the original X-ray (red), prediction of the position of the lesion using image based registration (green), and overlap between the two lesions (yellow) (right).

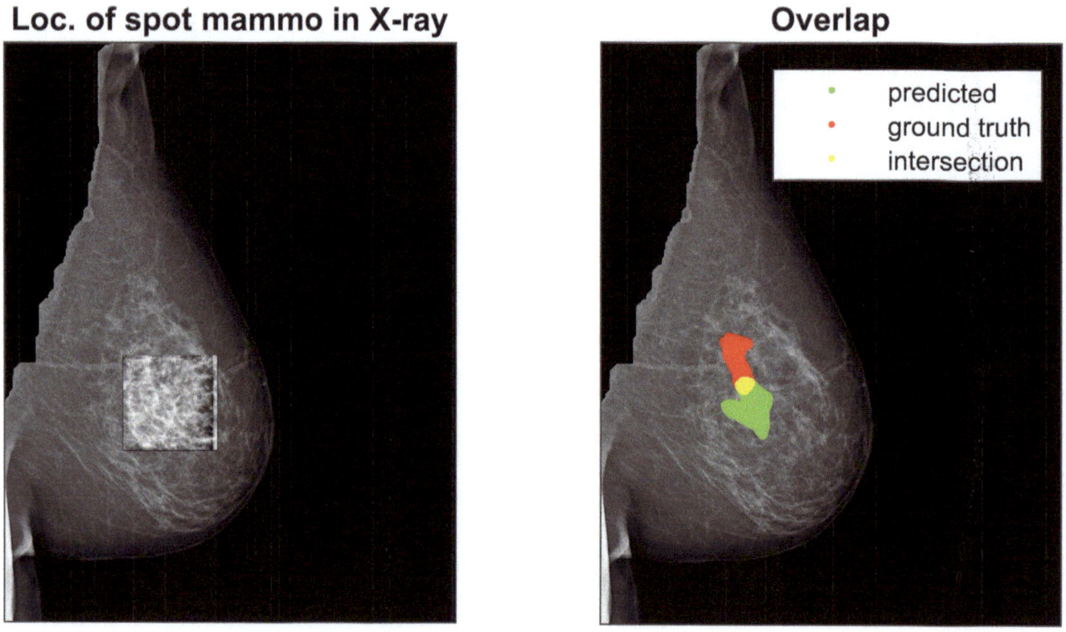

Figure 7.78: An example of registration between X-ray mammography and spot mammogram for the same showed case in the biomechanical model based registration (Patient ID 7): location of spot mammogram in full X-ray mammogram (left). Overlap between the lesion that has been annotated by radiologists in the original X-ray (red), prediction of the position of the lesion using image based registration (green), and overlap between the two lesions (yellow) (right).

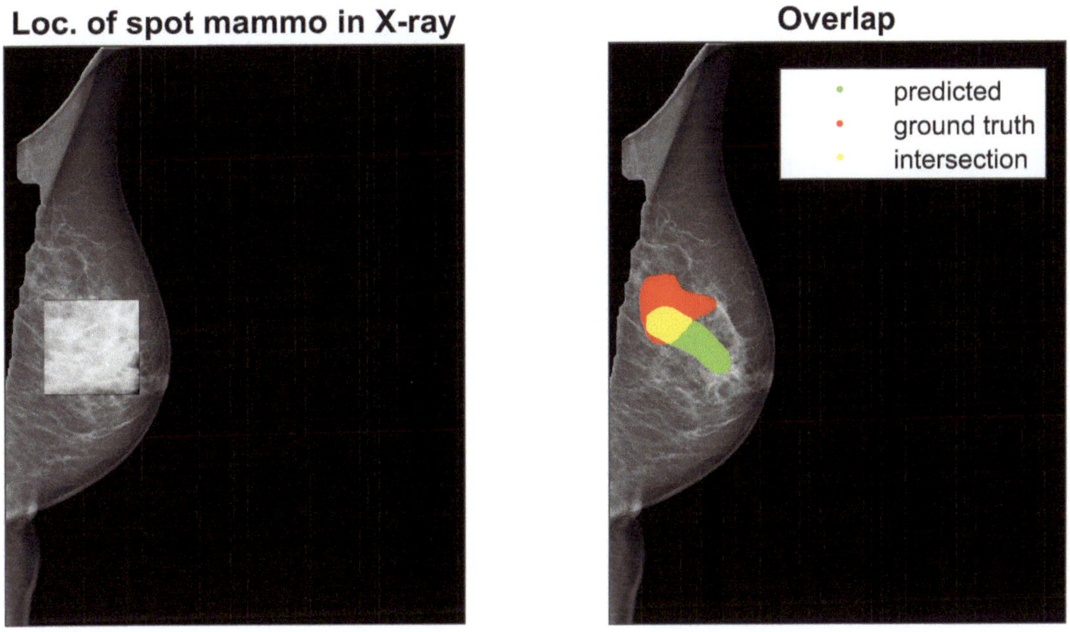

Figure 7.79: An example of registration between X-ray mammography and spot mammogram for the same showed case in the biomechanical model based registration (Patient ID 10): location of spot mammogram in full X-ray mammogram (left). Overlap between the lesion that has been annotated by radiologists in the original X-ray (red), prediction of the position of the lesion using image based registration (green), and overlap between the two lesions (yellow) (right).

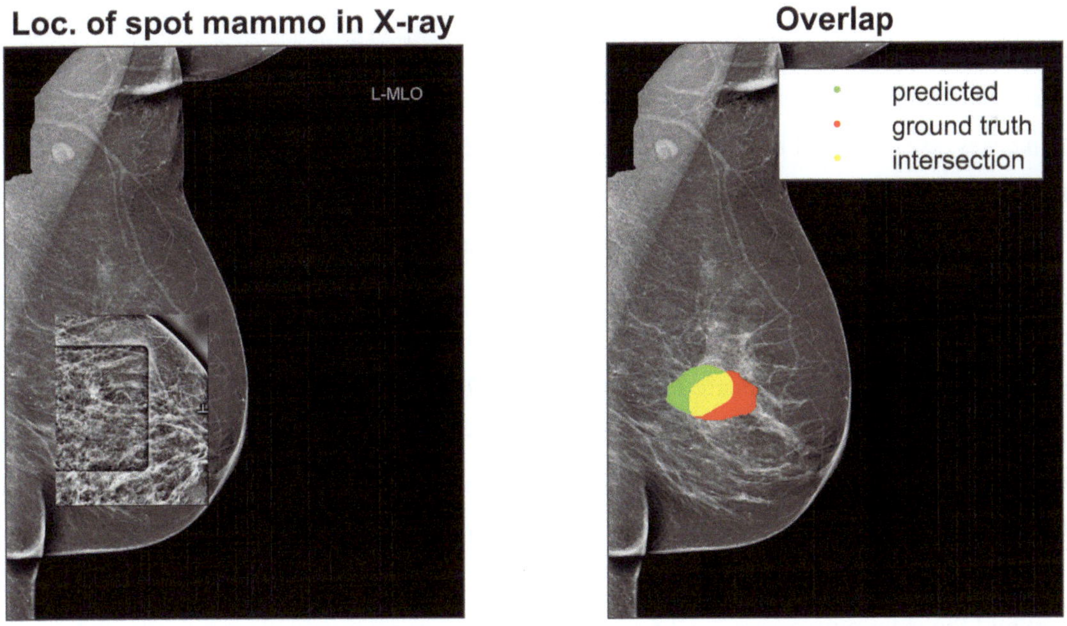

Figure 7.80: An example of registration between X-ray mammography and spot mammogram for the same showed case in the biomechanical model based registration (Patient ID 56): location of spot mammogram in full X-ray mammogram (left). Overlap between the lesion that has been annotated by radiologists in the original X-ray (red), prediction of the position of the lesion using image based registration (green), and overlap between the two lesions (yellow) (right).

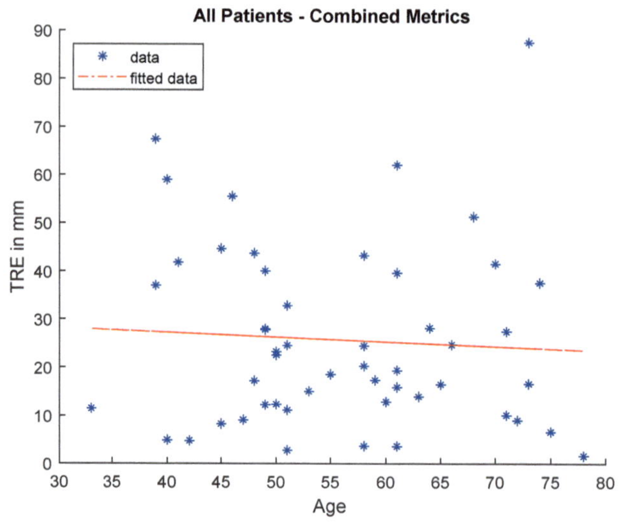

Figure 7.81: A very weak correlation between the accuracy of the image based registration method and the age of the patients with an r of -0.06.

because of the very small number of datasets in cases of ML and RL. Hence, for the three dominant views, boxplots have been used to evaluate the correlation between the accuracy of the method and the views of the full X-ray mammography as shown in Figure 7.82 (a). It was observed that error across the three main views differs considerably: CC/CC (sixteen datasets) have a median TRE of 12.6 mm, MLO/ML (eight datasets) have a median TRE of 17.6 mm, and MLO/ML (fourteen datasets) have a higher median TRE of 23.9 mm.

In addition, the number of intersections between the two annotations for CC/CC cases is eight cases out of sixteen cases, as shown in Figure 7.83, while the number of intersections between the two annotations for MLO/ML cases is four cases out of eight cases, as shown in Figure 7.84. For MLO/LM cases, the number of intersections between the two annotations is eight cases out of fourteen cases, as shown in Figure 7.85.

For CC/CC cases, the lower and the upper quarters are 4.7 mm and 35.15 mm, respectively. While the minimum and the maximum errors are 1.55 mm and 61.99 mm, respectively. The reason why one patient has a high error although, it is a CC and CC case is that the lesion is located at the very back of the breast. For the intersection between the two annotations as shown in Figure 7.83, the maximum DSC achieved is 0.73 and the maximum IoU achieved is 0.58, and the maximum TPR achieved is 0.63 in the same case.

For MLO/ML cases, the lower and the upper quarters are 8.6 mm and 40.1 mm, respectively. While the minimum and the maximum errors are 3.6 mm and 44.6 mm, respectively. Two reasons that could cause such a high error in this case, is that the data is from a "Sectra Imtec AB" device for which only six datasets in total are available. This machine has a different image contrast than the one that has been imaged from Siemens. Second, the lesion is located at the very back of the breast. For the intersection between the two annotations, as shown in Figure 7.84. The maximum DSC achieved is 0.56 and the maximum IoU achieved is 0.39 in the same case. The maximum TPR achieved is 0.49.

For MLO/LM cases, the lower and the upper quarters are $16.6mm$ and $39.6mm$, respectively. While the minimum and the maximum errors are $6.6mm$ and $67.3mm$, respectively. One reason that could cause such a high error in this case, is that the data is from a "FUJFILM Corporation" device for which only three datasets in total are available. Hence, due to the limitation of the number of datasets taken from this machine, it is likely too small for this conclusion. This could be optimized in the future. For the intersection between the two annotations as shown in Figure 7.85, the maximum DSC achieved is 0.43 and the maximum IoU achieved is 0.27. The maximum TPR achieved is 0.42 in the same case.

It was expected that MLO/LM has a higher error than the other two views since three difference occurs: imaging situation (one prone and one upright), and variance in projection angle (one MLO and the other is LM).

After splitting the datasets by projection angle in full and spot mammograms, these three classes of views were divided into left and right breasts, in which the number of datasets in each class is eight, eight, two, six, seven, and seven, respectively, as shown in Figure 7.82 (b). For MLO/ML, it is hard to conclude since one class (left breast) has only two patients. The other two categories are equally divided, but they are also considered a small dataset on each side. However, it could be assumed that right breasts work better than left ones.

In conclusion, the accuracy of the image based registration method correlates with the views of full X-ray mammography and spot mammograms regarding CC/ CC, MLO/ML, and MLO/LM cases and potentially with the laterality.

Table 7.19: Distribution of six patients (prone - other views) from the Medical University of Vienna regarding imaging situation for the X-ray guided biopsy

Modality / view	**X-ray**	**Spot mammogram**	*Patient ID*	*TRE*	*DSC*	*IoU*	*TPR*
left	CC	ML	19	$17.15mm$	0	0	0
right	CC	LM	1	$17.3mm$	0.19	0.1	0.15
right	CC	LM	14	$28.13mm$	0.0025	0.0012	0.0022
right	MLO	CC	29	$87.48mm$	0	0	0
right	MLO	CC	51	$15.83mm$	0	0	0
right	MLO	CC	56	$19.33mm$	0.14	0.075	0.14

Table 7.20: Distribution of four patients (prone - left breasts) from the Medical University of Vienna regarding imaging situation for the X-ray guided biopsy

Modality / view	**X-ray**	**Spot mammogram**	*Patient ID*	*TRE*	*DSC*	*IoU*	*TPR*
left	ML	ML	41	$41.8mm$	0	0	0
left	ML	LM	47	$16.4mm$	0.1	0.054	0.085
left	ML	CC	24	$43.7mm$	0	0	0
left	RL	ML	54	$13.9mm$	0.31	0.18	0.32

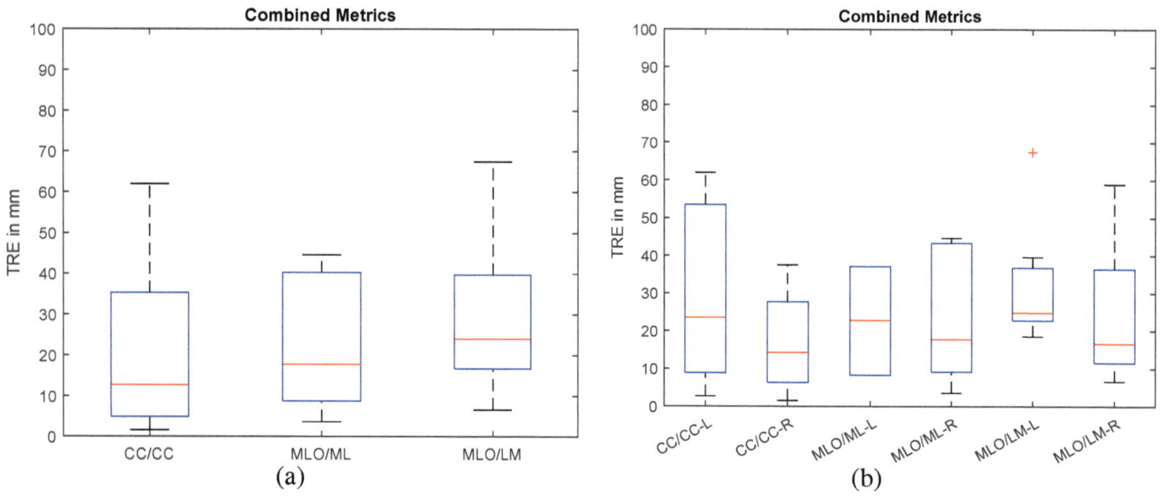

Figure 7.82: The three dominant views category: CC/CC, MLO/ML, and MLO/LM for the prone position (a). Analysis of TRE for the image based registration method regarding these categories by splitting into left (L) and right (R) breasts (b).

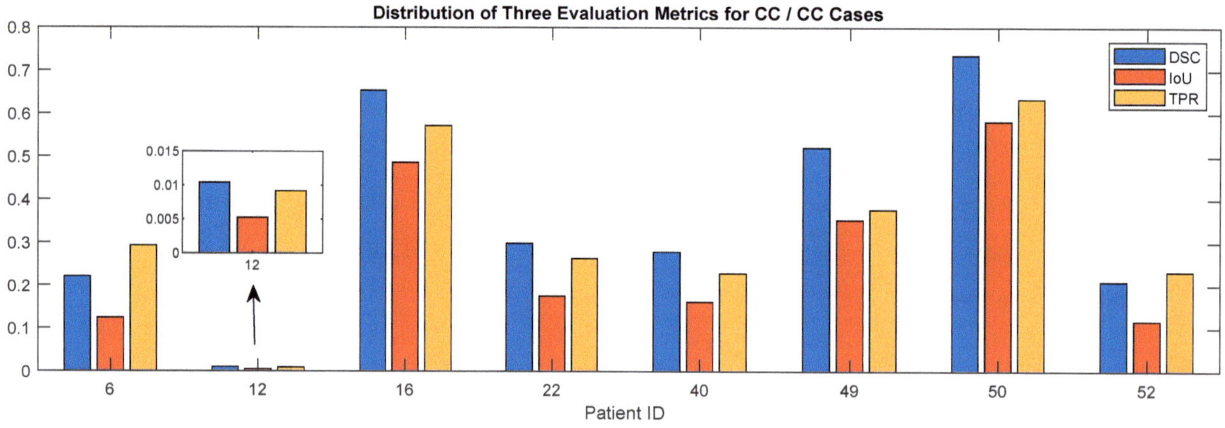

Figure 7.83: Distribution of the three evaluation metrics for the CC/CC cases that have been calculated for the intersection between the actual and the predicted lesion using image based registration method. Eight cases out of sixteen cases achieve an intersection.

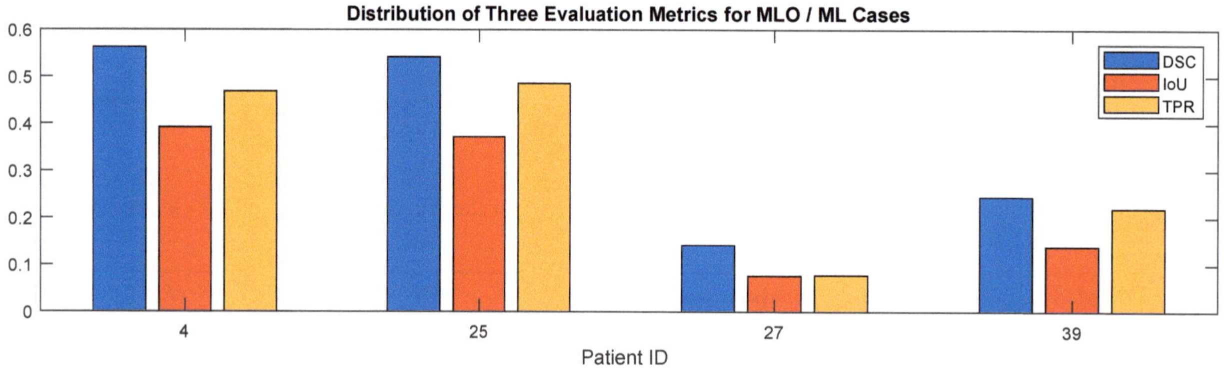

Figure 7.84: Distribution of the three evaluation metrics for the MLO/ML cases that have been calculated for the intersection between the actual and the predicted lesion using image based registration method. Four cases out of eight cases achieve an intersection.

Table 7.21: Distribution of three patients (upright) from the Medical University of Vienna regarding imaging situation for the X-ray guided biopsy

Modality / view	**X-ray**	**Spot mammogram**	*Patient ID*	*TRE*	*DSC*	*IoU*	*TPR*
left	CC	LM	26	24.42 *mm*	0	0	0
right	CC	ML	43	11.41 *mm*	0.034	0.018	0.033
right	MLO	LM	28	41.51 *mm*	0	0	0

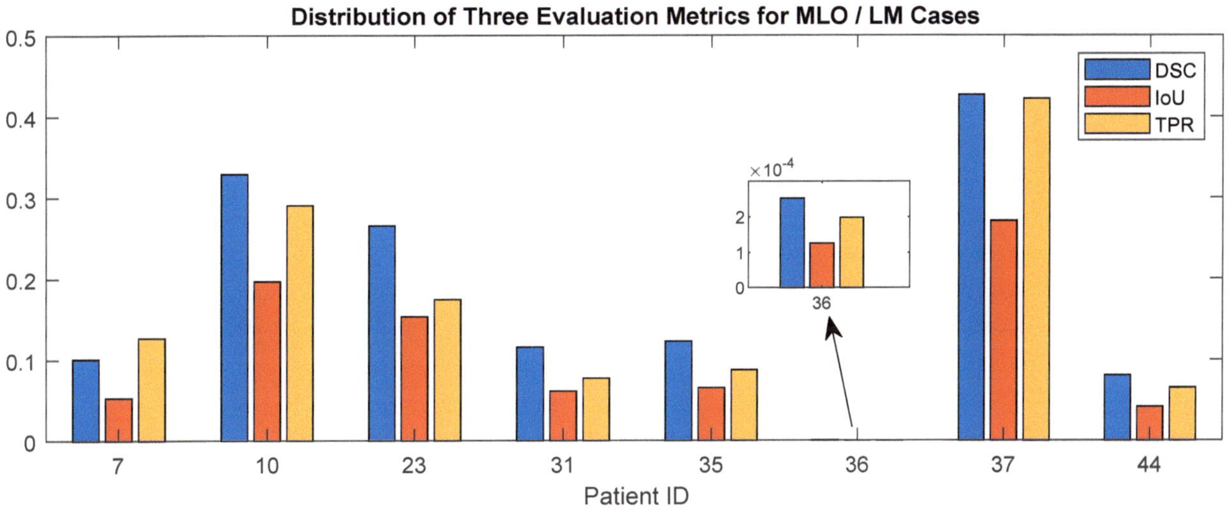

Figure 7.85: Distribution of the three evaluation metrics for the MLO/LM cases that have been calculated for the intersection between the actual and the predicted lesion using image based registration method. Eight cases out of fourteen cases achieve an intersection.

Histopathological Diagnosis

The third characteristic is in terms of the histopathological diagnosis of the breast needle biopsies as explained in Section 5.2.1. It has been observed that the median TRE is 22.5 *mm* and 24.7 *mm*, respectively, for B2 and B5a classes as shown in Figure 7.86. In addition, the minimum, maximum, lower quarter, and upper quarter of both classes, as shown in Table 7.22, are quite comparable. It could be concluded that there is no correlation with respect to histopathological diagnosis.

Table 7.22: Detailed analysis of boxplots of class B2 and B5a versus the accuracy of the image based registration

	min	max	lower quarter	upper quarter
B2	1.5 *mm*	61.9 *mm*	12.6 *mm*	48.9 *mm*
B5a	2.7 *mm*	67.4 *mm*	12.5 *mm*	40.1 *mm*

Breast Density

The breast density of clinical datasets is the fourth characteristic as detailed in Section 5.2.2. A polynomial fitting curve (first degree) has been used to evaluate how the datasets are distributed compared to the

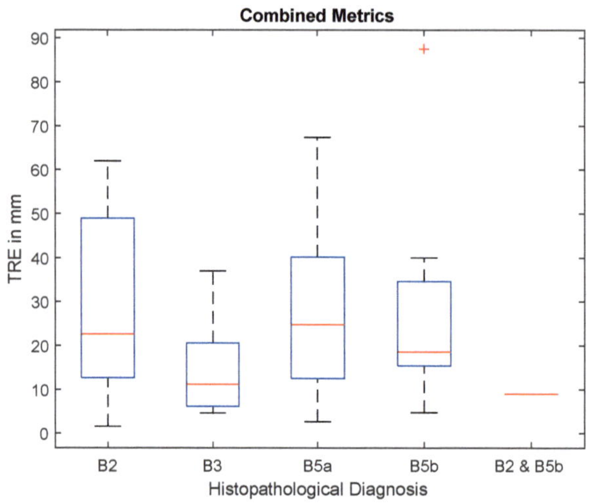

Figure 7.86: Analysis of TRE for the image based registration method regarding the breast needle biopsies into categories with no correlation: B2 (15 cases), B3 (5 cases), B5a (21 cases), B5b (9 cases), and B2 and B5b (1 case).

TRE of the image based registration method, as shown in Figure 7.87. It was observed that the glandular tissue related to the total breast volume and the ratio between glandular and fatty tissues in the breast as aspects have a very weak correlation with the TRE with an r of -0.16 and -0.16, respectively, and R^2 of 0.026 and 0.024, respectively.

Sizes of Breasts and Lesions

The size of breasts and lesions is the fifth characteristic from Section 5.2.3 that has been investigated. A polynomial fitting curve (first degree) has been used to evaluate how the datasets are distributed compared to the TRE of the image based registration method, as shown in Figures 7.88 and 7.89.

There is a very weak correlation with an r of 0.08 and -0.04 and with R^2 of 0.007 and 0.002 for full X-ray mammography CC and MLO, respectively, as shown in Figure 7.88. Since spot mammograms are a subset of the breast, this characteristic is not valid for this modality.

To analyze the impact of the sizes of lesions, a polynomial fitting curve (first degree) has been used to evaluate how the datasets are distributed compared to the TRE of the image based registration method. Also, it was observed in Section 5.2.3 that sizes of lesions in full X-ray mammography and spot mammograms are correlating. Hence, it is expected that sizes of lesions either full X-ray mammography or spot mammograms will have the same behavior in relation to the TRE of the image based registration method. There is a weak correlation with an r of -0.21, -0.23, and -0.24 and with R^2 of 0.045, 0.052, and 0.059 for full X-ray mammography CC and MLO, and spot mammograms, respectively, as shown in Figure 7.89.

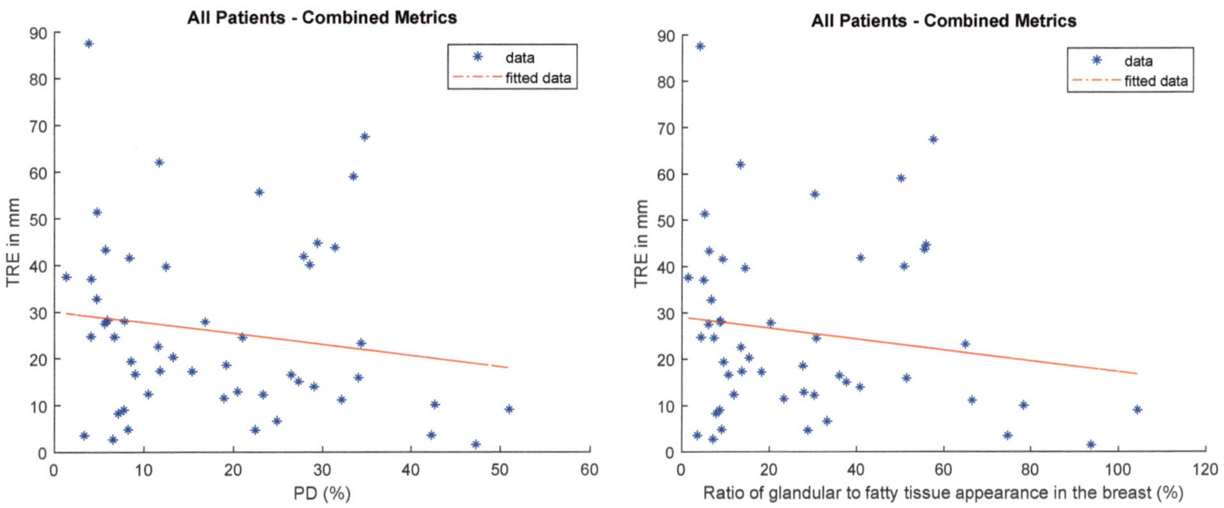

Figure 7.87: A very weak correlation between the accuracy of the image based registration method and the distribution of glandular tissue in the breast with an r of -0.16 (left). A very weak correlation between the accuracy of the image based registration method and the distribution of the ratio between glandular and fatty tissues in the breast with an r of -0.16 (right).

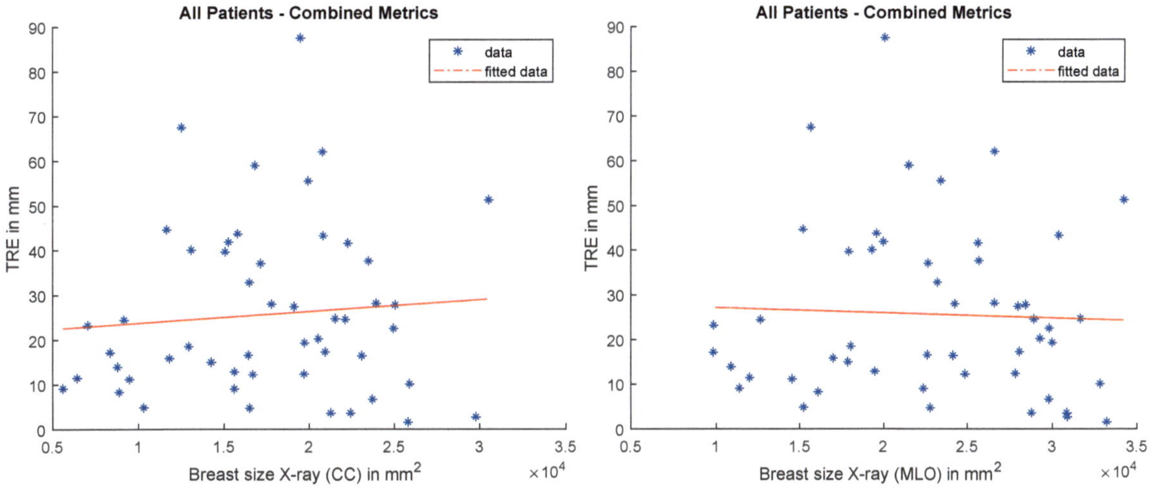

Figure 7.88: Relation between the accuracy of the image based registration method and distribution of the breast sizes in full X-ray mammograms (2D): CC (left) and MLO (right). A very weak correlation is found between the breast sizes and the TRE with an r of 0.084, and -0.041, respectively.

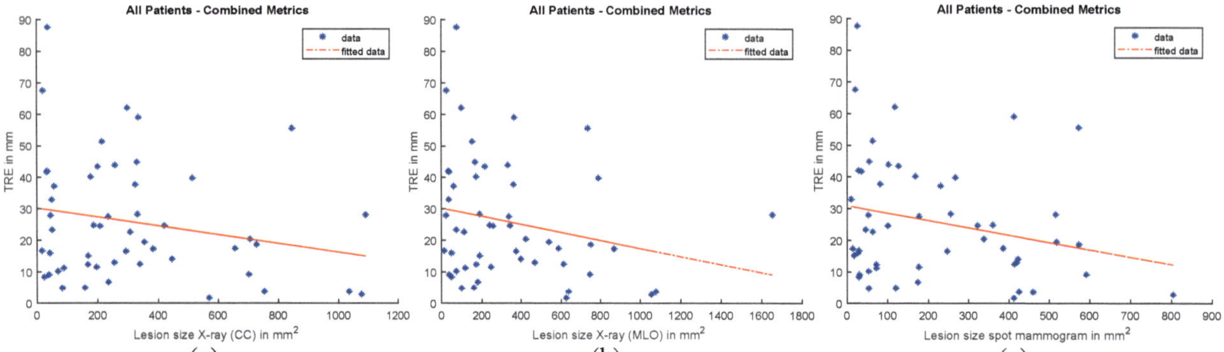

(a) (b) (c)

Figure 7.89: Relation between the accuracy of the image based registration method and distribution of the lesions sizes in full X-ray mammograms: CC (first) and MLO (middle) and spot mammograms (last). A weak correlation is found between the sizes of lesions and the TRE with an r of -0.21, -0.23, and -0.24, respectively.

Location of Lesions

The location of lesions in full X-ray mammography and the spot mammograms is the sixth characteristic that has been evaluated. Boxplots have been used for finding a correlation between locations of lesions and the TRE of image based registration. X-ray mammography was split into eight regions. Spot mammogram has been split into sixteen regions, as explained in Section 5.2.4.

For full X-ray mammography, some regions cannot be considered conclusive results, such as R4 (one case only) and R5 (five cases) for both views, and R3 (three cases) in case of CC view. It is clear that lesions at the back of the breast near the muscle in the posterior direction (R1, R2, and R6 in both views) are harder to register than the lesions at the center of the breast (R7), which achieves the least median of $16.5\,mm$ for eight cases in CC view and the median of $15.8\,mm$ for four cases in MLO view, as shown in Figure 7.90 (first and middle).

For spot mammograms, some regions do not have available datasets and some regions, such as R14 and R15 have only one case. R1 and R5 have two cases, and R3 and R8 have only 3 cases. Hence, it is hard to get a conclusive observation since the cases located in these regions are likely too small. While R2 though, has also a small number of cases (four cases only), it is expected that median TRE would be high $34\,mm$ since the lesions located in this region are at the corners of the spot mammogram. The highest number of cases belong to R6 (13 cases), in which the median TRE is $17.16\,mm$, as shown in Figure 7.90 (last). Thus, if the guidelines are followed well and the lesion is central in the spot mammogram, then they are better registered.

7.5 Accumulative Registration of the Two Methods

To evaluate the accuracy of the accumulative registration of the two methods, TRE, TPR, IoU, and DSC, have been calculated based on the predicted annotation position originating from MRI annotation from the first method. This was given as an input for the second method (image based registration) to predict

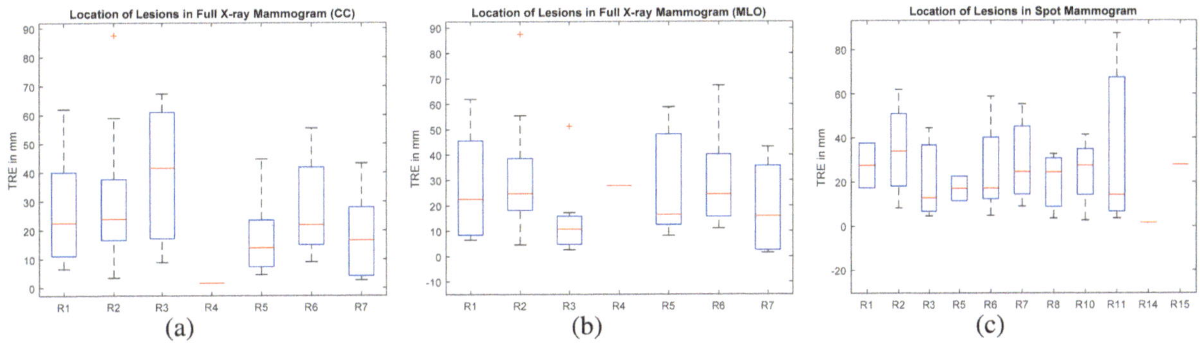

Figure 7.90: Relation between the location of lesions in full X-ray mammography in both views and spot mammograms and the accuracy of the image based registration method. R7 in both views of full X-ray mammograms (CC 8 cases (middle)) (MLO 4 cases (last)) is the least median error which corresponds to the area of the center of the breast.

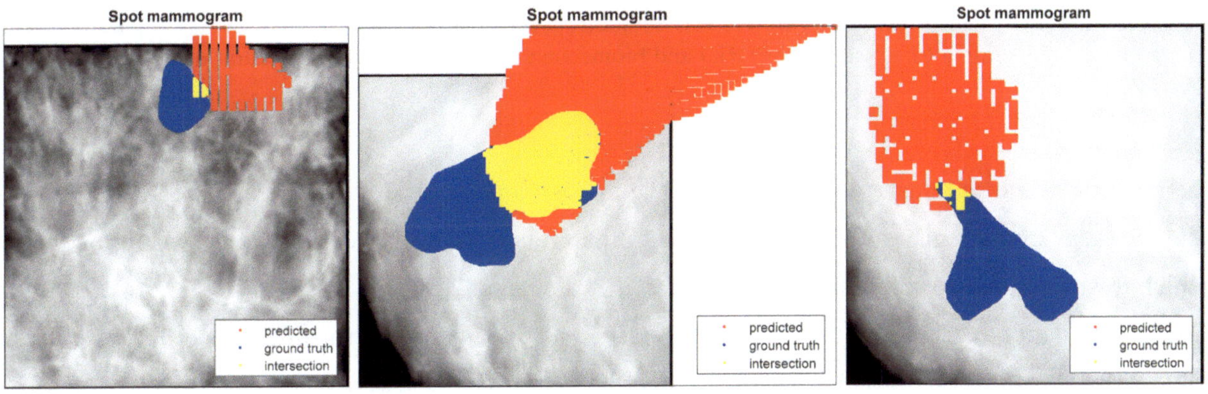

Figure 7.91: An example of registration between MRI and spot mammogram for the same showed cases of CC cases in the two methods of registration (Patient IDs: 40, 52, and 12): predicted lesion area based on the matching of the annotation in the MRI (red) and ground truth annotation in the spot mammogram (blue). The yellow color represents the overlap of both.

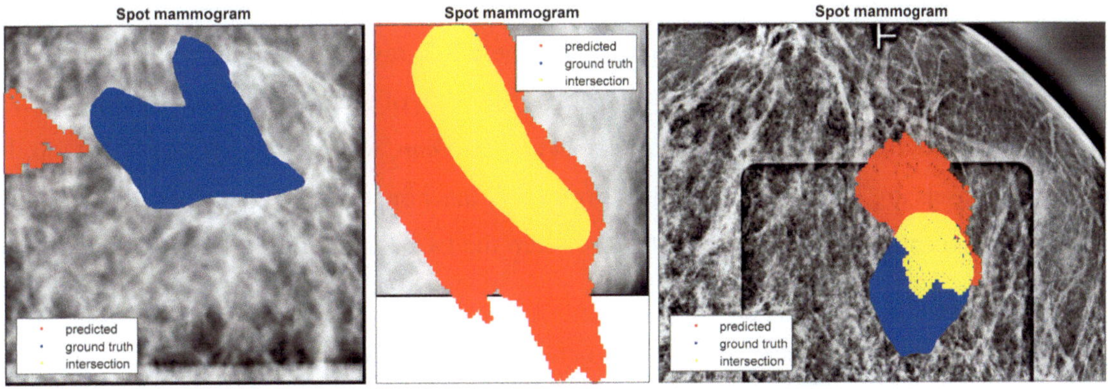

Figure 7.92: An example of registration between MRI and spot mammogram for the same showed cases of MLO cases in the two methods of registration (Patient IDs: 7, 10, and 56): predicted lesion area based on the matching of the annotation in the MRI (red) and ground truth annotation in the spot mammogram (blue). The yellow color represents the overlap of both.

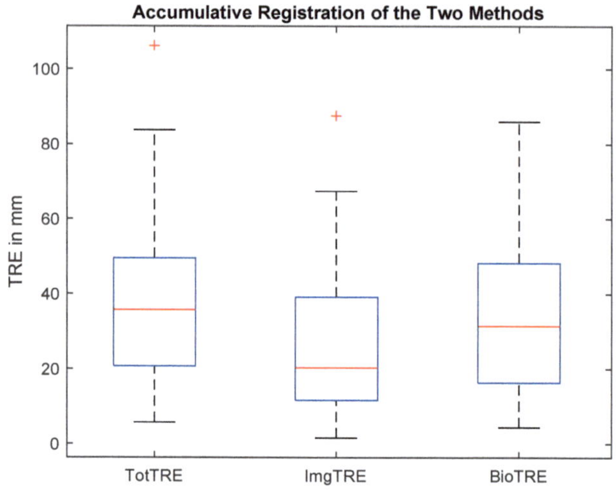

Figure 7.93: Analysis of TRE of the accumulative registration of the two methods (denoted as TotTRE) of 51 cases received from the Medical University of Vienna. In addition the analysis TRE of each method separately. The image based registration (denoted as ImgTRE) and biomechanical model based registration method (denoted as BioTRE).

the position in the spot mammogram image. The predicted position in the spot mammogram based on the two steps of the registration was compared to the ground truth marked in the spot mammogram in which it was possible to calculate the total TRE, TPR, IoU, and DSC.

The accumulative registration of the two methods achieves a median error of 35.6 mm for 51 clinical datasets as shown in Figure 7.93. The lower and the upper quarters are 20.7 mm and 49.5 mm, respectively. The minimum and the maximum errors are 5.7 mm and 106.1 mm, respectively. For such a high error for the case of 106.1 mm, the error in both methods is high, for the biomechanical model based registration, TRE is 43.2 mm and for the image based registration, TRE is 67.3 mm. Hence, the two errors are added together. For biomechanical model based registration, it is one of the MLO cases which could need more optimization in rotation angle and position of plates. For image based registration the reason behind that might be because of the data acquisition of the full X-ray. It is from the device of 'FUJIFILM Corporation', in which only three cases are available. The error of the image based registration method in the other two cases from the same machine is 55.4 and 39.6 mm, which are considered high too. In the future, more datasets are needed to be available from this machine to investigate the proposed methods and improve the accuracy. For the intersection between the two annotations, DSC, IoU, and TPR have

Table 7.23: Accuracy of accumulative registration of the two methods between MRI and spot mammograms

Accuracy in *mm*	mean	median	std
Biomechanical	32.9	31.3	18.94
Imagebased	25.6	20.2	18.95
Total	35.9	35.6	20.9

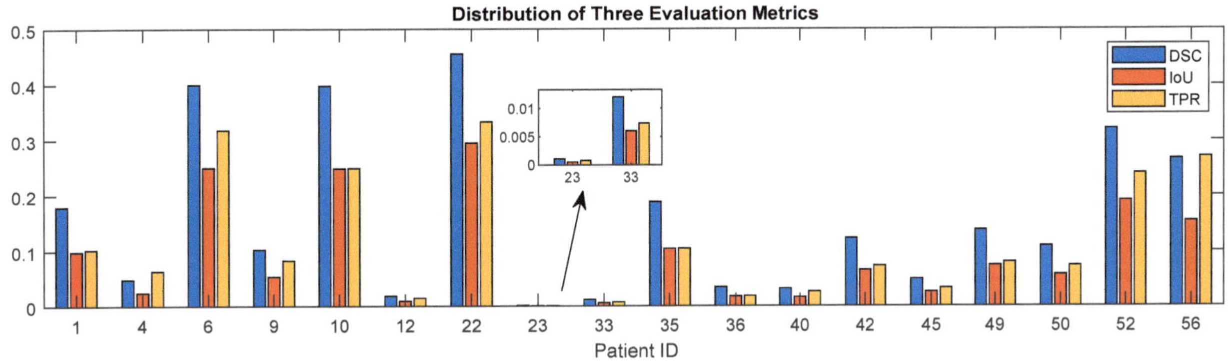

Figure 7.94: Distribution of the three evaluation metrics that have been calculated for the intersection between the actual and the predicted lesion for the total registration between MRI and spot mammograms. 18 out of 51 cases achieve an intersection between the two annotations.

been calculated. It is found that 18 cases out of the 51 cases achieve intersection, as shown in Figure 7.94. The maximum DSC achieved is 0.45, the maximum IoU achieved is 0.29, and the maximum TPR achieved is 0.33 in the same case.

The same six cases that have been visualized in both methods are shown for the accumulative registration of the two methods in Figure 7.91 and 7.92. One of the cases of MLO (first case), is a good case in both methods separately: biomechanical model based registration (12*mm* and has an overlap between the two annotations) and image based registration (24.6*mm* and has an overlap between the two annotations). However when combining the two methods, there is no overlap, and the error achieved is 35.6*mm*. It could be concluded even if the methods work well separately, it might not work for the final registration. Additionally, it could be concluded although the variance in views and image situation was partly compensated, the accumulative registration of the two methods could be clinically feasible which will be discussed in the next chapter 8.

7.5.1 Correlation with Features of the Breast

The accumulative registration of the two methods has been analyzed using seven characteristics that correspond to the variations in datasets as explained in Chapter 5 to investigate if the subgroups correlate with the accuracy of the method.

Patient Age

A polynomial fitting curve (first degree) has been used to evaluate how the datasets are distributed compared to the total TRE, as shown in Figure 7.95. It was observed that age has a very weak correlation with the TRE with an r of 0.09 and R^2 of 0.008.

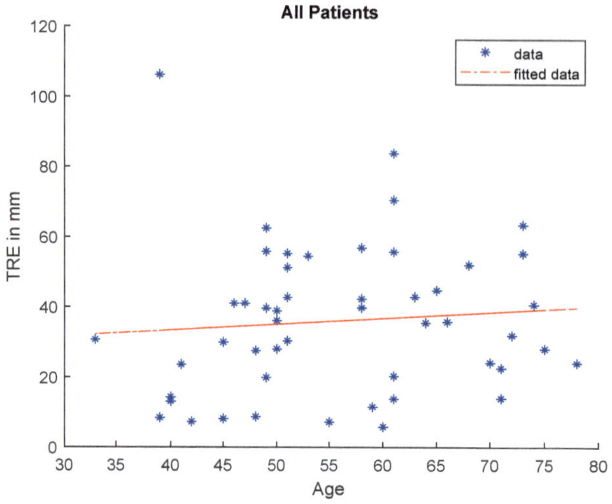

Figure 7.95: A very weak correlation between the accuracy of the accumulative registration of the two methods and the age of the patients with an r of -0.06.

Views

The second characteristic, which was investigated, is the views of full X-ray mammography that corresponds to the breast in MRI and spot mammograms, as explained in Section 5.1.2.

It was observed that the median total TRE of CC for full mammograms and CC for spot mammograms (sixteen datasets) is 33.85 mm, which is less than both the median TRE of MLO for full mammograms and ML for spot mammograms (eight datasets, 40.31 mm) and the median TRE of MLO for full mammograms and LM for spot mammograms (14 datasets, 37.21 mm). Then, these three classes of views were divided into left and right breasts, as shown in Figure 7.96 (b).

For the overlap between the two lesions, it has been observed that the CC and CC view for the prone position for spot mammogram has the highest number of overlapping cases, eight cases out of sixteen, as shown in Figure 7.98. MLO and ML cases have three overlapping cases out of eight, as shown in Figure 7.99. MLO and LM cases have five overlapping cases out of fourteen, as shown in Figure 7.100.

For all the other remaining views, there are only two datasets that have an overlap between the two lesions, which is shown in Table 7.24. That is why, the distribution of TRE for each method separately and the total error is only presented in Figure 7.97, in which it has been realized that TRE for views of MLO and CC for full X-ray mammography and spot mammogram, respectively are the worst for the proposed methods.

In conclusion, the total error of the accumulative registration of the two methods correlates with the views of full X-ray mammography and spot mammograms regarding CC/ CC, MLO/ML, and MLO/LM cases.

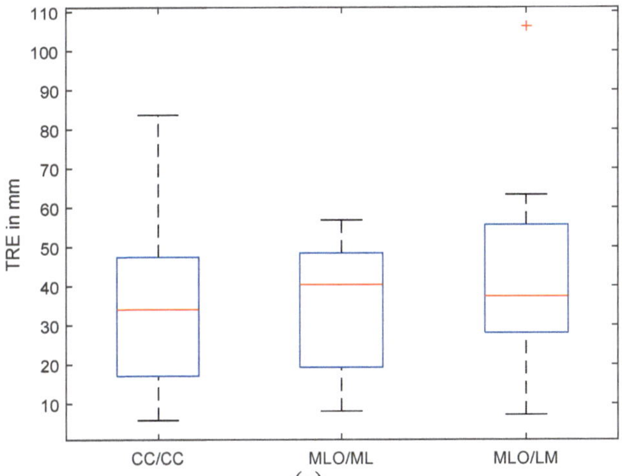

(a)

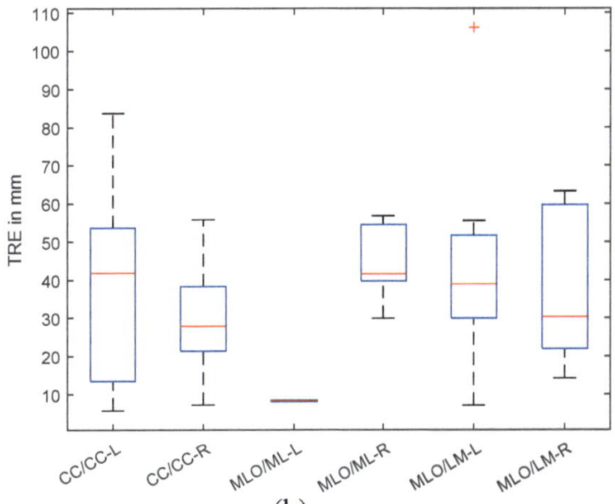

(b)

Figure 7.96: The three dominant views category: CC/CC, MLO/ML, and MLO/LM for the prone position (a). Analysis of TRE for the accumulative registration of the two methods regarding these categories by splitting into left (L) and right (R) breasts (b).

Table 7.24: Distribution of the total overlap of the remaining views (Only two patients)

Views	*Patient ID*	*DSC*	*IoU*	*TPR*
CC and LM	1	0.18	0.098	0.10
MLOand CC	56	0.36	0.22	0.37

Histopathological Diagnosis

The third characteristic is in terms of histopathological diagnosis of the breast needle biopsies, as explained in Section 5.2.1. The highest number of cases belongs to both B2 and B5a classes, in which it has been observed that median TRE is 38.8*mm* and 35.62*mm*, respectively, as shown in Figure 7.101. It could be concluded that there is no correlation with respect to the breast needle biopsies regarding histopathological appearance.

Breast Density

The breast density of clinical datasets is the fourth characteristic as detailed in Section 5.2.2. A polynomial fitting curve (first degree) has been used to evaluate how the datasets are distributed compared to the total TRE, as shown in Figure 7.102. It was observed that both aspects have a very weak correlation with the TRE with an r of -0.033 and -0.036, respectively, and R^2 of 0.0011 and 0.0013, respectively.

Sizes of Breasts and Lesions

The size of breasts and lesions is the fifth characteristic from Section 5.2.3 that has been investigated. A polynomial fitting curve (first degree) has been used to evaluate how the datasets are distributed compared to the total TRE, as shown in Figures 7.103 and 7.104.

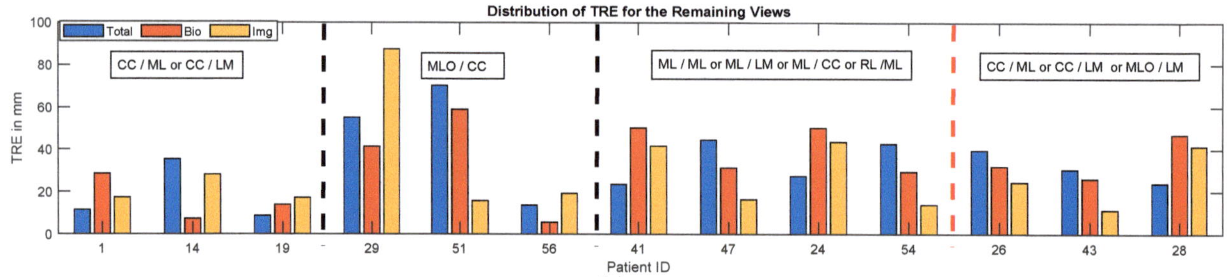

(a)

Figure 7.97: Distribution of the TRE for the remaining views that have a small number of datasets. The first three classes are cases of prone position (black dashes) and the last class is the cases of upright position. The first class belongs to the category of having either CC and ML or CC and ML (denoted as CC/ML or CC/LM) for full X-ray mammography and spot mammograms, respectively. Similarly, all the next classes are for the same two modalities. The second class belongs to MLO and CC (denoted as MLO/CC). The third class belongs to ML and ML or ML and CC or ML and LM or RL and ML (denoted as ML/ML or ML/CC or ML/LM or RL/ML). The fourth class belongs to CC and ML or CC and LM or MLO and LM (denoted as CC/ML or CC/LM or MLO/LM).

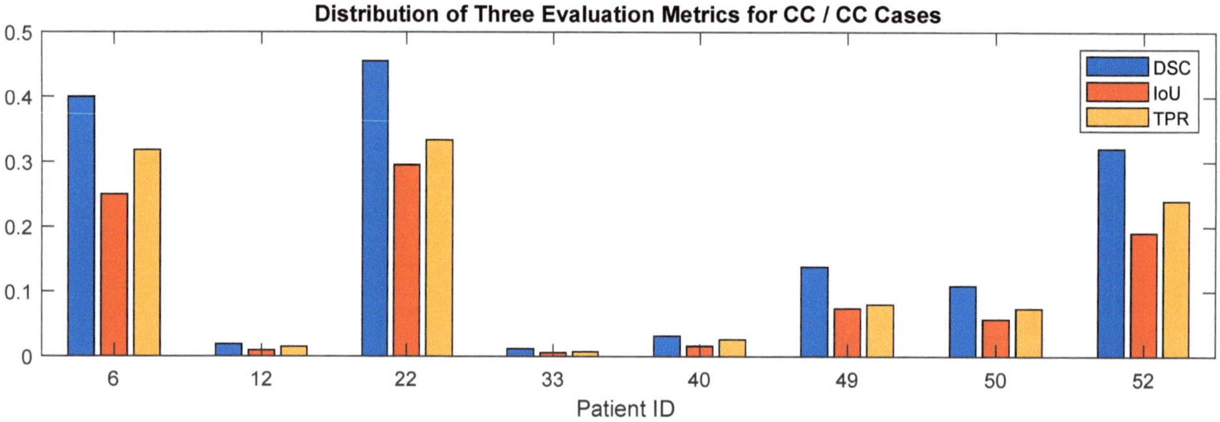

Figure 7.98: Distribution of the three evaluation metrics for the CC/CC cases that have been calculated for the intersection between the actual and the predicted lesion using accumulation registration of the two methods. Eight cases out of sixteen cases achieve an intersection.

There is a very weak correlation with an r of 0.041 and R^2 of 0.002 for MR images, as shown in Figure 7.103 (first). There is also a very weak correlation with an r of 0.11 and 0.02 and with R^2 of 0.011 and $6.02x10^{-4}$ for full X-ray mammography CC and MLO, respectively, as shown in Figure 7.103 (middle and last).

To analyze the impact of the sizes of lesions, a polynomial fitting curve (first degree) has been used to evaluate how the datasets are distributed compared to the total TRE. Also, it is observed in Section 5.2.3 that the sizes of lesions in full X-ray mammography and spot mammograms are strongly correlated. While the sizes of lesions in full X-ray mammogram and MRI intermediately correlate. Hence, it is expected that sizes of lesions either in MRI or full X-ray mammography or spot mammograms will have the same behavior in relation to the total TRE. There is a very weak correlation with an r of -0.18, -0.061, -0.14, and -0.15, and with R^2 of 0.032, 0.094, 0.021, and 0.022 for MRI, full X-ray mammography CC and MLO, and spot mammograms, respectively, as shown in Figure 7.104.

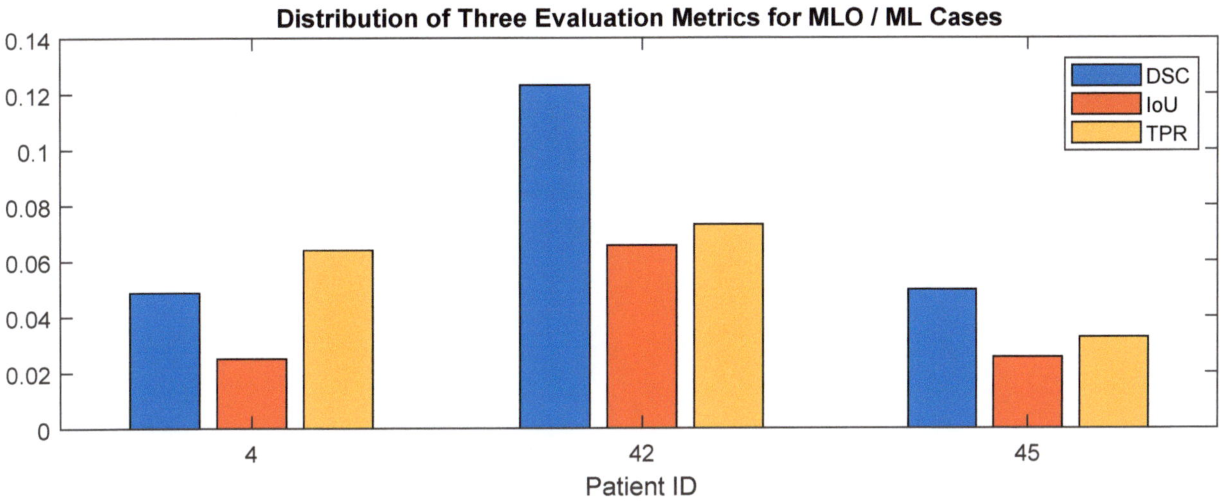

Figure 7.99: Distribution of the three evaluation metrics for the MLO/ML cases that have been calculated for the intersection between the actual and the predicted lesion using accumulative registration of the two methods. Four cases out of eight cases achieve an intersection.

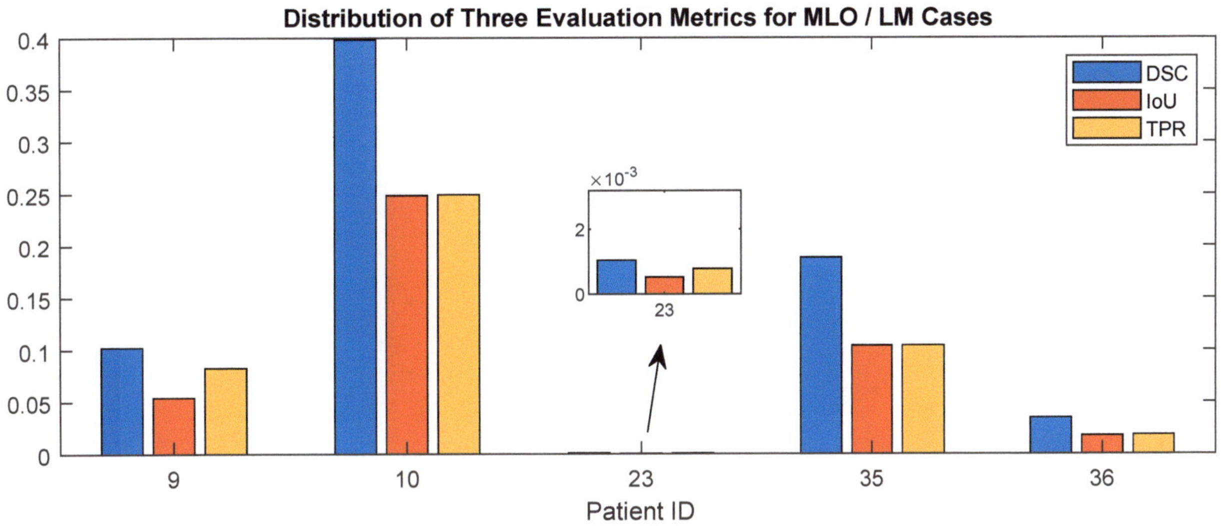

Figure 7.100: Distribution of the three evaluation metrics for the MLO/LM cases that have been calculated for the intersection between the actual and the predicted lesion using accumulative registration of the two methods. Eight cases out of fourteen cases achieve an intersection.

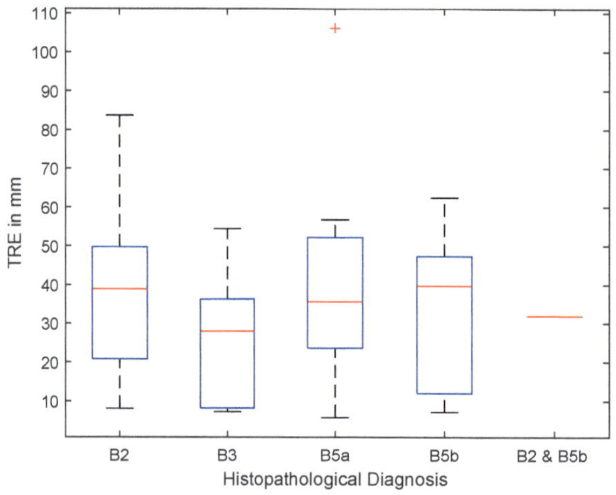

Figure 7.101: Analysis of the total TRE for the accumulative registration of the two methods regarding the breast needle biopsies into categories with no correlation: B2 (15 cases), B3 (5 cases), B5a (21 cases), B5b (9 cases), and B2 and B5b (1 case).

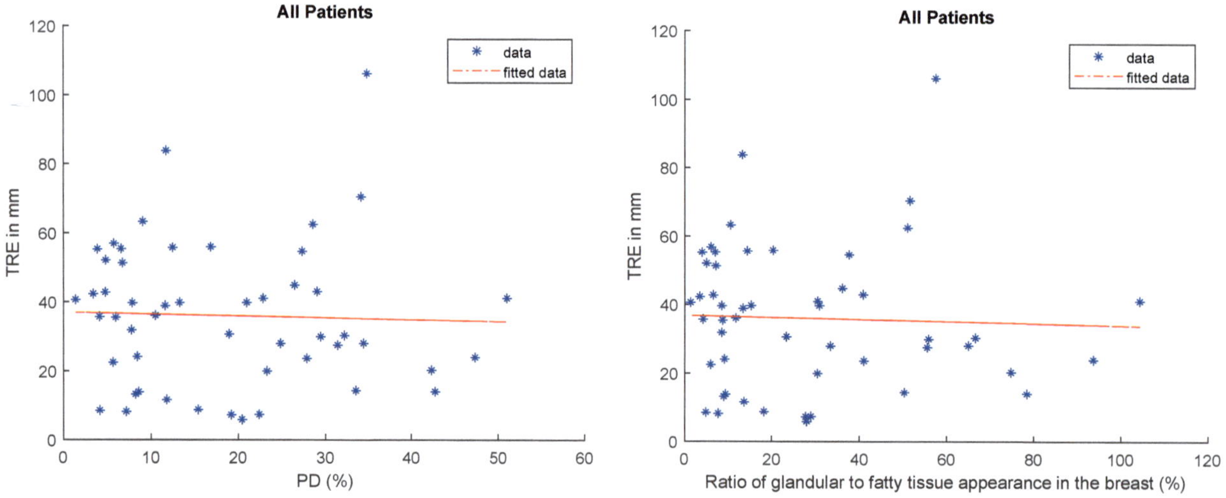

Figure 7.102: A very weak correlation between the accuracy of the accumulative registration of the two methods and the distribution of glandular tissue in the breast with an r of -0.033 (left). A very weak correlation between the accuracy of the accumulative registration of the two methods and the distribution of the ratio between glandular and fatty tissues in the breast with an r of -0.036 (right).

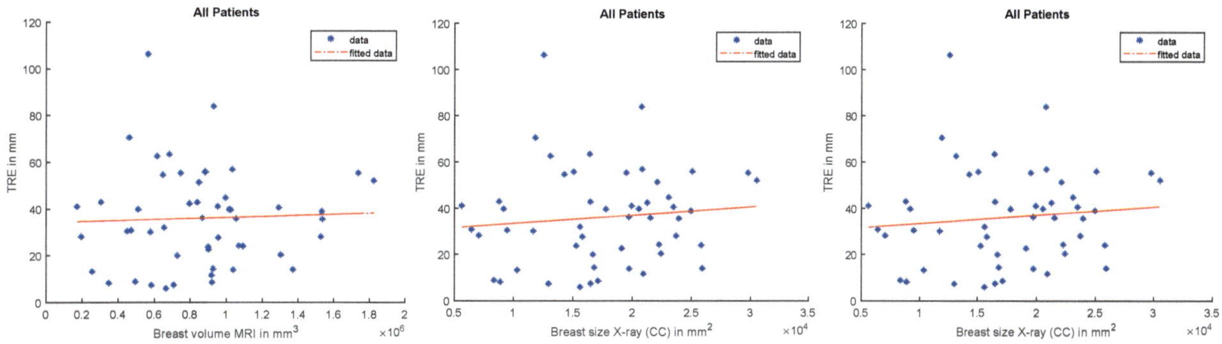

Figure 7.103: Relation between the accuracy of the accumulative registration of the two methods and distribution of the breast sizes in MRI (first) and full X-ray mammograms (2D): CC (middle) and MLO (last). A very weak correlation is found between the breast sizes and the TRE with an r of 0.041, 0.11, and 0.02, respectively.

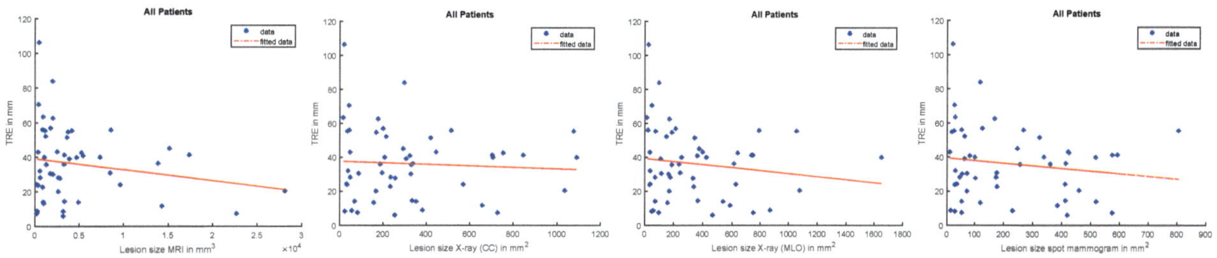

Figure 7.104: Relation between the accuracy of the accumulative registration of the two methods and distribution of the lesions sizes in MRI (first), full X-ray mammograms: CC (second) and MLO (third), and spot mammograms (fourth). A very weak correlation is found between the sizes of lesions and the TRE with an r of -0.18, -0.061, -0.14, and -0.15 respectively.

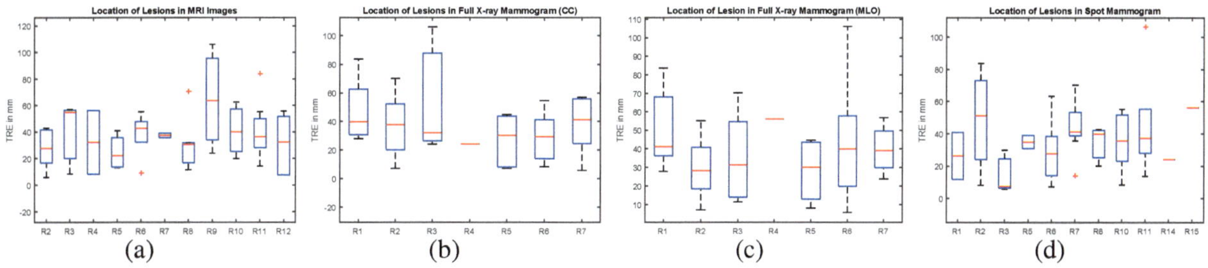

Figure 7.105: Relation between the location of lesions in MRI (first), full X-ray mammography in both views (second and third), and spot mammograms (fourth) and the accuracy of the total TRE of the accumulative registration of the two methods.

Location of Lesions

The location of lesions in MRI, full X-ray mammography, and spot mammogram is the sixth characteristic that has been evaluated. Boxplots have been used for finding a correlation between locations of lesions and the total TRE. MR images were split into twelve regions. X-ray mammograms were split into eight regions. Spot mammograms were split into sixteen regions as explained in Section 5.2.4.
For MRI, only two cases have annotated lesions in R4 and R7, and three cases have annotated lesions in R3, R9, and R10. A conclusive observation for these regions is not possible since the cases located in these regions are likely too small. R5, though, has a small number of cases (four cases only). Nevertheless, it is expected that median TRE would be small because the deformation in this region, which belongs to the lower inner middle slices, is relatively small compared to other areas. The median TRE achieved is 21.8 mm. The highest number of cases belongs to R11 (11 cases), in which the median TRE is 35.9 mm, as shown in Figure 7.105 (first), and it corresponds to lower outer middle slices. In addition, R11 in MRI corresponds to R2 in X-ray mammography which achieves a median of 37.2 mm in the case of CC and a median of 27.9 mm in the case of MLO.
For full X-ray mammography, some regions cannot be considered conclusive results, such as R4 for both views (one case only) and R3 (3 cases) in the case of the CC view. Lesions at the back of the breast near the muscle in the posterior direction (R1 in both views) are harder to register, as shown in Figure 7.105. For the lesions at the center of the breast (R7), it can be expected to achieve a lower error; however, it achieves a high error with a median of 40.9 mm for CC view and 38.7 mm for MLO. It could be concluded that each method separately works well in this region. It has a negative effect when they are added up together, as shown in one of the examples in Figures 7.41, 7.78, and 7.92 (first).
For spot mammograms, some regions do not have available datasets, and some regions, such as R14 and R15, have only one case, R1 and R5 have two cases, and R3 and R8 have only three cases. A conclusive observation is not possible for these cases since they are likely too small. R2, though, has also a small number of cases (four cases only). Nevertheless, it is expected that the median TRE would be high 50.9 mm because the lesions located in this region are at the corners of the spot mammogram. The highest number of cases belong to R6 (13 cases), in which the median TRE is 27.43 mm, as shown in Figure 7.52 (last). Thus, if the guidelines are followed well and the lesion is central in the spot mammogram since it is already a small window, then they are better registered.

7.6 Summary

This chapter presented the results from all the proposed methods. First, there are three approaches to segment MRI simultaneously for three tissue types: the whole 3D volume as a single object, 2D stacked volume as multiple objects, and a subset from the volume as a single object. Based on the analysis in this chapter, the segmentation based on the 3D volume is chosen from the three approaches for all the subsequent steps in the biomechanical model due to its best accuracy, despite being slower than the subset from the volume. It achieves a median of DSC of 0.89, 0.45, and 0.89 for fatty, glandular, and muscular tissues, respectively compared to a manually annotated ground truth.

Second, another novel algorithm was proposed to speed up the biomechanical model for mimicking the mammographic compression in case of CC view. The goal was to validate if machine learning algorithms could replace FEMs for accelerating computation time and keeping the same accuracy. The three models have been investigated separately but were not yet integrated into the registration process due to time constraints. However, two out of three models show very promising results. The estimation of the unloaded state and the compression simulation achieves an average RMSE of 3.4 and 4.7 mm, respectively for XGBoost. While for Att-BLSTM, the average RMSE achieved is 3.0 and 4.9 mm for the unloaded state estimation and the compression simulation, respectively.

Third, the proposed biomechanical model based registration method achieves a median TRE of 31.76 mm for 51 clinical datasets received from the Medical University of Vienna. 22 cases (43.1%) have an overlap between the two annotations.

Fourth, the proposed image based registration method achieves a median TRE of 20.2 mm and the number of overlapping annotations in 26 datasets (51.0%) of the available data. Based on the predicted annotation position resulting from the MRI annotation from the biomechanical model based registration, the image based registration is carried out. By that, it is possible to calculate the accumulative registration of the two methods. The median total TRE achieved is 35.6 mm and 18 cases have an overlap between the two annotations, which means 35.4% of the available data.

The three registration methods has been analyzed using six characteristics that correspond to the variations in datasets: age, views, breast needle biopsies in categories, dense breasts, breast sizes, lesion sizes, and location of lesions. Overall, in the three analyses, the same observation is similar. The methods are significantly dependent on the views of full X-ray mammograms and spot mammograms and also on where the lesion is located in the breast. Moreover, the methods are independent of age, breast needle biopsies, dense breasts, breast sizes, and lesion sizes.

8 Clinical Use

In this chapter, the aim is to stimulate what would happen if a biopsy needle is inserted at the predicted position of the lesion based on the proposed methods. The goal is to evaluate if this biopsy needle hits the actual lesion. Hence, the question, if the methods could be clinically used, will be answered. This investigation is done by modeling the size and dimensions of the needle that is used at the Medical University of Vienna. Then, it estimates approximately the area which tissue is biopsied. These areas are named artificial areas based on the view of the biopsy compression. Then afterward, it evaluates if this area overlaps with the predicted lesion annotation.

8.1 Dimensions of the Needle

There are several dimensions of needles such as 8, 9, 11, and 14-needle gauges for taking a biopsy using stereotactic X-ray guidance. From the literature, an analysis has been done to check how much specimen is enough to be taken during a biopsy based on the size of the needle. It was found that for the needle size of 9 gauge, 14 specimens are enough (145). This article stated the advantages of a larger size of the needle which consequently has larger sample volumes which leads to the increasing accuracy of histological diagnosis (145). Another article presented the difference between dimensions 8 and 11 needle gauges. It was found that there was no big difference in terms of diagnostic accuracy, rate of complications, and subjective experience of pain between the 8 and 11-needle gauges. It was recommended that lesions with a maximum diameter of up to 18 mm to be taken using an 8 gauge needle, and those with a maximum diameter of up to 11 mm to be removed with an 11 gauge needle. However, the assumption that a large tissue volume can be removed using a large needle was only partly correct. The limitation could be caused by one of these reasons: the size of the lesion if it is close to the skin or surface, the size of the breast, and the bleeding rate of the patient (69).

Another article showed that a 9-needle gauge was even better than an 11-needle gauge regarding three points: the average weight of the biopsy specimen, the core length, and the degree of fragmentation (144). The specimens were extracted by rotating the entire system about its longitudinal axis. Geometrically, the longitudinal axis of the needle was located concentrically within the device with its tip opening at the center, the distance was 8 mm. The specimens obtained with needle 9 were significantly longer than the others. The mean length of biopsy cores was 26.6 mm. Although the 9 needle gauge showed better performance than 8 or 11-needle gauges, the removal of specimens from lesions closer to the chest wall was not good.

Table 8.1: Dimensions of the needles

Part number refer to Figure 8.1	length (mm)	width (mm)
1	1	3.75
2	2	3.75
3	1	3.75
4	1	2.99
5	15	2.99
6	1	2.99
7	0.5	3.75
8	1	3.75
9	0.5	3.75

The proposed method is examined on clinical data using a simulation model for stereotactic guided biopsy in order to assess the applicability of the clinical application. The purpose of this simulation is to ascertain the success rate (sampling of the target lesion) when using the model-predicted location to center a stereotactic guided biopsy. According to radiologists from the Medical University of Vienna, they employ a 9-gauge vacuum-assisted biopsy equipment. The outer diameter and the inner diameter are $3.75\,mm$ and $2.99\,mm$, respectively. The dimensions of the inner details of the needle are shown in Figure 8.1 and Table 8.1. 12 or 24 specimens were typically acquired. An area of the biopsy of approximately $15\,mm$ for 12 specimens and $25\,mm$ for 24 specimens was estimated.

The position of inserting the needle inside the breast is decided based on the shortest distance to the surface of the breast. Based on the position of the compression view of the biopsy, the area view differs from one position to another, as shown in Figure 8.1. There are two shapes for the simulated area consequently. For the craniocaudal position of inserting the needle, as shown in Figure 8.2 (left), the area is a circle. The craniocaudal position is applicable for CC, ML, and LM views. The area of the lateral position of inserting the needle is assumed to be a rectangle, as shown in Figure 8.6 (left). This position is only applicable for the CC view because the breast will not be slipped like in the cases of ML and LM views. That is why this approach (lateral position) is only investigated for CC cases (129). The dimensions of 12 (denoted as small area) and 24 specimens (denoted as large area) are mentioned in Table 8.2.

The center of gravity of the annotation in the MRI volume is transferred to the spot mammogram and a circle around this center point mimics the area of excised tissue for CC, ML, and ML cases, as shown in Figure 8.2 and similarly a rectangle shape only for CC cases, as shown in Figure 8.6.

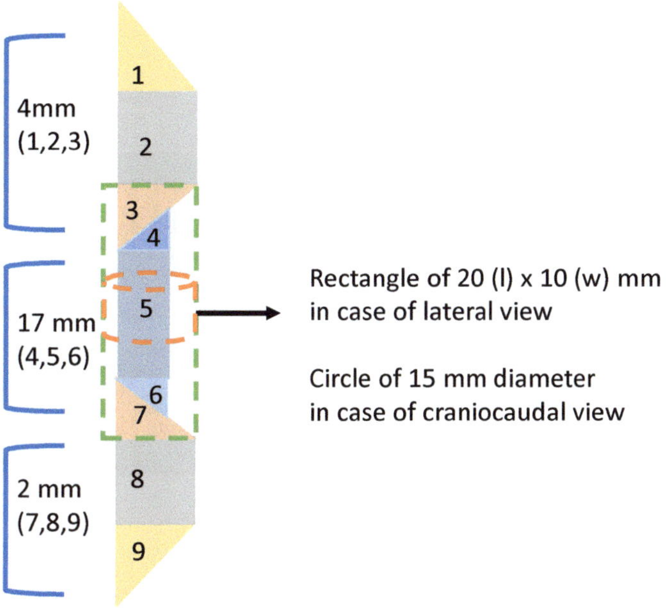

Figure 8.1: The dimensions of the needle. Each part is numbered from 1 to 9 and the dimensions of each part are stated in Table 8.1 based on literature for needle gauges of size 9. Additionally, the dimensions of the artificial areas are given: rectangle in case of lateral view and circle in case of craniocaudal view are estimated by expert radiologists to verify the applicability of clinical use.

Table 8.2: Dimensions of the simulated needle

Simulated Needle shape	12 specimen	24 specimens
Craniocaudal (circle)	$15\,mm$ diameter	$25\,mm$ diameter
Lateral (rectangle)	20×10 mm	30×20 mm

8.1.1 Craniocaudal Approach

Small Area

For a craniocaudal approach of 12 specimens, it is found that eleven patients could be clinically used, which is 21.6% of the available data. All of them are from the prone imaging situation of stereotactically guided biopsy. Different views are included as shown in Table 8.3 which could be considered a positive property since the proposed method could work on a variety of views.

Big Area

For a craniocaudal view of 24 specimens, it is found that more patients could be clinically used. In this case, fourteen patients could be used which is 27.5% of the available data. All of them are from the prone imaging situation of stereotically guided biopsy as well as shown in Table 8.4.

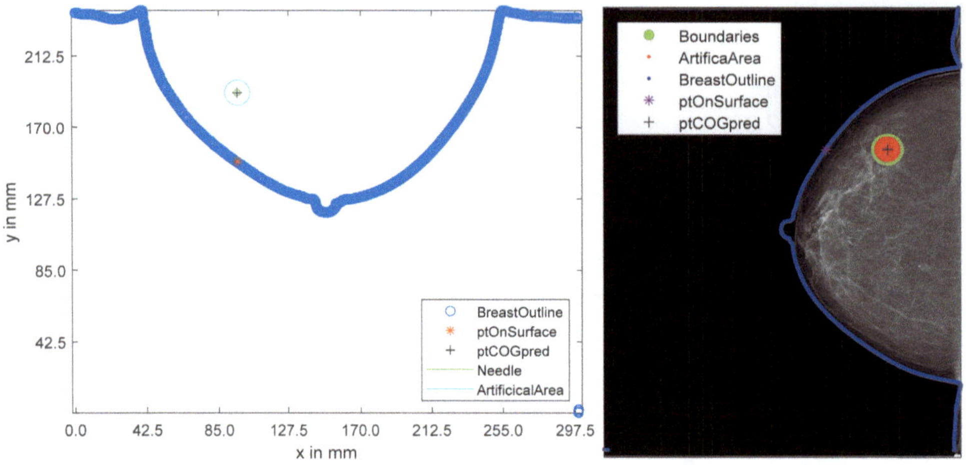

Figure 8.2: The shortest distance to the surface for the craniocaudal approach (left). The artificial area in this case is a circle based on the COG of the predicted annotation of MRI (right).

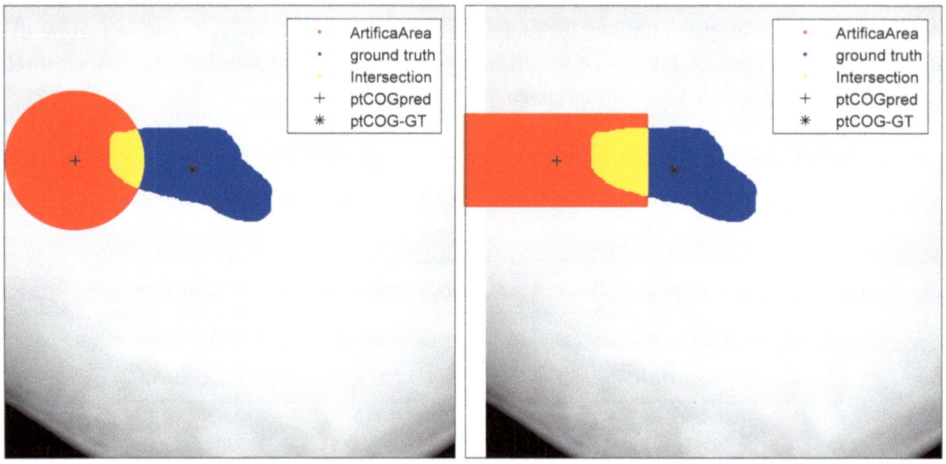

Figure 8.3: Predicted area (red) which a vacuum biopsy would cover if the biopsy was performed around the predicted lesion position in the spot mammogram (+ sign) in cranio-caudal direction and tissue will be excised in an area of 15 *mm* diameter (left) and in lateral direction and tissue will be excised in an area of 20×10 mm (right). Blue represents the annotation of the lesion area in the spot mammogram, and yellow represents the overlap of both. In this case, a part of the lesion would be included in the specimens taken during X-ray guided biopsy and thus the biopsy could be considered successful.

Table 8.3: Distribution of the eleven patients with an overlap of the predicted biopsied area of 15 *mm* diameter for the craniocaudal view in views category

Views	CC + CC	MLO + ML	MLO + LM	CC + ML	CC + LM	MLO + CC
No. of Patients	4	2	2	1	1	1

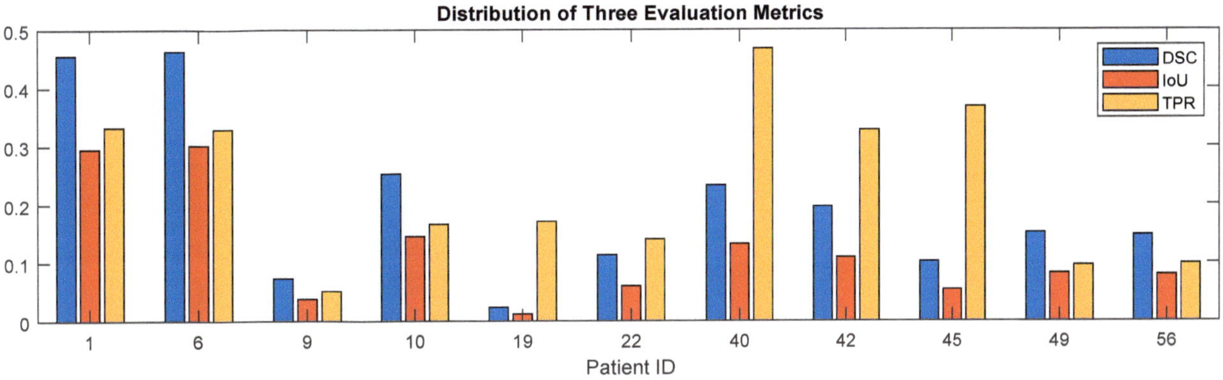

Figure 8.4: Distribution of the three evaluation metrics for the small area of craniocaudal view of 15 *mm* diameter applied. Eleven patients could be clinically used by this artificial area.

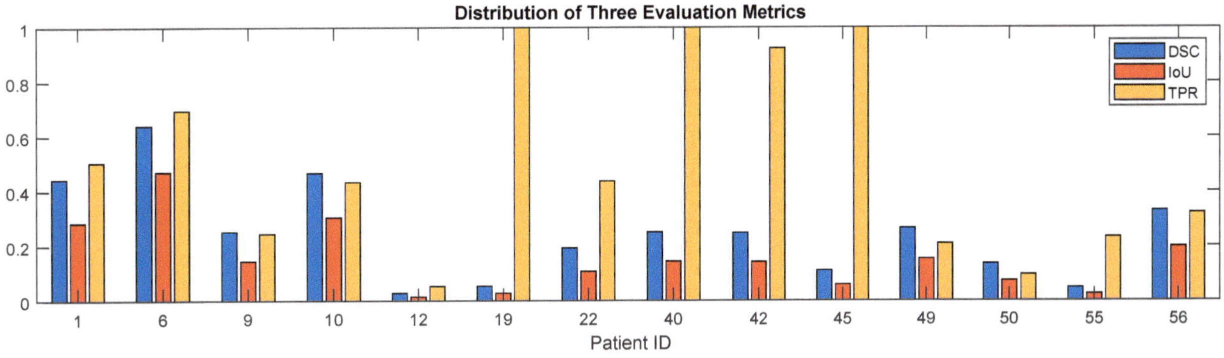

Figure 8.5: Distribution of the three evaluation metrics for the big area of craniocaudal view of 25 *mm* diameter applied. Fourteen patients could be clinically used by this artificial area.

Table 8.4: Distribution of the fourteen patients with an overlap of the predicted biopsied area of 25 *mm* diameter for the craniocaudal view in views category

Views	CC + CC	MLO + ML	MLO + LM	CC + ML	CC + LM	MLO + CC
No. of Patients	7	2	2	1	1	1

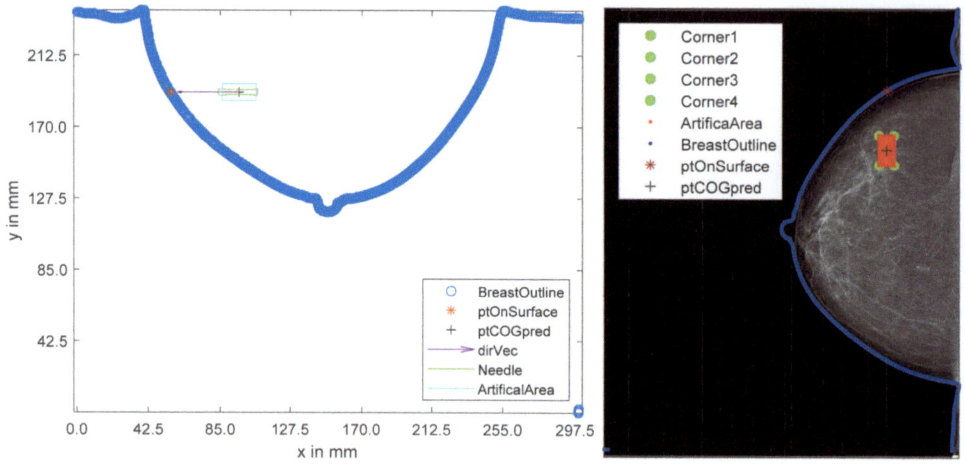

Figure 8.6: The shortest distance to the surface for the lateral approach (left). The artificial area in this case is a rectangle based on the COG of the predicted annotation of MRI (right).

Table 8.5: Distribution of the six patients with an overlap of the predicted biopsied area of 20×10 mm for the lateral view in views category

Views	CC + CC	MLO + CC
No. of Patients	5	1

8.1.2 Lateral Approach

Small Area

For a lateral approach of 12 specimens, it is found that six patients could be clinically used which is 30% of the available data. All of them are from the prone imaging situation of stereotically guided biopsy. Different views are included as shown in Table 8.5 which could be considered a positive property since the proposed method could work on a variety of views.

Big Area

Similarly, a lateral approach of 24 specimens is only investigated for CC cases. It is found that more patients could be clinically used. In this case, eight patients could be used which is 40% of the available data. Similarly, all of them are from the prone imaging situation of stereotically guided biopsy and include different views as shown in Table 8.6.

Two examples have been visualized for the cranial-caudal approach and lateral approach for small and big areas, as shown in Figures 8.3 and 8.7. In the first example, based on the evaluation presented in the results, there is overlap already in the two methods separately, and in the accumulative registration of the

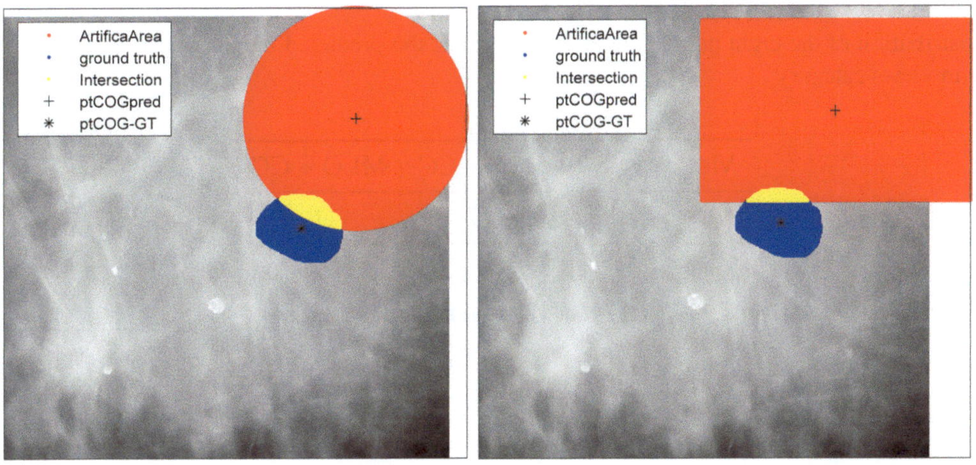

Figure 8.7: Predicted area (red) which a vacuum biopsy would cover if the biopsy was performed around the predicted lesion position in the spot mammogram (+ sign) in cranio-caudal direction and tissue will be excised in an area of 25 *mm* diameter (left) and in lateral direction and tissue will be excised in an area of 30×20 mm (right). Blue represents the annotation of the lesion area in the spot mammogram, and yellow represents the overlap of both. In this case, a part of the lesion would be included in the specimens taken during X-ray guided biopsy and thus the biopsy could be considered successful.

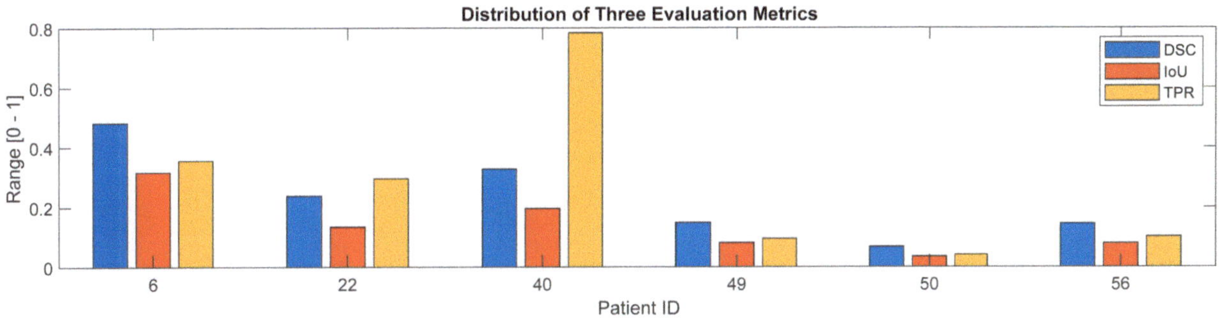

Figure 8.8: Distribution of the three evaluation metrics for the small area of lateral view of 20×10 mm applied. Six patients could be clinically used by this artificial area

two methods. There is also an overlap based on the clinical evaluation. In the second example, there is only overlap based on the clinical evaluation. Both of them are considered clinically applicable.

8.1.3 Summary

In conclusion, although 18 patients have a total overlap as shown in Section 7.5 and 17 patients have a total TRE of less than 25 *mm*, the proposed methods could be clinically applicable for fourteen patients in total. In other words, the results have clinical relevance for fourteen patients, i.e. biopsy would be successful. From the 18 patients with an overlap annotation, 12 patients could be clinically applicable and from 17 patients with a TRE less than 25 *mm*, 14 patients could be clinically applicable. One reason might be because the artificial area is centered around the center of gravity of annotation of MRI which might be deviated from the actual annotation because it is a free hand shape, which might be irregular.

Table 8.6: Distribution of the eight patients with an overlap of the predicted biopsied area of 30 × 20 mm for the lateral view in views category

Views	CC + CC	MLO + CC
No. of Patients	7	1

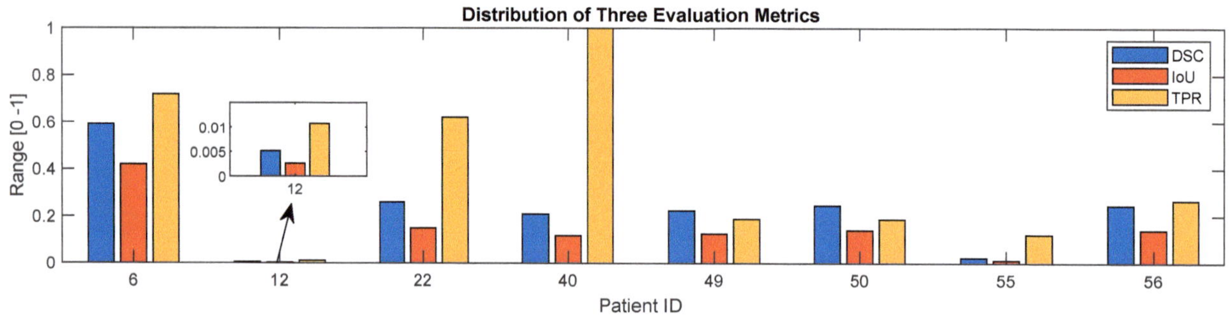

Figure 8.9: Distribution of the three evaluation metrics for the big area of lateral view of 30 × 20 mm applied. Eight patients could be clinically used by this artificial area

One way to improve and avoid this type of problem, instead of selecting COG, selecting a target position based on the lesion shape would be better.

9 Discussion, Conclusions, and Outlook

An image registration method between MRI, full X-ray mammograms, and spot mammograms for X-ray-guided stereotactic breast biopsy has been presented in this thesis. The aim is that lesions only seen in MRI can have a widely accessible, affordable lesion workup with X-ray-guided biopsies. Compared to literature, aligning MRI volumes with spot mammograms has not been implemented before. The method is based on two registration steps: a biomechanical model based registration between MRI and full X-ray mammograms followed by an image based registration between full X-ray and spot mammograms.
The biomechanical model based registration originates from an earlier work presented by Ruiter; Hopp (156; 76). It estimates a breast configuration relative to its shape in X-ray mammography using a patient-specific breast geometry that was imaged using MRI in 3D. A biomechanical model simulates the deformation applied virtually to the MRI to take into consideration non-linear deformation and the behavior of incompressible tissue throughout the deformation process for the breast. Two main blocks in this method have been developed in this thesis using machine learning algorithms, which are segmentation and deformation simulation. In order to transfer the lesion position from the full X-ray mammography to the spot mammogram, which is obtained during X-ray-guided biopsy, an image based registration has been developed and evaluated.

9.1 Discussion and Outlook

9.1.1 Biomechanical Model Based Registration

The proposed biomechanical model based registration method achieves a median TRE of $31.8\,mm$ for 51 clinical datasets received from the Medical University of Vienna. It is found that 22 cases, which is approximately 43.1%, have an overlap between the two annotations of lesions in both modalities. These annotations serve as landmarks for evaluation. Compared to literature in Table 3.4, the error is larger. High errors such as $85.9\,mm$ in Figure 7.30 are more probably for MLO cases than for CC cases. One possible reason is that MLO X-ray mammograms tend to vary considerably more with respect to the rotation angle of the muscle around the longitudinal axis in the transverse plane and rotation around the sagittal axis in the frontal plane for applying compression. Additionally, inaccurate segmentation might add to the error. A novel approach based on unsupervised NNs was proposed for this reason but did not improve for this particular case. Optimizing parameters like rotation angles and compression thickness could be a solution to decrease the error for one patient (78), but this will increase the computation time to two or three days. Hence, a novel approach to replace FEM was proposed. It makes use of deep learning

and machine learning algorithms. It enables the deformation simulation of X-ray mammography in a couple of seconds. Further work can apply this method in the registration process.
To generalize robustness and accelerate computing time, two main processing blocks in the biomechanical model based registration have been researched. The two main blocks are segmentation and deformation simulation.

Segmentation

A novel automatic method has been proposed using unsupervised NNs for segmenting three tissue types: fatty, glandular, and muscular tissues. Compared to literature in Table 7.3, it is the first time to classify three tissues simultaneously. This method was initially developed for the T2-weighted MRI series. One of the challenges was that 30% of the available datasets do not have this series included. Therefore, the proposed method has been improved in the postprocessing to also function for T1-weighted MRI. This thesis proposes three approaches for training the model: the full 3D volume as a single object, 2D stacked volume as multiple objects, and a subset from the volume as a single object while adapting the same trained parameters to the whole volume. The segmentation of full 3D volume is chosen from the three approaches for all the subsequent steps in the biomechanical model due to the best achieved accuracy and despite being slower than the segmentation of the subset from the volume.
The proposed method shows very promising results with a mean DSC of 0.87, 0.45, and 0.88 for fatty, glandular, and muscular tissues, respectively. While the segmentation of fatty and muscular tissues performs similarly to values found in the literature as shown in Table 7.3, the mean DSC for glandular tissue is noticeably lower. The reason for this might be partially attributed to a statistical impact, as errors are overemphasized in this class due to the lower number of voxels. Furthermore, the ground truth segmentation misses tiny details of glandular structures, which are partially segmented with the proposed method which is considered a positive sign property. As a consequence, this may result in a lower DSC score, despite really producing excellent results. In contrast to a patient's X-ray mammography, the concordance between the glandular structures identified by the segmentation was demonstrated, as shown in Figure 9.1. Unsupervised learning will enable us to segment a breast MRI, without the need for labeled training data. As a result, it serves as a fundamental, reliable tool for the automatic creation of biomechanical models unique to each patient.
One of the limitations of the segmentation algorithm is that the datasets have only been tested quantitatively only for T2-weighted MR images. In the future, the database of labeled datasets could be increased for more quantitative validation of the method. The method could be extended for more different MRI series to generalize robustness. Interestingly, the segmentation of a subset from the volume is from the computational point of view 45 times faster than the segmentation of the full 3D volume though the accuracy is slightly lower e.g. for fatty tissue, mean DSC is 0.82 instead of 0.87. In the future, it would be interesting to investigate a broader range of slice numbers from the subset of volume, including e.g.

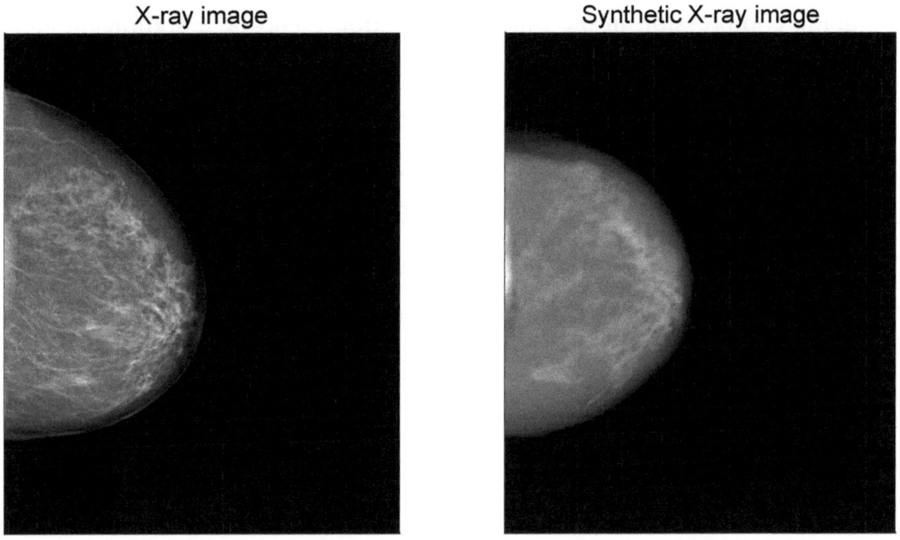

Figure 9.1: An example showing how segmentation using NNs will help in creating the deformed MR image from the biomechanical model: X-ray image (left) and synthetic X-ray (deformed MRI) (right) (162).

three, five, and ten slices, rather than just one slice, and compare the results to the segmentation of the full 3D volume.

Machine Learning Algorithms for Simulation of Biomechanical Model Deformation

In this thesis, estimation of the unloaded state from a gravity-loaded MRI and execution of a mammographic breast compression based on two machine learning algorithms (ERT and XGBoost), and one deep learning algorithm (Att-BLSTM) has been proposed as a novel method. In the literature, machine learning algorithms have not been implemented to simulate the unloaded state. Furthermore, this makes a significant contribution by testing the machine learning based deformation simulation with realistic compression ratios and a large number of clinical datasets, covering a variety of breast shapes from daily clinical practice. The estimation of the unloaded state step and the compression simulation step using the proposed method both exhibit highly encouraging results with an average RMSE of 3.4 and 4.7 mm, respectively for XGBoost. For the machine learning based simulation of the mammographic compression, the investigation of the correlation between error and compression ratio leads us to the conclusion that the XGBoost and Att-BLSTM models are stable up to a compression ratio of approximately 53% with 5.8 mm. The model was tested up to a 76% compression ratio. In comparison to literature where the compression ratio was only 20% as shown in Table 7.7, the proposed models are significantly better for the usually applied compression ratios in clinical practice. The normal range of compression ratios is from 50 to 65% (see Appendix B.3). Additionally, it has been shown that there is no dependence on breast volume, which is a positive sign since it is typical for large breasts to distort significantly when compressed for mammography. To replace FEM simulations that describe the deformation of the breast during mammographic compression in the image registration could be a significant step. This would result in a huge time reduction, from almost 20 minutes to only 5 seconds which equals a speedup of 240 and makes the computation time clinically more relevant. Furthermore, it will give rise to the opportunity to greatly expand the optimization space for biomechanically informed image registration, as some image registration techniques, such as those in (77), need an optimization of some variables like rotation angle around body axes, compression thickness, and Young's moduli of various tissues. The optimization of parameters in such a high dimensional space requires iterative execution of the biomechanical simulation. With the help of machine learning and deep learning models, it is possible to test out many rotations, compression thicknesses, and Young's moduli in a matter of seconds now, potentially increasing the registration accuracy dramatically.

There are several restrictions on the models, despite the fact that the proposed machine learning algorithms produced promising results. The proposed deep learning and machine learning models have, in general, been trained on a specific biomechanical model. Although this model is rather complex and has undergone testing in clinical use cases, it is evident that the machine learning model cannot be broadly applied to significant modifications in the biomechanical model. A certain resolution of the finite element mesh, for instance, was used to train the model. Results may vary depending on whether the deformation is predicted for finer or coarser meshes. Furthermore, meshes that do not meet specific quality requirements as they are, for example, evaluated by ABAQUS, may produce varied outcomes. Similar restrictions have been applied to training and testing for the material properties of tissues, where constant values from the literature have been employed. In order to improve generalization for other FEM use cases, these material attributes could be introduced to the model in the future as a free parameter. In order to prevent overfitting issues, a cross-validation analysis was done on a total of more than 500 cases. The patient cohort contained breasts of various shapes and sizes obtained from clinical routines and images taken at two different clinical sites. It must be taken into account that the machine learning and deep learning models have been trained for the realistic compression applied to the specific patient during X-ray mammography, so the overall number of cases with a high compression ratio of more than 65 is 14.5% in the available patient cohort. At very high compression ratios of more than 65%, this might have caused the error to grow exponentially. It would be intriguing to extend the datasets that have a higher compression ratio in the future. In order to enhance statistics and generalization, the suggested strategy might also be evaluated with more unknown datasets in the future. For optimizing the hyperparameters and examining various loss functions for the three models, extra trials such as a mesh grid search could be added. A breast mesh evaluation rather than a nodal evaluation is another possibility. Currently, the proposed method is not integrated into the registration method due to time constraints. However, it will be advantageous to incorporate it in the future, reducing registration time by a factor of 240.

9.1.2 Image Based Registration

Image based registration between full X-ray mammography and spot mammograms has not been presented in literature. The proposed image based registration method achieves a median TRE of 20.2 *mm*, and annotated landmarks overlapped in 26 datasets, which is 50.1% of the total datasets available at this time.

While the median error of image based registration is already relevant for clinical application and could be feasible to be used in clinical use, more research for potential improvements of the case, e.g. with large errors, is needed. More optimal parameters, such as the number of iterations or restricting the search of the ROI to anatomically plausible regions, may improve such outliers. In addition, more research in combining image similarity metrics may be added. Moreover, another image contrast enhancement method from the proposed ones could be used since the cases that have high errors are from a Philips device, of which only four cases were included in the available dataset. Thus, increasing and diversifying the datasets would be very helpful for validating the optimization. Also, it would be interesting to consider the deformation state of the two images using affine transformation or more complex registration methods. One limitation is that this method only allows rigid transformation which simplifies the problem in a less complex way. Three main parameters have been neglected: the different repositioning of the patient, the difference in the gravity forces applied, and the difference in the projection angle. In the future, more information about these parameters could be collected to explore other transformations.

9.1.3 Accumulative Registration of the Two Methods

Based on the predicted annotation position resulting from the MRI annotation from the biomechanical model based registration, the position of the lesion in the spot mammography was predicted. The median total TRE achieved is 35.6 *mm* and 18 cases have an overlap between the annotated landmarks, which means 35.4% of the available data. The presented methods provide the ability to use X-ray-guided biopsy instead of MRI-guided interventions for those cases. There is still a large error in some clinical datasets, which could be investigated in the future. One reason for these large errors is that the errors of the two methods separately are likely added up when concatenating the registration results, although each method separately works well.

One way to solve this is to allow the image based registration to use information from the biomechanical model based registration. The ROI in image based registration might be selected based on the predicted lesion position from the biomechanical model based registration (see the appendix in section B.2). A window of 7.5×7.5 cm around the predicted lesion has been selected to apply image based registration. This method has been tested until now for three patients which initially had a high TRE and it provides initially promising results with a median TRE of 21.9 *mm* instead of 56.2 *mm*. One disadvantage of this method is that it depends on the biomechanical model based registration. Hence, if biomechanical model based registration does not work, then the accumulative registration of methods will not work as well. Enhancing the number of overlaps between annotations in each method individually, is necessary for cases with large errors, as this evaluation demonstrates greater reliability on this metric than solely assessing TRE.

In the beginning, it was mentioned that the registration between MRI and spot mammograms could be done using three possible approaches. There are five parameters that have a huge impact on the three approaches: the number of modalities used, the resolution of the modalities, dependency of errors, accuracy, and computation time needed.

The presented methods used two steps of registration involving three modalities. It mostly overcomes the restriction to different resolutions in each modality. The error of the second registration method does not depend on the error of the first registration method. Thus, the two individual registration methods, are not dependent. They could work separately. High errors are however possible since the errors are added up as shown in one of the examples in Section 7.5. In terms of computation time, around 30 minutes in total are required using a fixed parameter set.

The requirement to obtain the patient's full mammogram is one disadvantage of this method. While this is typically available because the majority of patients who undergo MRI have recently undergone a screening or diagnostic mammogram, or a mammogram is acquired prior to X-ray guided biopsy for planning purposes, it could potentially be avoided in case of lesions only available in MRI, such as lesions detected in young, high-risk women undergoing MRI screening only. This could be accomplished with the next two approaches.

The second approach of registration consists of two steps, involving two modalities. First, a biomechanical model based registration is used for creating a projection of the deformed MRI, which should be similar to full X-ray mammogram. This is then followed by image based registration between projection of the deformed MRI and the spot mammogram. The proposed segmentation method using NNs, shows that the tiny structure of glandular and fatty tissues are similar in the full X-ray mammogram and the projection of the deformed MRI, as shown in Figure 9.1. The proposed deformation simulation of the breast using machine learning methods could also be integrated into the biomechanical registration method to create the projection of the deformed MRI. The proposed image based registration in this thesis could also be used to register between the synthetic X-ray and spot mammograms. Hence, all the methods are available to investigate this approach in the future. In terms of the five parameters mentioned, it would be similar to the first approach regarding dependency of errors, expected accuracy, and computation time. One advantage is that the number of modalities involved in the registration process has been reduced to two. One limiting factor might be the low resolution of the MRI which may affect the image based registration because it will be applied based on the synthetic X-ray mammogram, not the original X-ray mammogram.

The third approach of registration consists of one step, involving two modalities. A simulation of artificial spot mammograms from MRI datasets was initially developed to align them with real spot mammograms directly. Based on this proposed biomechanical model, the artificial spot mammography could be con-

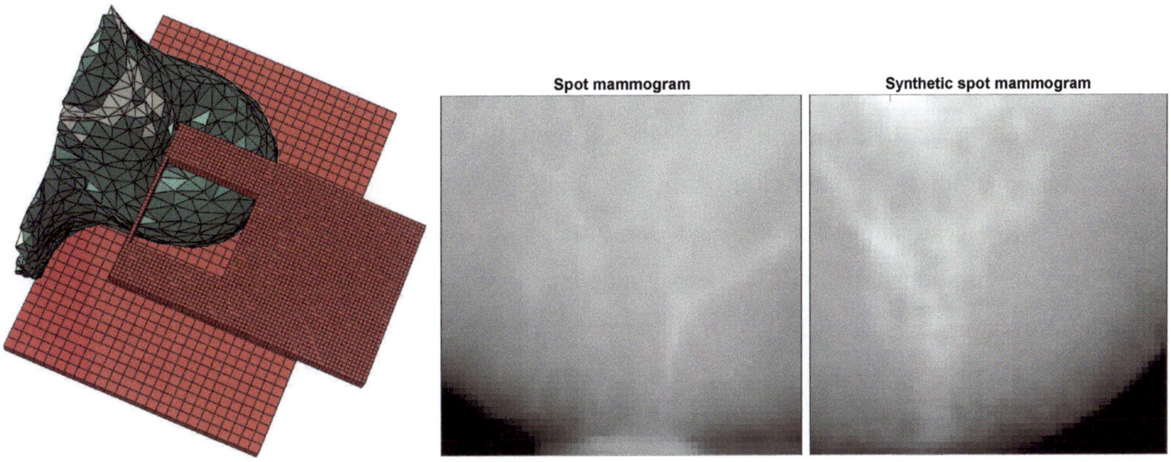

Figure 9.2: An example showing how a synthetic spot mammogram is created from the biomechanical model (left): spot mammogram (middle) and synthetic spot mammogram (right).

structed by simulating the prone patient or upright positioning on a biopsy table with small compression paddles used for stereotactic biopsy. Similarly, synthetic spot mammograms using a ray-casting algorithm could be produced. One of the challenges in this approach is that the compression plate has to be positioned on the breast surface so that the artificial spot mammogram shows the lesion visible in the MRI volume as centrally as possible. An iterative search, which would require up to three hours, is needed. Preliminary results have been achieved with an average TRE of $6\,mm$ for eight cases. Despite, the promising results the following challenges need to be resolved in the future: the lesion's location on the actual spot mammography heavily influences the approach. It is yet unknown, however, if actual spot mammograms can be obtained at the correct anatomical position without the planning based on the full X-ray mammography, which is standard procedure. The iterative search approach for the location of the plates in the simulation can offer general guidelines for positioning the plates clinically, and it may be tested in the future. One drawback of this approach is that this method might fail when breast structures are not visible due to low or extremely high breast densities, or because inner breast structures could not be distinguished due to the low resolution of MR images in the artificial spot mammograms. Regarding the five parameters, the number of modalities used is two, the method will be dependent on the resolution and contrast of inner structures of the breast, as shown in Figure 9.2, and there will be no dependency of errors since it is only one method. The accuracy shows promising results but needs to be tested in the future with more datasets. One drawback is that it currently requires around three hours to compute. In future, using the proposed machine learning algorithm to replace this proposed biomechanical model of simulating artificial spot mammograms from MRI might be helpful for this reason.

In conclusion, the two-step matching method is used currently since it offers more reliable performance and would diverge less from the standard clinical workflow.

9.1.4 General Limitations

Each method separately has been analyzed using seven characteristics that correspond to the variations in datasets: age, views, histopathology, density of breasts, breast sizes, lesion sizes, and location of lesions. The accumulative registration of the two methods has been analyzed using the same seven characteristics to investigate if the subgroups correlate with the method's accuracy. Overall, in the three analyses, the observations are similar. The methods are significantly dependent on the views of the full X-ray mammograms and spot mammograms and also on where the lesion is located in the breast. Some limitations are that the method is working better with CC views and with the location of lesions in the center of the breast. The methods are however independent of age, histopathology, breast density, breast sizes, and lesion sizes. This is considered a positive property since the method can be applied to a lot of variances in the characteristics of the patients and pathologies, e.g. also with dense breasts, which are usually hard to register in case of the biomechanical model based registration and easier to register in case of image based registration. For the compression simulation in the biomechanical model, dense breasts are hard to register due to more non-linear deformation and the heterogeneity of the tissues. However, dense breasts are easier to register in the image based registration due to the clearly defined inner structure of glandular tissue which makes it easier to be identified using image similarity.

One of the general limitations of the proposed method is that it has been tested on the available datasets, which is considered a small one of only 51 patients. Though there is already a lot of variance in the datasets. In the future, it would be interesting to test it on a larger number of datasets.

Currently, the proposed methods transfer the lesions from MRI (3D) to spot mammograms (2D), it would be interesting to enhance the methods for tomosynthesis-guided biopsy (3D) to eliminate the projection to 2D, opting instead for a direct transition from 3D to 3D. In tomosynthesis-guided biopsy, some limitations may be overcome by this, such as the breast being too thin and the target abnormality being located directly behind the nipple or near the chest wall. In the future, it would be intriguing to test tomosynthesis-guided biopsy instead, since there has been already research done for registering between MRI and tomosynthesis images by Smole et al. (176). Additionally, Rochat et al. (149) concluded that tomosynthesis-guided biopsy could be better than stereotactically-guided biopsy.

Another general limitation is that the available datasets only contained cases with lesions visible in all modalities. While not being used for registration explicitly, a clinical use case would be different since the purpose is to apply this method for patients with lesions not visible in X-ray mammography and spot mammograms. It remains to be investigated if the proposed methods are affected by that, for example in solving the image based registration.

9.1.5 Applicability for Clinical Use

In order to determine whether the proposed procedure could be applied clinically, it was tested on clinical data using a simulation model for stereotactic guided biopsy. This simulation was done to evaluate the success rate when a stereotactic guided biopsy was centered using the model's projected location. In order to simulate a needle, radiologists mentioned to have used a 9-gauge vacuum-assisted biopsy equipment to collect 12 or 24 specimens. For 12 specimens and 24 specimens, the estimated area of the biopsy was $15mm$ and $25mm$, respectively. This is considered a limitation since these dimensions originate from personal communication with two experienced radiologists. They are estimating these dimensions using the particular needle used in practice at the Medical University of Vienna. Based on this analysis, 11 and 14 datasets could be clinically used if 12 respectively 24 specimens are taken during the biopsy. The methods thereby show the feasibility of clinical use especially for cases of lesions located centrally in the breast and the views of CC and CC for full X-ray mammography and spot mammograms, respectively.

9.2 Conclusion and Contributions

One of the biggest barriers to a wider use of CE-MRI in clinical practice is the lack of MRI-guided breast interventions (96). The developed methods provide the ability to transfer lesions found by MRI to spot mammography, or close-up stereotactic biopsy, which is utilized for X-ray guided biopsy, in order to make a low-cost and broadly accessible lesion workup with X-ray-guided biopsy possible for lesions only seen in MRI. It is a novel approach for the clinical workflow, by matching MRI and spot mammograms. Having such a method would increase the probability of detecting lesions, reduce the costs by an estimated 50%, decrease the time needed for taking a biopsy by approximately 67% from 90 minutes (for MRI-guided biopsy) to 30 minutes (for X-ray guided biopsy), and solve the issue of shortage of MRI guided breast interventions.

The following points summarize the main contributions of this thesis:

- A novel method for segmenting MR images was developed and tested on clinical cases
- A novel method for replacing FEM using machine learning algorithms was investigated with a large number of datasets and clinically realistic deformations
- For the first time, an image based registration between full X-ray and spot mammograms was developed and tested with clinical data
- Accumulative registration between MRI and spot mammograms was explored, which allows a novel clinical workflow and evaluation of the applicability of clinical use

Acknowledgments

I would like to express my gratitude to my supervisor Dr. Torsten Hopp for his continuous support, technically and personally. He understands and relates to all stresses any PhD student could have. I am thankful for all the conferences we attended together because they allowed us to have more conversations on a professional and personal level. I would also like to thank PD. Dr. Nicole Ruiter for her assistance and supervision. I would like to express my gratitude to both of them for the revision of my thesis.

My aspiration as a scientist is that the research done in this thesis will be of clinical use and help patients in the future.

I would like to say it was such a pleasure to be in the USCT group for creating a suitable environment to work in. Although I was working on a different project, it has been a great pleasure to get to know them closer, Martin, Olga, Zewei, Michael, Patrick, Nima, and Hartmut. I would like to thank my students who greatly affected this project: Dana, Michael, Yu, Zeyu, Rémi, Oussama, and Luit.

I would like to thank my father who is not on earth anymore. I would not have been able to continue this journey without his support to me during his life and now through his continuous prayers and whispering in God's ear for me. I believe that the memories that we shared are the key that makes me pursue my dreams. I would like to thank my mother for always supporting and encouraging me to keep moving forward. She has always been there for me no matter how long the distance was. I would like to thank my siblings as well. Having them around gives me the energy to continue.

I would like to thank my family members in Egypt and some of which are now in Heaven.

I would like to thank my second family in Karlsruhe; Sherif, Hassan, Donna, Shady, Feby, Amir, and Michael who were always there to help and support me. I would like to thank my family in Dresden who inspired me to travel to Germany from the beginning, and whom I considered family: Mina, Michael, Manuel, and Mirna. I would like to thank my friends in Egypt, knowing that if I call them they would never be late to listen or support my family in Egypt when I am not there, Suzan, Bahkhoum, Marina, Phoebe, Youssef, and the list goes on. I would like to thank my friends from church in Egypt who were always asking and kept sending courageous quotes and verses.

Last but not least, I would like to thank God for all the blessings, He granted to me.

A Publications and Supervision

A.1 Journal

S. Said, Z. Yang, P. Clauser, N. V. Ruiter, P. A. T. Baltzer, T. Hopp, "Estimation of the biomechanical mammographic deformation of the breast using machine learning models," in Clinical Biomechanics, Volume 110, 2023, 106117, ISSN 0268-0033.

A.2 Proceedings

S. Said, P. Clauser, N. V. Ruiter, P. A. T. Baltzer, T. Hopp, "Image registration between MRI and spot mammograms for X-ray guided stereotactic breast biopsy: preliminary results," Proc. SPIE 11598, Medical Imaging 2021: Image-Guided Procedures, Robotic Interventions, and Modeling, 115981C (15 February 2021).

Maul, J., **Said, S.**, Ruiter, N., Hopp, T. (2021). X-ray Synthesis Based on Triangular Mesh Models Using GPU-Accelerated Ray Tracing for Multi-modal Breast Image Registration. In: Svoboda, D., Burgos, N., Wolterink, J.M., Zhao, C. (eds) Simulation and Synthesis in Medical Imaging. SASHIMI 2021. Lecture Notes in Computer Science, vol 12965. Springer, Cham.

S. Said, P. Clauser, N. Ruiter, P. A. T. Baltzer, T. Hopp, "Image based registration between full x-ray and spot mammograms for x-ray guided stereotactic breast biopsy," Proc. SPIE 12034, Medical Imaging 2022: Image-Guided Procedures, Robotic Interventions, and Modeling, 120342L (4 April 2022).

S. Said, M. Meyling, R. Huguenot, M. Horning, P. Clauser, N. V. Ruiter, P. A. T. Baltzer, T. Hopp, "MRI breast segmentation using unsupervised neural networks for biomechanical models," Proc. SPIE 12286, 16th International Workshop on Breast Imaging (IWBI2022), 122860C (13 July 2022).

S. Said, P. Clauser, N. Ruiter, P. A. T. Baltzer, T. Hopp, "Image based registration between full X-ray and spot mammograms: Analysis of registration accuracy in subgroups," Proc. SPIE 12466, Medical Imaging 2023: Image-Guided Procedures, Robotic Interventions, and Modeling, 124661V (3 April 2023).

T. Hopp, I. Tabet, **S. Said**, P. Clauser, P.A.B. Baltzer, and N.V. Ruiter, "Image registration of diffusion weighted and conventional breast MRI," Proc. SPIE 12464, Medical Imaging 2023: Image Processing, 124641T (3 April 2023).

A.3 Supervision of students

Supervision of Dana Grund, who studied mathematics in Münster. She did an internship investigating mass scaling, compression plate models and quality of meshes. (July - August, 2020)

Supervision of Micheal Meyling, who studies electrical engineering at KIT. He did his bachelor thesis entitled "Breast Tissue Segmentation of MRI images using Unsupervised Neural Networks." (October, 2020 - April, 2021)

Supervision of Gao Yu, who studies electrical engineering at the KIT. He did his internship by implementing ML position in biomechanical model followed by doing his master thesis with a title "Simulation of spot mammograms from magnetic resonance images based on biomechanical modeling." (December, 2020 - October, 2021)

Supervision of Rémi Huguenot, who studies engineer telecom at Saint-Etienne University. He started his internship by making the segmentation more robust for all patients and solving the problem of interpolation for MRI images. (April, 2021 - August, 2021)

Supervision of Oussama Kadmi, who studies image processing and electronics systems at telecom Saint Etienne University. He started his internship by analyzing the segmentation data for MRI images. (August, 2022)

Supervision of Zeyu Yang, who studied communication engineering at the KIT. He started his master thesis with a title "Modeling the Deformation of Breast Tissue Using Machine Learning Approaches." (February, 2021 - September, 2022)

Supervision of Luit Düringer, who is a student at DHBW. He did his practical part in robustifying the segmentation methods for T1 - weighted MRI images. (January - March, 2023)

B Appendix

B.1 Interpolation

Interpolation is needed for two purposes. First, MRI data is not isotropic when received from the Medical University of Vienna to save time during data acquisition. Hence, MRI has a small number of slices and that is why interpolation is required. Interpolation is needed for the original data of MRI and the segmented data of MRI. Upsampling increases the number of sampling points. While downsampling, the resolution is reduced. For example, if there are two images, A (high resolution) and B (low resolution) whose resolutions are to be adjusted to each other, downsampling produces fewer artifacts than upsampling due to the higher density of the existing data points. Upsampling artifacts can be huge due to the large gaps and missing information where interpolation is required.

It has been tested to apply linear interpolation (classical interpolation) before applying segmentation algorithms. However, it was observed that it negatively affects the quality of segmentation due to missing information and the big gap between slices. To avoid this issue, interpolation was applied after segmenting MRI data using morphological merge (26) (see equation B.1).

$$C = [(As)((AB)s)](AB)(26) \tag{B.1}$$

The main difference between the classical interpolation and this morphological method is that classical interpolation sticks to an intensity range and the interpolation is done by keeping the overall variation of this range. The morphological merge method is based on binary interpolation. It is applied to each tissue separately and between slices, in which it is repeated as many times as necessary to obtain the right number of slices.

For example as in equation B.1, A and B are considered two original slices, S is a morphological object of small size, and C is considered to be the interpolated slice. The required number of slices is computed based on the number of slices that exist in the original data using this equation

$$\alpha = N_z * \frac{Z}{X} \tag{B.2}$$

where α and N_z are the number of slices after interpolation and before interpolation, respectively. Z and X are the resolutions of each coordinate.

Second, the annotation of the MRI segmentation, which serves as ground truths, was performed based on the T2-weighted images. The T1-weighted images have a lower physical resolution, i.e. the centers of the pixels of the two images have different distances. If the number of sampling points of an image object is smaller while the object size remains the same, this results in a smaller number of voxels, i.e. a smaller volume. Accordingly, the resolutions must be adapted to each other in order to be able to make an accurate comparison using the DSC. This can be done either by downsampling the ground truth or upsampling the T1-weighted images.

B.2 Select ROI Using Window 7.5 x 7.5

The proposed algorithm here instead of taking the full view of the X-mammogram, only takes a small part 7.5×7.5 cm or 10×10 cm of it based on the predicted lesion from the biomechanical model registration. It will not only cause a speeding up for image based registration, but it will also affect the TRE of image based registration as it is concentrated on a region of interest. This method has been tested for three cases as shown in table B.1. It has been realized that some $TRE_{Imagebased}$ which were $71.5mm$ decreased to $21.9mm$. One drawback is if the biomechanical registration did not work, then as consequence, image based registration will not work as well.

Table B.1: Comparison between the results of full size and window 7.5×7.5 of three patients for image based registration

Window	**P1**	**P2**	**P3**
Full size	$14.9mm$	$56.2mm$	$71.5mm$
W 7.5X7.5 cm	$12.7mm$	$34.68mm$	$21.9mm$

B.3 Distribution of Patients Regarding Compression Ratio

This figure shows the distribution of the available datasets that have been tested using machine learning algorithms relative to the compression ratio read from the metadata. The normal range from 50% to 65% compression ratio, as shown in Figure B.1. Only 33% out of the total patients have a compression ratio up to 53%.

Table B.2: Accumulative registration using window 7.5×7.5 of the same three patients

TRE/mm	**P1**	**P2**	**P3**
Biomechanical	6.10	35.7	24.6
Imagebased (W 7.5x7.5)	12.7	34.68	21.9
Total	15.5	13.7	4.48

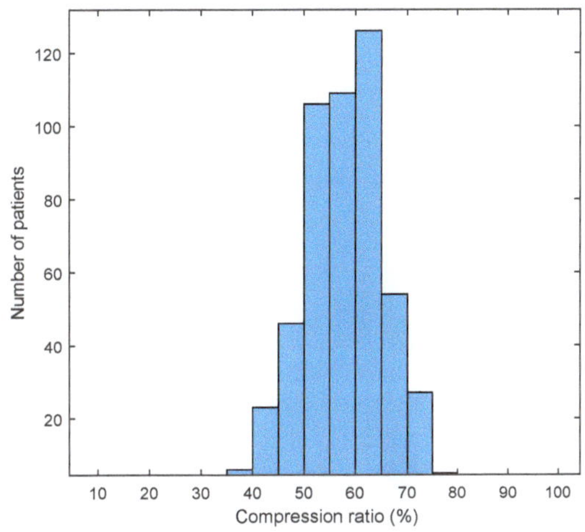

Figure B.1: Histogram of patients regarding the compression ratio read from the metadata for 516 patients that have been tested using machine learning algorithms for the compression step.

B.4 More Details about Data Characteristics as in Chapter 5

B.4.1 Resolution

The figures B.2, B.3, and B.4 have been sorted based on the minimum resolution available regarding MRI images in x and y coordinates denoted as $resMRI_{x,y}$ and showing the spacing between slices which are denoted as $resMRI_z$ and the variance of resolutions of the other modalities: full X-ray mammogram denoted as resX-ray and spot mammogram denoted as *resSpot* that have been used in the registration methods for the same dataset.

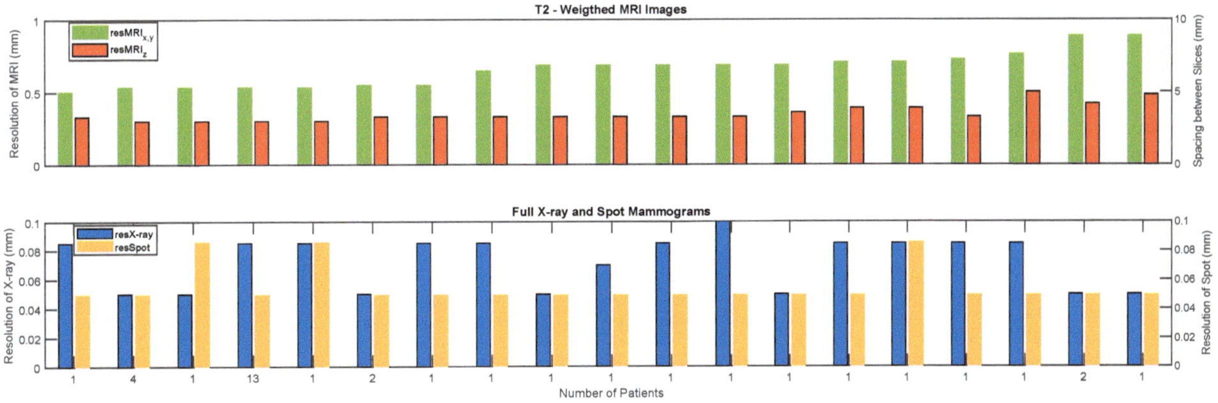

Figure B.2: Combination of resolution of the three modalities: MRI, full X-ray mammogram, and spot mammograms of the available T2- weighted MRI images in the clinical datasets received from the Medical University of Vienna. The green bar indicates the resolution of the MRI modality in the x and y directions. The red bar indicates the resolution of the MRI modality in the z direction. The blue bar indicates the resolution of full X-ray mammograms. The yellow bar indicates the resolution of the spot mammogram.

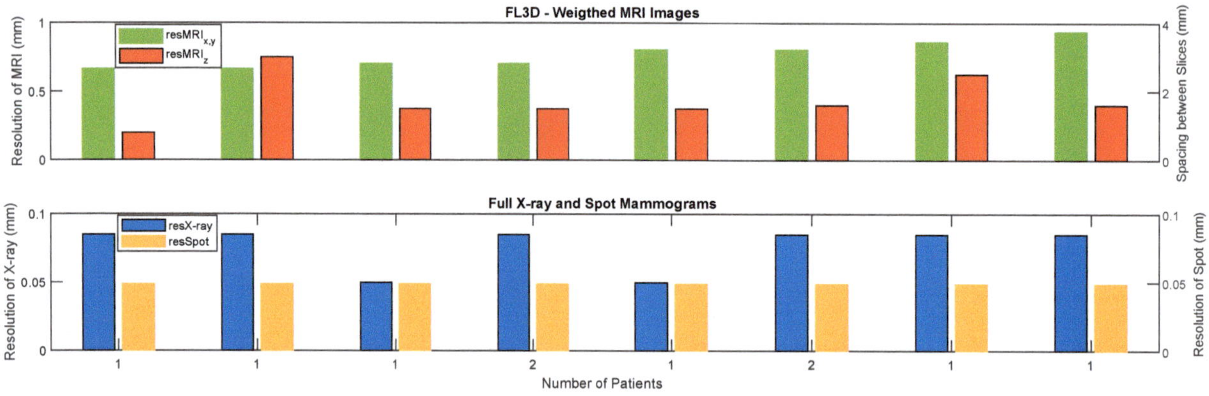

Figure B.3: Combination of resolution of the three modalities: MRI, full X-ray mammogram, and spot mammograms of the available FL3D - weighted MRI images in the clinical datasets received from the Medical University of Vienna. The green bar indicates the resolution of the MRI modality in the x and y directions. The red bar indicates the resolution of the MRI modality in the z direction. The blue bar indicates the resolution of full X-ray mammograms. The yellow bar indicates the resolution of the spot mammogram.

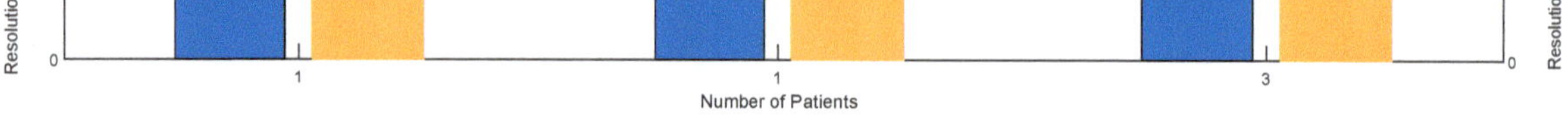

Figure B.4: Combination of resolution of the three modalities: MRI, full X-ray mammogram, and spot mammograms of the available T1 - weighted MRI images in the clinical datasets received from the Medical University of Vienna. The green bar indicates the resolution of the MRI modality in the x and y directions. The red bar indicates the resolution of the MRI modality in the z direction. The blue bar indicates the resolution of full X-ray mammograms. The yellow bar indicates the resolution of the spot mammogram.

C List of Figures

1.1 The proposed clinical workflow using image registration 2
1.2 Registration between MRI and spot mammograms 4

2.1 Breast anatomy 8
2.2 Imaging situation of each modality 9
2.3 Examples of the artifacts available in MRI. 10
2.4 Description of neuron from two points of view 19

3.1 The components needed to build the proposed biomechanical model 28
3.2 The block diagram of the biomechanical based registration 32
3.3 Three views MRI volume in one of the clinical cases 34
3.4 The block diagram of the segmentation process 35
3.5 The steps of preprocessing in one of the slices 35
3.6 Histogram of voxels intensities in MRI volume 36
3.7 Concepts of loss functions 37
3.8 The first approach: the same MRI volume from NN of ten iterations - transverse view . . 39
3.9 Experiment of selecting the number of iterations 40
3.10 The first step in postprocessing 40
3.11 Histogram of the unassigned segmented labels 41
3.12 The second step in the postprocessing 41
3.13 The result of segmentation 44
3.14 Correlation between certain rations in images and meshes 45
3.15 Solving the tilting problem using more restricted nodes 48
3.16 Simulation of biomechanical model for CC view 48
3.17 Simulation of biomechanical model for MLO view 49
3.18 Problems of bad quality meshes that could lead to non-convergence 52
3.19 Diagram of a tetrahedron mesh 53
3.20 Monitoring of mesh quality - first scenario 56
3.21 Monitoring of mesh quality - second scenario 56
3.22 Monitoring of mesh quality - third scenario 57
3.23 Monitoring of mesh quality - fourth scenario 57

3.24 Two algorithms for the estimation of the position of lower plate in the inferior direction . 58
3.25 Analysis of TRE for different scenarios and estimations for the position of lower plate . . 59
3.26 Effect of skin thickness 60
3.27 Magnitude of specific nodes for the full scenario using different mass scaling factors and time increment values 62
3.28 Magnitude of specific nodes for the full scenario using different mass scaling factors . . 62
3.29 Magnitude of specific nodes for the unloaded state and moving plates (gravity+plates) scenario using different mass scaling factors 63
3.30 Magnitude of specific nodes for the full scenario using increasing total time 63
3.31 Manually changing D_1 and C_{10} 64
3.32 The total kinetic energy and internal energy 64
3.33 A clinical case solved by FEM in two steps of simulation 66
3.34 Workflow of the proposed machine learning algorithm for optimization 66
3.35 Simulation of compression step in case of CC view 67
3.36 Validation of removing some intermediate steps from the compression step 67
3.37 Data structure of the features 69

4.1 Auotmatically annotated atructures 74
4.2 The block diagram of the image based registration 80
4.3 Preprocessing block of image based registration method for full X-ray mammogram . . . 81
4.4 Preprocessing block of image based registration method for spot mammogram 81
4.5 Rotation of spot mammogram 82
4.6 Six techniques for image contrast enhancement 84
4.7 Evaluation of image quality 85
4.8 Removing muscle area from X-ray mammograms 86

5.1 Flow diagram of clinical datasets regarding views 92
5.2 Number of patients: CC for full X-ray and LM or ML or CC for spot mammograms . . . 92
5.3 Number of patients: MLO for full X-ray and LM or ML or CC for spot mammograms . . 93
5.4 Distribution of machines and models - MR images 93
5.5 Distribution of machines and models - full X-ray mammograms 94
5.6 Breast needle biopsy categories 94
5.7 Histogram - percentage density (PD%) 97
5.8 Distribution of breast sizes 98
5.9 Lesion sizes MRI vs X-ray mammogram 98
5.10 Lesion sizes X-ray vs spot mammogram 99
5.11 Location of lesions for the three modalities 99
5.12 Histogram of location of lesions based on the defined regions for the three modalities . . 99

6.1 Classifying fatty and glandular tissues using niftySeg 102
6.2 Annotation using Slicer 3D 103
6.3 Evaluation methods 104

7.1 Seven common problems used for qualitative evaluation 108
7.2 Mean DSC for the three categories of slices 109
7.3 Segmentation of T1-weighted MRI series 110
7.4 Mean DSC of segmentation 110
7.5 Mean Jaccard Index of segmentation 111
7.6 One clinical case using segmentation of 2D stacked volume 111
7.7 DSC for each patient 112
7.8 Jaccard Index for each patient 113
7.9 Segmentation of three tissues using the three approach 113
7.10 Computation Time in CPU 114
7.11 An Example of segmentation of the proposed method compared to Fuzzy C-means . . . 114
7.12 Comparison of percentage of nodes in each category of the error 117
7.13 Probability density function versus RMSE 119
7.14 Relation between the breast volume and the RMSE 120
7.15 Relation between the compression ratio and the RMSE for the three models 121
7.16 Comparison between 1-fold vs 10-fold of percentages of nodes into categories - unloaded state estimation 122
7.17 Comparison between 1-fold vs 10-fold of percentages of nodes into categories - compression step 123
7.18 Clinical case with a visually good result in the sagittal plane (unloaded state estimation) . 123
7.19 Clinical case with a visually good result in the transversal plane (unloaded state estimation) 124
7.20 Clinical case with a visually good result in the coronal plane (unloaded state estimation) 124
7.21 Clinical case with a visually good result in the sagittal plane (compression step) 124
7.22 Clinical case with a visually good result in the transversal plane (compression step) . . . 125
7.23 Clinical case with a visually good result in the coronal plane (compression step) 125
7.24 Clinical case with visually non-optimal result in the sagittal plane (unloaded state estimation) 125
7.25 Clinical case with visually non-optimal result in the transversal plane (unloaded state estimation) 125
7.26 Clinical case with visually non-optimal result in the coronal plane (unloaded state estimation) 126
7.27 Clinical case with visually non-optimal result in the sagittal plane (compression step) . . 126
7.28 Clinical case with visually non-optimal result in the transversal plane (compression step) 126

7.29 Clinical case with visually non-optimal result in the coronal plane (compression step) . . 127
7.30 An example of registration between MRI and X-ray mammography - bad case 128
7.31 Tracking the lesion in MR image before deformation 128
7.32 Tracking the lesion in MR image after deformation 129
7.33 Accuracy of the biomechanical model based registration method 129
7.34 Results of the three evaluation metrics for biomechanical model based registration 129
7.35 Histogram of percentage of distorted elements and total loss volume 131
7.36 Relation between the accuracy and the quality mesh or loss volume - method 1 132
7.38 An example of registration between MRI and X-ray mammograhy - case 1 133
7.39 An example of registration between MRI and X-ray mammograhy - case 2 133
7.40 An example of registration between MRI and X-ray mammograhy - case 3 133
7.41 An example of registration between MRI and X-ray mammograhy - case 4 134
7.42 An example of registration between MRI and X-ray mammograhy - case 5 134
7.43 An example of registration between MRI and X-ray mammograhy - case 6 134
7.44 Relation between the accuracy and the age of clinical datasets - method 1 135
7.45 Distribution of the three evaluation metrics for the CC cases - method 1 136
7.46 Distribution of the three evaluation metrics for the MLO cases - method 1 136
7.47 Relation between the accuracy and the views of clinical datasets - methods 1 137
7.48 Relation between the accuracy and the breast needle biopsies into categories - method 1 . 138
7.49 Relation between the accuracy and the dense breasts of clinical datasets - method 1 . . . 139
7.50 Relation between the accuracy and the sizes of breasts - method 1 139
7.51 Relation between the accuracy and the sizes of lesions - method 1 139
7.52 Relation between the accuracy and the location of lesions - method 1 140
7.53 Boxplots of the four properties extracted for contrast Enhancement of spot mammograms 142
7.54 The Euclidean distance for contrast and energy parameters 142
7.55 Distribution of the contrast property for all the available datasets 143
7.56 Distribution of the energy property for all the available datasets 143
7.57 Analysis of different image similarity metrics using method 2 of image contrast 143
7.58 Analysis of different image similarity metrics using method 3 of image contrast 143
7.59 Analysis of different image similarity metrics using method 4 of image contrast 144
7.60 Manual annotation for the rotation angle of the muscle area 144
7.61 Distribution of TRE for all image similarity metrics for full size of X-ray mammogram . 146
7.62 Distribution of TRE for all image similarity metrics for ROI of X-ray mammogram using multi-resolution approach 146
7.63 Distribution of TRE for all image similarity metrics for ROI of X-ray mammogram using pixelwise approach 146

7.64 Distribution of three evaluation metrics for all image similarity metrics for full size of X-ray mammogram . . . 147
7.65 Distribution of three evaluation metrics for all image similarity metrics for ROI of X-ray mammogram using a multi-resolution approach . . . 147
7.66 Distribution of three evaluation metrics for all image similarity metrics for ROI of X-ray mammogram using normal convoution . . . 147
7.67 Number of datasets with an overlap of lesion annotations for three evaluation metrics . . 148
7.68 Comparison between computation time of getting ROI . . . 149
7.69 Comparison between the pixelwise vs multi-resolution approach visually . . . 150
7.70 Experiment of testing the most robust rescaling factor . . . 150
7.71 Comparison of computation time of applying all the image similarity metrics using the full size vs ROI . . . 151
7.72 Analysis of TRE for the image based registration method using the five proposed combined metrics . . . 155
7.73 Analysis of TRE for the image based registration method . . . 156
7.74 Distribution of the three evaluation metrics - method 2 . . . 156
7.75 An example of registration between X-ray mammogram and spot mammogram - case 1 . 157
7.76 An example of registration between X-ray mammogram and spot mammogram - case 2 . 157
7.77 An example of registration between X-ray mammogram and spot mammogram - case 3 . 158
7.78 An example of registration between X-ray mammogram and spot mammogram - case 4 . 158
7.79 An example of registration between X-ray mammogram and spot mammogram - case 5 . 159
7.80 An example of registration between X-ray mammogram and spot mammogram - case 6 . 159
7.81 Relation between the accuracy and the age of clinical datasets - method 2 . . . 160
7.82 Relation between the accuracy and the views of clinical datasets - method 2 . . . 162
7.83 Distribution of the three evaluation metrics for the CC/CC cases - method 2 . . . 162
7.84 Distribution of the three evaluation metrics for the MLO/ML cases - method 2 . . . 162
7.85 Distribution of the three evaluation metrics for the MLO/LM cases - method 2 . . . 163
7.86 Relation between the accuracy and the breast needle biopsies into categories - method 2 . 164
7.87 Relation between the accuracy and the dense breasts of clinical datasets - method 2 . . . 165
7.88 Relation between the accuracy and the sizes of breasts - method 2 . . . 165
7.89 Relation between the accuracy and the sizes of lesions - method 2 . . . 166
7.90 Relation between the accuracy and the location of lesions - method 2 . . . 167
7.91 An example of registration between MRI and spot mammogram - CC cases . . . 167
7.92 An example of registration between MRI and spot mammogram - MLO cases . . . 167
7.93 Total TRE of accumulative registration of the two methods . . . 168
7.94 Distribution of the three evaluation metrics of accumulative registration of the two methods 169
7.95 Relation between the accuracy and the age of clinical datasets - accumulative registration 170

7.96 Relation between the accuracy and the views of clinical datasets - accumulative registration 171
7.97 Relation between the accuracy and the views of clinical datasets - accumulative registration 172
7.98 Distribution of the three evaluation metrics for the CC/CC cases - accumulative registration 172
7.99 Distribution of the three evaluation metrics for the MLO/ML cases - accumulative registration . . . 173
7.100 Distribution of the three evaluation metrics for the MLO/LM cases - accumulative registration . . . 173
7.101 Relation between the accuracy and the breast needle biopsies into categories - accumulative registration . . . 174
7.102 Relation between the accuracy and the dense breasts of clinical datasets - accumulative registration . . . 174
7.103 Relation between the accuracy and the sizes of breasts - accumulative registration 175
7.104 Relation between the accuracy and the sizes of lesions - accumulative registration 175
7.105 Relation between the accuracy and the location of lesions - accumulative registration . . 175

8.1 Needle dimensions . . . 181
8.2 Applicability of clinical use . . . 182
8.3 Applicability of clinical use - small area . . . 182
8.4 Clinical use - craniocaudal view - margin 0 . . . 183
8.5 Clinical use - craniocaudal view - margin 1 . . . 183
8.6 Applicability of clinical use - big area . . . 184
8.7 Applicability of clinical use - big area . . . 185
8.8 Clinical use - lateral view - margin 0 . . . 185
8.9 Clinical use - lateral view - margin 1 . . . 186

9.1 An example of synthetic X-ray mammogram . . . 189
9.2 An example of synthetic spot mammogram . . . 193

B.1 Distribution of patients regarding compression ratio . . . 203
B.2 Combination of resolutions - T2 weighted MRI . . . 203
B.3 Combination of resolutions - FL3D weighted MRI . . . 204
B.4 Combination of resolutions - T1 - weighted MRI . . . 204

D List of Tables

2.1 Types of needle . . . 12
2.2 Difference between modalities . . . 15

3.1 Literature review of finite element methods for breast deformation . . . 27
3.2 Literature of biomechanical models . . . 27
3.3 Literature of biomechanical models - geometry definitions . . . 29
3.4 Evaluation of clinical datasets available from literature . . . 31
3.5 Literature of skin model . . . 31
3.6 The range and the step size that have been tried for creating a tetrahedron mesh (first column). The best parameters were found with the least distorted elements (second column) 54
3.7 Tissue labels . . . 69

5.1 Distribution of combination views of the three patients available from the Medical University of Vienna in case of upright position regarding imaging situation for the X-ray guided biopsy . . . 91
5.2 Distribution of combination views of the four patients available from the Medical University of Vienna in the case of prone position (left breasts) regarding imaging situation for the X-ray guided biopsy . . . 95
5.3 Image resolution of X-ray and spot mammogram for image based registration . . . 95
5.4 Categories of breast needle biopsies regarding histopathological appearances . . . 96
5.5 Regions specified for location of lesion in MRI . . . 100

7.1 Percentage of datasets in the group score of each category of slices . . . 108
7.2 Accuarcy in terms of DSC of three proposed approaches for segmentation (mean, median) $\pm$ standard deviation RMSE (mm) in the eight datasets . . . 112
7.3 Comparison of the mean DSC for three tissues compared to literature review . . . 115
7.4 Unloaded state estimation: (mean, median) $\pm$ standard deviation RMSE (mm) of the validation set. . . . 116
7.5 Compression step: (mean, median) $\pm$ standard deviation RMSE (mm) of the validation set 118
7.6 Average RMSE for tissues in mm . . . 119
7.7 Literature review of machine learning algorithms for breast deformation . . . 121

7.8 RMSE ((mean, median) ± standard deviation) of 10-fold cross-validation for the unloaded state estimation and the compression step in the last step for XGBoost and Att-BLSTM . . . 122
7.9 Accuracy of biomechanical model based registration between MRI and X-ray mammography . . . 128
7.10 The correlation strength into categories based on Wuensch and Evans (209) . . . 130
7.11 The number of elements in the initial mesh exceeded the limit of the minimum angle (less than <10°) . . . 130
7.12 TRE of biomechanical model based registration regarding views . . . 136
7.13 Evaluation of biomechanical model based registration regarding ML and RL cases . . . 137
7.14 Detailed analysis of boxplots of class B2 and B5a versus the accuracy of the biomechanical model based registration . . . 137
7.15 Rotation angle of the muscle area in the full X-ray mammography . . . 145
7.16 The second criteria for selecting a good cluster . . . 153
7.17 The third criteria for selecting a good cluster . . . 154
7.18 Accuracy of image based registration between full X-ray mammography and spot mammograms . . . 155
7.19 Distribution of six patients (prone - other views) from the Medical University of Vienna regarding imaging situation for the X-ray guided biopsy . . . 161
7.20 Distribution of four patients (prone - left breasts) from the Medical University of Vienna regarding imaging situation for the X-ray guided biopsy . . . 161
7.21 Distribution of three patients (upright) from the Medical University of Vienna regarding imaging situation for the X-ray guided biopsy . . . 163
7.22 Detailed analysis of boxplots of class B2 and B5a versus the accuracy of the image based registration . . . 163
7.23 Accuracy of accumulative registration of the two methods between MRI and spot mammograms . . . 168
7.24 Distribution of the total overlap of the remaining views (Only two patients) . . . 171

8.1 Dimensions of the needles . . . 180
8.2 Dimensions of the simulated needle . . . 181
8.3 Distribution of the eleven patients with an overlap of the predicted biopsied area of 15 mm diameter for the craniocaudal view in views category . . . 182
8.4 Distribution of the fourteen patients with an overlap of the predicted biopsied area of 25 mm diameter for the craniocaudal view in views category . . . 183
8.5 Distribution of the six patients with an overlap of the predicted biopsied area of 20×10 mm for the lateral view in views category . . . 184

8.6 Distribution of the eight patients with an overlap of the predicted biopsied area of 30×20 mm for the lateral view in views category 186

B.1 Comparison between the results of full size and window 7.5×7.5 of three patients for image based registration 202

B.2 Accumulative registration using window 7.5×7.5 of the same three patients 202

E Bibliography

[1] Breast Anatomy. URL https://www.learnanatome.com/breast. Accessed: November 10, 2023.

[2] Aspect Ratio. URL https://docs.salome-platform.org/latest/gui/SMESH/aspect_ratio_3d.html. Accessed: May 20, 2022.

[3] Neuron. URL https://neuroscientificallychallenged.com/glossary/neuron. Accessed: June 4, 2023.

[4] Systematic review: Using Magnetic Resonance Imaging to Screen Women at High Risk for Breast Cancer. *Annals of Internal Medicine*, 148(9):671–679, 2008. doi: 10.7326/0003-4819-148-9-200805060-00007. URL https://www.acpjournals.org/doi/abs/10.7326/0003-4819-148-9-200805060-00007. PMID: 18458280.

[5] Latest global cancer data: Cancer burden rises to 19.3 million new cases and 10.0 million cancer deaths in 2020. *International Agency for Research on Cancer*, 2020.

[6] A. Karpathy and J. Johnson. CS231n Convolutional Neural Networks for Visual Recognition, 2023. URL http://cs231n.github.io/convolutional-networks/. Accessed: July 28, 2023.

[7] M. N. Ahmed, S. M. Yamany, N. Mohamed, A. A. Farag, and T. Moriarty. A modified Fuzzy C-means algorithm for bias field estimation and segmentation of MRI data. *IEEE Transactions on Medical Imaging*, 21(3):193–199, 2002.

[8] R. Alcantara, M. Posso, M. Pitarch, N. Arenas, B. Ejarque, V. Iotti, and G. Besutti. Contrast-enhanced mammography-guided biopsy: technical feasibility and first outcomes. *European radiology*, 33, 07 2022. doi: 10.1007/s00330-022-09021-w.

[9] O. Alonzo-Proulx, N. Packard, J. M. Boone, A. Al-Mayah, K. K. Brock, S. Z. Shen, and M. J. Yaffe. Validation of a method for measuring the volumetric breast density from digital mammograms. *Physics in Medicine and Biology*, 55(11):3027, may 2010. doi: 10.1088/0031-9155/55/11/003. URL https://dx.doi.org/10.1088/0031-9155/55/11/003.

[10] V. Ames and P. D. Britton. Stereotactically guided breast biopsy: a review. *Insights into Imaging*, 2:171 – 176, 2011.

[11] A. AMoghe and S. Jyoti. Image Registration: A Review of Elastic Registration Methods Applied to Medical Imaging. *International Journal of Computer Applications*, 70:6–11, 05 2013. doi: 10.5120/11972-7827.

[12] K. Anjyo and H. Ochiai. *Rigid Transformation*, pages 5–29. Springer International Publishing, Cham, 2017. ISBN 978-3-031-79561-9. doi: 10.1007/978-3-031-79561-9_2. URL https://doi.org/10.1007/978-3-031-79561-9_2.

[13] A. G. Asuero, A. Sayago, and A. G. González. The Correlation Coefficient: An Overview. *Critical Reviews in Analytical Chemistry*, 36(1):41–59, 2006. doi: 10.1080/10408340500526766. URL https://doi.org/10.1080/10408340500526766.

[14] A. Athanasiou, A. Tardivon, M. Tanter, B. Sigal-Zafrani, J. Bercoff, T. Deffieux, J.-L. Gennisson, M. Fink, and S. Neuenschwander. Breast Lesions: Quantitative Elastography with Supersonic Shear Imaging—Preliminary Results. *Radiology*, 256(1):297–303, 2010. doi: 10.1148/radiol.10090385. URL https://doi.org/10.1148/radiol.10090385. PMID: 20505064.

[15] F. Azar, D. Metaxas, and M. Schnall. A finite element model of the breast for predicting mechanical deformations during biopsy procedures. In *Proceedings IEEE Workshop on Mathematical Methods in Biomedical Image Analysis. MMBIA-2000 (Cat. No.PR00737)*, pages 38–45, 2000. doi: 10.1109/MMBIA.2000.852358.

[16] F. S. Azar, D. N. Metaxas, and M. D. Schnall. A Deformable Finite Element Model of the Breast for Predicting Mechanical Deformations under External Perturbations. *Academic Radiology*, 8 (10):965–975, 2001. ISSN 1076-6332. doi: https://doi.org/10.1016/S1076-6332(03)80640-2. URL https://www.sciencedirect.com/science/article/pii/S1076633203806402.

[17] P. R. Bakic. *Breast tissue description and modeling in mammography*. Lehigh University, 2001.

[18] J. R. Beckett and C. J. Kotre. Dosimetric implications of age related glandular changes in screening mammography. *Physics in Medicine and Biology*, 45(3):801–813, feb 2000.

[19] C. P. Behrenbruch, K. Marias, P. A. Armitage, M. Yam, N. Moore, R. E. English, J. Clarke, and M. Brady. Fusion of contrast-enhanced breast MR and mammographic imaging data. *Medical Image Analysis*, 7(3):311–340, 2003. ISSN 1361-8415. doi: https://doi.org/10.1016/S1361-8415(03)00015-X. URL https://www.sciencedirect.com/science/article/pii/S136184150300015X. Functional Imaging and Modeling of the Heart.

[20] M. Berger, Q. Yang, and A. Maier. *X-ray Imaging*, pages 119–145. Springer International Publishing, Cham, 2018. ISBN 978-3-319-96520-8. doi: 10.1007/978-3-319-96520-8_7. URL https://doi.org/10.1007/978-3-319-96520-8_7.

[21] J. Beutel, M. Sonka, J. Fitzpatrick, H. Kundel, R. Van Metter, and S. of Photo-optical Instrumentation Engineers. *Handbook of Medical Imaging, Volume 2: Medical Image Processing and Analysis*. Handbook of medical imaging. SPIE Press, 2009. ISBN 9780819477606. URL https://books.google.de/books?id=qVDFSAAACAAJ.

[22] M. H. Bharati, J. Liu, and J. F. MacGregor. Image texture analysis: methods and comparisons. *Chemometrics and Intelligent Laboratory Systems*, 72(1):57–71, 2004. ISSN 0169-7439. doi: https://doi.org/10.1016/j.chemolab.2004.02.005. URL https://www.sciencedirect.com/science/article/pii/S0169743904000528.

[23] U. Bick, R. Trimboli, A. Athanasiou, C. Balleyguier, P. Baltzer, M. Bernathova, K. Borbély, B. Brkljacic, L. Carbonaro, P. Clauser, E. Cassano, C. Colin, G. Esen, A. Evans, E. Fallenberg, M. Fuchsjäger, F. Gilbert, T. Helbich, S. Heywang-Köbrunner, and F. Sardanelli. Image-guided breast biopsy and localisation: recommendations for information to women and referring physicians by the European Society of Breast Imaging. *Insights into Imaging*, 11, 12 2020. doi: 10.1186/s13244-019-0803-x.

[24] F. Bookstein. Principal warps: thin-plate splines and the decomposition of deformations. *IEEE Transactions on Pattern Analysis and Machine Intelligence*, 11(6):567–585, 1989. doi: 10.1109/34.24792.

[25] V. B. Bora, A. G. Kothari, and A. G. Keskar. Robust Automatic Pectoral Muscle Segmentation from Mammograms Using Texture Gradient and Euclidean Distance Regression. *Journal of Digital Imaging*, 29(1):115—125, February 2016. ISSN 0897-1889. doi: 10.1007/s10278-015-9813-5. URL https://europepmc.org/articles/PMC4722023.

[26] A. Bors, L. Kechagias, and I. Pitas. Binary morphological shape-based interpolation applied to 3-D tooth reconstruction. *IEEE Transactions on Medical Imaging*, 21(2):100–108, 2002. doi: 10.1109/42.993129.

[27] A. Bower. *Applied Mechanics of Solids*. 10 2009. ISBN 9780429193323. doi: 10.1201/9781439802489.

[28] M. Brahim, K. Westerkamp, L. Hempel, R. Lehmann, D. Hempel, and P. Philipp. Automated Assessment of Breast Positioning Quality in Screening Mammography. *Cancers*, 14(19), 2022. ISSN 2072-6694. doi: 10.3390/cancers14194704. URL https://www.mdpi.com/2072-6694/14/19/4704.

[29] L. G. Brown. A Survey of Image Registration Techniques. *ACM Comput. Surv.*, 24(4):325–376, dec 1992. ISSN 0360-0300. doi: 10.1145/146370.146374. URL https://doi.org/10.1145/146370.146374.

[30] A. Bumberger, P. Clauser, M. Kolta, P. Kapetas, M. Bernathova, T. H. Helbich, K. Pinker, and P. A. Baltzer. Can we predict lesion detection rates in second-look ultrasound of MRI-detected breast lesions? a systematic analysis. *European Journal of Radiology*, 113:96–100, 2019. ISSN 0720-048X. doi: https://doi.org/10.1016/j.ejrad.2019.02.008. URL https://www.sciencedirect.com/science/article/pii/S0720048X19300592.

[31] J. Burkardt. Computational Geometry Lab: Tetrahedrons, 2010. URL https://people.sc.fsu.edu/~jburkardt/presentations/cg_lab_tetrahedrons.pdf. Accsessed: March 3, 2022.

[32] D. Camacho, R. Hopper, G. Lin, and B. Myers. An improved method for finite element mesh generation of geometrically complex structures with application to the skullbase. *Journal of biomechanics*, 30(10):1067—1070, October 1997. ISSN 0021-9290. doi: 10.1016/s0021-9290(97)00073-0. URL https://doi.org/10.1016/s0021-9290(97)00073-0.

[33] G. Capobianco, C. Cerrone, A. Di Placido, D. Durand, L. Pavone, D. D. Russo, and F. Sebastiano. Image Convolution: A Linear Programming Approach for Filters Design. *Soft Comput.*, 25(14): 8941–8956, jul 2021. ISSN 1432-7643. doi: 10.1007/s00500-021-05783-5. URL https://doi.org/10.1007/s00500-021-05783-5.

[34] M. Cardoso, M. Clarkson, M. Modat, and S. Ourselin. Niftyseg: open-source software for medical image segmentation, label fusion and cortical thickness estimation. In *IEEE International Symposium on Biomedical Imaging, Barcelona, Spain*, 2012.

[35] J. J. Carr, P. F. Hemler, P. W. Halford, R. I. Freimanis, R. H. Choplin, and M. Y. M. Chen. Stereotactic Localization of Breast Lesions: How It Works and Methods to Improve Accuracy. *RadioGraphics*, 21(2):463–473, 2001. doi: 10.1148/radiographics.21.2.g01mr11463. URL https://doi.org/10.1148/radiographics.21.2.g01mr11463. PMID: 11259709.

[36] T. Carter, C. Tanner, N. Beechey-Newman, D. Barratt, and D. Hawkes. MR Navigated Breast Surgery: Method and Initial Clinical Experience. volume 11, pages 356–63, 02 2008. ISBN 978-3-540-85989-5. doi: 10.1007/978-3-540-85990-1_43.

[37] G. B. C. Centre. Mammography Test. URL https://gangabreastcare.com/mammography.php. Accessed: June 10, 2023.

[38] M. Cetin and N. Musaoglu. Merging hyperspectral and panchromatic image data: qualitative and quantitative analysis. *International Journal of Remote Sensing*, 30(7):1779 – 1804, 2009. doi: 10.1080/01431160802639525. URL https://doi.org/10.1080/01431160802639525.

[39] CGAL Editorial Board. *CGAL User and Reference Manual*, 3.5 edition, 2009. URL `http://www.cgal.org/Manual/3.5/doc_html/cgal_manual/packages.html`. Accessed: February 25, 2022.

[40] T. Chen and C. Guestrin. XGBoost: A Scalable Tree Boosting System. In *Proceedings of the 22nd ACM SIGKDD International Conference on Knowledge Discovery and Data Mining*, KDD '16, page 785–794, New York, NY, USA, 2016. Association for Computing Machinery. ISBN 9781450342322. doi: 10.1145/2939672.2939785. URL `https://doi.org/10.1145/2939672.2939785`.

[41] J. Chung, V. Rajagopal, P. M. F. Nielsen, and M. Nash. A biomechanical model of mammographic compressions. *Biomechanics and modeling in mechanobiology*, 7(1):43—52, February 2008. ISSN 1617-7959. doi: 10.1007/s10237-006-0074-6. URL `https://doi.org/10.1007/s10237-006-0074-6`.

[42] J.-H. Chung, V. Rajagopal, T. A. Laursen, P. M. Nielsen, and M. P. Nash. Frictional contact mechanics methods for soft materials: Application to tracking breast cancers. *Journal of Biomechanics*, 41(1):69–77, 2008. ISSN 0021-9290. doi: https://doi.org/10.1016/j.jbiomech.2007.07.016. URL `https://www.sciencedirect.com/science/article/pii/S002192900700334X`.

[43] P. Clauser, R. Mann, A. Athanasiou, H. Prosch, K. Pinker, M. Dietzel, T. H. Helbich, M. Fuchsjäger, J. Camps-Herrero, F. Sardanelli, G. Forrai, and P. A. T. Baltzer. A survey by the European Society of Breast Imaging on the utilisation of breast MRI in clinical practice. *European Radiology*, 28(5):1909—1918, May 2018. ISSN 0938-7994. doi: 10.1007/s00330-017-5121-4. URL `https://europepmc.org/articles/PMC5882636`.

[44] A. M. F. Collignon, F. Maes, D. Delaere, D. Vandermeulen, P. Suetens, and G. Marchal. Automated multi-modality image registration based on information theory. 1995.

[45] I. Despotovic, B. Goossens, and W. Philips. MRI Segmentation of the Human Brain: Challenges, Methods, and Applications. *Computational and Mathematical Methods in Medicine*, 2015, 2015.

[46] M. Dietzel, P. A. Baltzer, T. Hopp, N. V. Ruiter, and W. A. Kaiser. Co-registration of MR-mammography and X-ray mammography. *European Journal of Radiology*, 81:S27–S29, 2012. ISSN 0720-048X. doi: https://doi.org/10.1016/S0720-048X(12)70011-6. URL `https://www.sciencedirect.com/science/article/pii/S0720048X12700116`. Extended abstracts and Abstracts of the Sixth International Congress on MR-Mammography.

[47] M. Dietzel, T. Hopp, N. V. Ruiter, C. G. Kaiser, W. A. Kaiser, and P. A. Baltzer. 4D co-registration of X-ray and MR-mammograms: initial clinical results and potential incremental diagnostic value. *Clinical Imaging*, 39(2):225–230, 2015. ISSN 0899-7071. doi: https:

//doi.org/10.1016/j.clinimag.2014.11.003. URL https://www.sciencedirect.com/science/article/pii/S089970711400271X.

[48] F. Documentation. FEM Mesh calculix. URL https://wiki.freecad.org/FEM_Mesh_CalculiX. Accessed: June 14, 2022.

[49] B. Dolly and D. Raj. Texture Based Image Retrieval Using GLCM and LBP. In S. Bhattacharyya, J. Nayak, K. B. Prakash, B. Naik, and A. Abraham, editors, *International Conference on Intelligent and Smart Computing in Data Analytics*, pages 35–45, Singapore, 2021. Springer Singapore. ISBN 978-981-33-6176-8.

[50] O. Dössel. *Magnetresonanz-Tomographie*, chapter 11, page 285 ff. Springer Vieweg, 2016. ISBN 978-3-642-54406-4. doi: 10.1007/978-3-642-54407-1.

[51] B. Eiben, V. Vavourakis, J. Hipwell, S. Kabus, C. Lorenz, T. Buelow, and D. Hawkes. Breast deformation modelling: Comparison of methods to obtain a patient specific unloaded configuration. volume 9036, page 903615, 03 2014. doi: 10.1117/12.2043607.

[52] B. Eiben, V. Vavourakis, J. Hipwell, S. Kabus, T. Buelow, C. Lorenz, T. Mertzanidou, S. Reis, N. Williams, M. Keshtgar, and D. Hawkes. Symmetric Biomechanically Guided Prone-to-Supine Breast Image Registration. *Annals of biomedical engineering*, 44, 11 2015. doi: 10.1007/s10439-015-1496-z.

[53] L. Erasmus, D. Hurter, M. Naude, H. Kritzinger, and S. Acho. A short overview of MRI artefacts. *South African Journal of Radiology*, 8(2):13, 2004. ISSN 2078-6778. doi: 10.4102/sajr.v8i2.127. URL https://sajr.org.za/index.php/sajr/article/view/127.

[54] Q. Fang. Iso2mesh, 2018. URL https://iso2mesh.sourceforge.net/cgi-bin/index.cgi/. Accessed: April 20, 2020.

[55] Q. Fang and D. A. Boas. Tetrahedral Mesh Generation from Volumetric Binary and Gray-Scale Images. In *Proceedings of the Sixth IEEE International Conference on Symposium on Biomedical Imaging: From Nano to Macro*, ISBI'09, page 1142–1145. IEEE Press, 2009. ISBN 9781424439317.

[56] A. Fedorov, R. Beichel, J. Kalpathy-Cramer, J. Finet, J.-C. Fillion-Robin, S. Pujol, C. Bauer, D. Jennings, F. Fennessy, M. Sonka, J. Buatti, S. Aylward, J. V. Miller, S. Pieper, and R. Kikinis. 3D Slicer as an image computing platform for the Quantitative Imaging Network. *Magnetic Resonance Imaging*, 30(9):1323–1341, 2012. ISSN 0730-725X. doi: https://doi.org/10.1016/j.mri.2012.05.001. URL https://www.sciencedirect.com/science/article/pii/S0730725X12001816. Quantitative Imaging in Cancer.

[57] J. Ferlay, M. Ervik, F. Lam, M. Colombet, L. Mery, M. Piñeros, A. Znaor, I. Soerjomataram, and F. Bray. Global Cancer Observatory: Cancer Today. *International Agency for Research on Cancer*, 2020.

[58] J. Fitzpatrick and J. West. The Distribution of Target Registration Error in Rigid-Body, Point-Based Registration. *IEEE transactions on medical imaging*, 20:917–27, 10 2001. doi: 10.1109/42.952729.

[59] E. García, Y. Diez, O. Diaz, X. Lladó, R. Martí, J. Martí, and A. Oliver. A step-by-step review on patient-specific biomechanical finite element models for breast MRI to X-ray mammography registration. *Medical Physics*, 45(1):e6–e31, 2018. doi: https://doi.org/10.1002/mp.12673. URL `https://aapm.onlinelibrary.wiley.com/doi/abs/10.1002/mp.12673`.

[60] E. García, Y. Diez, O. Diaz, X. Lladó, A. Gubern-Mérida, R. Martí, J. Martí, and A. Oliver. Breast MRI and X-ray mammography registration using gradient values. *Medical Image Analysis*, 54:76–87, 2019. ISSN 1361-8415. doi: https://doi.org/10.1016/j.media.2019.02.013. URL `https://www.sciencedirect.com/science/article/pii/S1361841519300179`.

[61] D. J. Gavaghan, J. P. Whiteley, S. J. Chapman, J. M. Brady, and P. Pathmanathan. Predicting Tumor Location by Modeling the Deformation of the Breast. *IEEE Transactions on Biomedical Engineering*, 55(10):2471–2480, 2008. doi: 10.1109/TBME.2008.925714.

[62] A. Gefen and B. Dilmoney. Mechanics of the normal woman's breast. *Technology and health care : official journal of the European Society for Engineering and Medicine*, 15 4:259–71, 2007.

[63] P. Geurts, D. Ernst, and L. Wehenkel. Extremely Randomized Trees. *Mach. Learn.*, 63(1):3–42, apr 2006. ISSN 0885-6125. doi: 10.1007/s10994-006-6226-1. URL `https://doi.org/10.1007/s10994-006-6226-1`.

[64] S. González, S. García, J. Del Ser, L. Rokach, and F. Herrera. A practical tutorial on bagging and boosting based ensembles for machine learning: Algorithms, software tools, performance study, practical perspectives and opportunities. *Information Fusion*, 64:205–237, 2020. ISSN 1566-2535. doi: https://doi.org/10.1016/j.inffus.2020.07.007. URL `https://www.sciencedirect.com/science/article/pii/S1566253520303195`.

[65] S. Govindjee and P. A. Mihalic. Computational methods for inverse finite elastostatics. *Computer Methods in Applied Mechanics and Engineering*, 136(1):47–57, 1996. ISSN 0045-7825. doi: https://doi.org/10.1016/0045-7825(96)01045-6. URL `https://www.sciencedirect.com/science/article/pii/0045782596010456`.

[66] A. Gubern-Mérida, M. Kallenberg, R. Martí, and N. Karssemeijer. Segmentation of the Pectoral Muscle in Breast MRI Using Atlas-Based Approaches. In N. Ayache, H. Delingette, P. Golland,

and K. Mori, editors, *Medical Image Computing and Computer-Assisted Intervention – MICCAI 2012*, pages 371–378, Berlin, Heidelberg, 2012. Springer Berlin Heidelberg.

[67] A. Gubern-Mérida, M. Kallenberg, R. M. Mann, R. Martí, and N. Karssemeijer. Breast Segmentation and Density Estimation in Breast MRI: A Fully Automatic Framework. *IEEE Journal of Biomedical and Health Informatics*, 19(1):349–357, 2015.

[68] R. Ha, P. Chang, E. Mema, S. Mutasa, J. Karcich, R. T. Wynn, M. Z. Liu, and S. Jambawalikar. Fully Automated Convolutional Neural Network Method for Quantification of Breast MRI Fibroglandular Tissue and Background Parenchymal Enhancement. *Journal of Digital Imaging*, 32 (1):141–147, Feb 2019. ISSN 1618-727X.

[69] M. Hahn, S. Okamgba, P. Scheler, K. Freidel, G. Hoffmann, B. Kraemer, D. D. Wallwiener, and U. Krainick-Strobel. Vacuum-assisted breast biopsy: A comparison of 11-gauge and 8-gauge needles in benign breast disease. *World journal of surgical oncology*, 6:51, 02 2008. doi: 10.1186/1477-7819-6-51.

[70] L. Han, J. H. Hipwell, C. Tanner, Z. Taylor, T. Mertzanidou, J. Cardoso, S. Ourselin, and D. J. Hawkes. Development of patient-specific biomechanical models for predicting large breast deformation. *Physics in Medicine and Biology*, 57(2):455–472, dec 2011. doi: 10.1088/0031-9155/57/2/455. URL `https://doi.org/10.1088/0031-9155/57/2/455`.

[71] L. Han, J. H. Hipwell, B. Eiben, D. Barratt, M. Modat, S. Ourselin, and D. J. Hawkes. A nonlinear biomechanical model based registration method for aligning prone and supine MR breast images. *IEEE transactions on medical imaging*, 33(3):682—694, March 2014. ISSN 0278-0062. doi: 10.1109/tmi.2013.2294539. URL `https://doi.org/10.1109/TMI.2013.2294539`.

[72] S. Heywang-Köbrunner, D. Hahn, H. Schmidt, I. Krischke, W. Eiermann, and R. Bassermann. MR imaging of the breast using gadolinium-dtpa. *Journal of computer assisted tomography*, 10: 199–204, 03 1986. doi: 10.1097/00004728-198603000-00005.

[73] D. L. G. Hill, P. G. Batchelor, M. Holden, and D. J. Hawkes. Medical image registration. *Physics in Medicine and Biology*, 46(3):R1, mar 2001. doi: 10.1088/0031-9155/46/3/201. URL `https://dx.doi.org/10.1088/0031-9155/46/3/201`.

[74] J. H. Hipwell, C. Tanner, W. R. Crum, J. A. Schnabel, and D. J. Hawkes. A New Validation Method for X-ray Mammogram Registration Algorithms Using a Projection Model of Breast X-ray Compression. *IEEE Transactions on Medical Imaging*, 26(9):1190–1200, 2007. doi: 10.1109/TMI.2007.903569.

[75] J. H. Hipwell, V. Vavourakis, L. Han, T. Mertzanidou, B. Eiben, and D. J. Hawkes. A review of biomechanically informed breast image registration. *Physics in Medicine and Biology*, 61(2):

R1–R31, jan 2016. doi: 10.1088/0031-9155/61/2/r1. URL https://doi.org/10.1088/0031-9155/61/2/r1.

[76] T. Hopp. *Multimodal Registration of X-Ray Mammograms with 3D Volume Datasets.* Dr. Hut, 2012. ISBN 9783843905497. URL https://books.google.de/books?id=ND9BMwEACAAJ.

[77] T. Hopp, P. Baltzer, M. Dietzel, W. A. Kaiser, and N. V. Ruiter. 2D/3D image fusion of X-ray mammograms with breast MRI: visualizing dynamic contrast enhancement in mammograms. *International journal of computer assisted radiology and surgery*, 7(3):339—348, May 2012. ISSN 1861-6410. doi: 10.1007/s11548-011-0623-z. URL https://doi.org/10.1007/s11548-011-0623-z.

[78] T. Hopp, M. Dietzel, P. Baltzer, P. Kreisel, W. Kaiser, H. Gemmeke, and N. Ruiter. Automatic multimodal 2D/3D breast image registration using biomechanical FEM models and intensity-based optimization. *Medical Image Analysis*, 17(2):209–218, 2013. ISSN 1361-8415.

[79] T. Hopp, B. Neupane, and N. V. Ruiter. Automated Multimodal Computer Aided Detection Based on a 3D-2D Image Registration. In A. Tingberg, K. Lång, and P. Timberg, editors, *Breast Imaging*, pages 400–407, Cham, 2016. Springer International Publishing. ISBN 978-3-319-41546-8.

[80] T. Hopp, P. Cotič Smole, and N. Ruiter. Automated Multimodal Breast CAD Based on Registration of MRI and Two View Mammography. pages 365–372, 09 2017. ISBN 978-3-319-67557-2. doi: 10.1007/978-3-319-67558-9_42.

[81] N. Houssami, S. Ciatto, P. Macaskill, S. J. Lord, R. M. Warren, J. M. Dixon, and L. Irwig. Accuracy and Surgical Impact of Magnetic Resonance Imaging in Breast Cancer Staging: Systematic Review and Meta-Analysis in Detection of Multifocal and Multicentric Cancer. *Journal of Clinical Oncology*, 26(19):3248–3258, 2008. doi: 10.1200/JCO.2007.15.2108.

[82] M. L. Huang, B. E. Adrada, R. Candelaria, D. Thames, D. Dawson, and W. T. Yang. Stereotactic Breast Biopsy: Pitfalls and Pearls. *Techniques in Vascular and Interventional Radiology*, 17(1): 32–39, 2014. ISSN 1089-2516. doi: https://doi.org/10.1053/j.tvir.2013.12.006. URL https://www.sciencedirect.com/science/article/pii/S1089251613000929. Breast Interventions.

[83] C. F. D. IMAGING. Stereotactic Breast Biopsy, 2016. URL https://www.cdimiami.com/stereotactic-breast-biopsy/. Accessed: June 10, 2023.

[84] T. Jeon, S. Lee, Y.-S. Kim, and H. Son. Tips for finding magnetic resonance imaging-detected suspicious breast lesions using second-look ultrasound: a pictorial essay. *Ultrasonography*, 41, 02 2022. doi: 10.14366/usg.21219.

[85] Y. Ji, B. Li, R. Zhao, Y. Zhang, J. Liu, and H. Lu. The relationship between breast density, age, and mammographic lesion type among chinese breast cancer patients from a large clinical dataset. *BMC Medical Imaging*, 21, 03 2021. doi: 10.1186/s12880-021-00565-9.

[86] W. Kaiser and E. Zeitler. MR imaging of the breast: fast imaging sequences with and without Gd-DTPA. Preliminary observations. *Radiology*, 170:681–6, 04 1989. doi: 10.1148/radiology.170.3.2916021.

[87] S. Kalli, P. Freer, and E. Rafferty. Lesions of the Skin and Superficial Tissue at Breast MR Imaging. *Radiographics : a review publication of the Radiological Society of North America, Inc*, 30:1891–913, 11 2010. doi: 10.1148/rg.307105064.

[88] A. Kanezaki. Unsupervised Image Segmentation by Backpropagation. In *2018 IEEE International Conference on Acoustics, Speech and Signal Processing (ICASSP)*, pages 1543–1547, 2018.

[89] D. Kang, S. Y. Shin, C. O. Sung, J. Y. Kim, J.-K. Pack, and H. D. Choi. An improved method of breast mri segmentation with simplified k-means clustered images. In *Proceedings of the 2011 ACM Symposium on Research in Applied Computation*, RACS '11, page 226–231, New York, NY, USA, 2011. Association for Computing Machinery. ISBN 9781450310871. doi: 10.1145/2103380.2103425. URL `https://doi.org/10.1145/2103380.2103425`.

[90] S. R. Kannan, S. Ramathilagam, R. Devi, and A. Sathya. Robust kernel FCM in segmentation of breast medical images. *Expert Syst. Appl.*, 38:4382–4389, 2011.

[91] B. Keller, D. Nathan, Y. Wang, Y. Zheng, J. Gee, E. Conant, and D. Kontos. Estimation of breast percent density in raw and processed full field digital mammography images via adaptive fuzzy c-means clustering and support vector machine segmentation. *Medical physics*, 39:4903–17, 08 2012. doi: 10.1118/1.4736530.

[92] B. Keller, J. Chen, D. Daye, E. Conant, and D. Kontos. Preliminary evaluation of the publicly available laboratory for breast radiodensity assessment (LIBRA) software tool: Comparison of fully automated area and volumetric density measures in a case-control study with digital mammography. *Breast cancer research : BCR*, 17:117, 08 2015. doi: 10.1186/s13058-015-0626-8.

[93] W. Kim, A. Kanezaki, and M. Tanaka. Unsupervised Learning of Image Segmentation Based on Differentiable Feature Clustering. *IEEE Transactions on Image Processing*, 29:8055–8068, 2020. ISSN 1941-0042.

[94] Z. Knops, J. Maintz, M. Viergever, and J. Pluim. Normalized mutual information based registration using k-means clustering and shading correction. *Medical Image Analysis*, 10(3):432–439, 2006. ISSN 1361-8415. doi: https://doi.org/10.1016/j.media.2005.03.009. URL `https:`

//www.sciencedirect.com/science/article/pii/S1361841505000435. Special Issue on The Second International Workshop on Biomedical Image Registration (WBIR'03).

[95] T. Krouskop, T. Wheeler, F. Kallel, B. Garra, and T. Hall. The Elastic Moduli of Breast and Prostate Tissues Under Compression. *Ultrasonic imaging*, 20:260–74, 11 1998. doi: 10.1177/016173469802000403.

[96] C. K. Kuhl, K. Strobel, H. Bieling, E. Wardelmann, W. Kuhn, N. Maass, and S. Schrading. Impact of Preoperative Breast MR Imaging and MR-guided Surgery on Diagnosis and Surgical Outcome of Women with Invasive Breast Cancer with and without DCIS Component. *Radiology*, 284(3): 645–655, 2017.

[97] U. Kumbhar, V. Patil, and S. Rudrakshi. Enhancement of Medical Images Using Image Processing in Matlab. *International journal of engineering research and technology*, 2, 2013.

[98] Y. Lecun, L. Bottou, Y. Bengio, and P. Haffner. Gradient-based learning applied to document recognition. *Proceedings of the IEEE*, 86(11):2278–2324, 1998. doi: 10.1109/5.726791.

[99] A. W. Lee, V. Rajagopal, T. P. Babarenda Gamage, A. J. Doyle, P. M. Nielsen, and M. P. Nash. Breast lesion co-localisation between X-ray and MR images using finite element modelling. *Medical Image Analysis*, 17(8):1256–1264, 2013. ISSN 1361-8415. doi: https://doi.org/10.1016/j.media.2013.05.011. URL https://www.sciencedirect.com/science/article/pii/S1361841513000856.

[100] T. Lei, R. Wang, Y. Wan, B. Zhang, H. Meng, and A. K. Nandi. Medical Image Segmentation Using Deep Learning: A Survey, 2020.

[101] H. Lester and S. R. Arridge. A survey of hierarchical non-linear medical image registration. *Pattern Recognition*, 32(1):129–149, 1999. ISSN 0031-3203. doi: https://doi.org/10.1016/S0031-3203(98)00095-8. URL https://www.sciencedirect.com/science/article/pii/S0031320398000958.

[102] B. Li, G. Yang, Z. Liu, J. L. Coatrieux, and H. Shu. Multimodal Medical Image Registration Based on an Information-Theory Measure with Histogram Estimation of Continuous Image Representation. *Mathematical Problems in Engineering*, 2018:1–12, 2018. URL https://EconPapers.repec.org/RePEc:hin:jnlmpe:2135453.

[103] A. Liu and B. Joe. Relationship between tetrahedron shape measures. *BIT Numerical Mathematics*, 34:268–287, 1994.

[104] Y. Liu, W. Lin, and Y. Cheng. Density Calculation of Pseudo Breast MRI Based on Adversarial Generative Network. In R. Kountchev, R. Mironov, and K. Nakamatsu, editors, *New Approaches*

for Multidimensional Signal Processing, pages 175–185, Singapore, 2022. Springer Singapore. ISBN 978-981-16-8558-3.

[105] Y.-L. Liu, P.-Y. Liu, M.-L. Huang, J.-T. Hsu, R.-P. Han, and J. Wu. Simulation of breast compression in mammography using finite element analysis: A preliminary study. *Radiation Physics and Chemistry*, 140:295–299, 2017. ISSN 0969-806X. doi: https://doi.org/10.1016/j.radphyschem.2017.01.017. URL `https://www.sciencedirect.com/science/article/pii/S0969806X17300774`. 2nd International Conference on Dosimetry and its Applications (ICDA-2) University of Surrey, Guildford, United Kingdom, 3-8 July 2016.

[106] J. Lorenzen, R. Sinkus, M. Lorenzen, M. Dargatz, C. Leussler, P. Röschmann, and G. Adam. MR elastography of the breast: Preliminary clinical results. *RöFo : Fortschritte auf dem Gebiete der Röntgenstrahlen und der Nuklearmedizin*, 174:830–4, 08 2002. doi: 10.1055/s-2002-32690.

[107] F. Lugauer and J. Wetzl. *Magnetic Resonance Imaging*, pages 91–118. Springer International Publishing, Cham, 2018. ISBN 978-3-319-96520-8. doi: 10.1007/978-3-319-96520-8_6. URL `https://doi.org/10.1007/978-3-319-96520-8_6`.

[108] M. Luong, H. Pham, and C. D. Manning. Effective Approaches to Attention-based Neural Machine Translation. *CoRR*, abs/1508.04025, 2015. URL `http://arxiv.org/abs/1508.04025`.

[109] F. Maes, A. Collignon, D. Vandermeulen, G. Marchal, and P. Suetens. Multimodality Image Registration by Maximization of Mutual Information. *IEEE Transactions on Medical Imaging*, 16 (2):187–198, 1997. doi: 10.1109/42.563664.

[110] F. Maes, D. Loeckx, D. Vandermeulen, and P. Suetens. *Image registration using mutual information*, pages 295–308. Springer US, Boston, MA, 2015. ISBN 978-0-387-09749-7. doi: 10.1007/978-0-387-09749-7_16. URL `https://doi.org/10.1007/978-0-387-09749-7_16`.

[111] J. Maintz and M. A. Viergever. A survey of medical image registration. *Medical Image Analysis*, 2(1):1–36, 1998. ISSN 1361-8415. doi: https://doi.org/10.1016/S1361-8415(01)80026-8. URL `https://www.sciencedirect.com/science/article/pii/S1361841501800268`.

[112] L. E. Malvern. Introduction to the mechanics of a continuous medium. 1969.

[113] R. Mann, C. Balleyguier, P. Baltzer, U. Bick, C. Colin, E. Cornford, A. Evans, E. Fallenberg, G. Forrai, M. Fuchsjäger, F. Gilbert, T. Helbich, S. Heywang-Köbrunner, J. Herrero, C. Kuhl, L. Martincich, F. Pediconi, P. Panizza, L. Pina, and F. Sardanelli. Breast MRI: EUSOBI recommendations for women's information. *European radiology*, 25, 05 2015. doi: 10.1007/s00330-015-3807-z.

[114] R. M. Mann, N. Cho, and L. Moy. Breast MRI: State of the Art. *Radiology*, 292(3):520–536, 2019. doi: 10.1148/radiol.2019182947. URL https://doi.org/10.1148/radiol.2019182947. PMID: 31361209.

[115] G. E. Martin. *Affine Transformations*, pages 167–181. Springer New York, New York, NY, 1982. ISBN 978-1-4612-5680-9. doi: 10.1007/978-1-4612-5680-9_15. URL https://doi.org/10.1007/978-1-4612-5680-9_15.

[116] F. Martínez-Martínez, M. Rupérez-Moreno, M. Martínez-Sober, J. Solves-Llorens, D. Lorente, A. Serrano-López, S. Martínez-Sanchis, C. Monserrat, and J. Martín-Guerrero. A finite element-based machine learning approach for modeling the mechanical behavior of the breast tissues under compression in real-time. *Computers in Biology and Medicine*, 90:116–124, 2017. ISSN 0010-4825. doi: https://doi.org/10.1016/j.compbiomed.2017.09.019. URL https://www.sciencedirect.com/science/article/pii/S0010482517303177.

[117] J. Maul, S. Said, N. Ruiter, and T. Hopp. X-ray synthesis based on triangular mesh models using gpu-accelerated ray tracing for multi-modal breast image registration. In D. Svoboda, N. Burgos, J. M. Wolterink, and C. Zhao, editors, *Simulation and Synthesis in Medical Imaging*, pages 87–96, Cham, 2021. Springer International Publishing. ISBN 978-3-030-87592-3.

[118] medgadget. Siemens rolls for Breast Visualization, 2009. URL https://www.medgadget.com/2009/02/siemens_rolls_out_15_tesla_mri_for_breast_visualization.html. Accessed: August 11, 2020.

[119] A. Melbourne, G. Ridgway, and D. J. Hawkes. Image similarity metrics in image registration. In B. M. Dawant and D. R. Haynor, editors, *Medical Imaging 2010: Image Processing*, volume 7623, page 762335. International Society for Optics and Photonics, SPIE, 2010. doi: 10.1117/12.840389. URL https://doi.org/10.1117/12.840389.

[120] A. Mendizabal, E. Tagliabue, J.-N. Brunet, D. Dall'Alba, P. Fiorini, and S. Cotin. Physics-based Deep Neural Network for Real-Time Lesion Tracking in Ultrasound-guided Breast Biopsy. In *Computational Biomechanics for Medicine XIV*, Shenzhen, China, Oct. 2019. URL https://hal.inria.fr/hal-02311277.

[121] T. Mertzanidou, J. Hipwell, M. J. Cardoso, X. Zhang, C. Tanner, S. Ourselin, U. Bick, H. Huisman, N. Karssemeijer, and D. Hawkes. MRI to X-ray mammography registration using a volume-preserving affine transformation. *Medical Image Analysis*, 16(5):966–975, 2012. ISSN 1361-8415. doi: https://doi.org/10.1016/j.media.2012.03.001. URL https://www.sciencedirect.com/science/article/pii/S1361841512000461.

[122] T. Mertzanidou, J. Hipwell, S. Johnsen, L. Han, B. Eiben, Z. Taylor, S. Ourselin, H. Huisman, R. Mann, U. Bick, N. Karssemeijer, and D. Hawkes. MRI to X-ray mammography intensity-based registration with simultaneous optimisation of pose and biomechanical transformation parameters. *Medical Image Analysis*, 18(4):674–683, 2014. ISSN 1361-8415. doi: https://doi.org/10.1016/j.media.2014.03.003. URL `https://www.sciencedirect.com/science/article/pii/S1361841514000358`.

[123] K. Miller, G. Joldes, D. Lance, and A. Wittek. Total Lagrangian Explicit Dynamics Finite Element Algorithm for computing soft tissue deformation. *Communications in Numerical Methods in Engineering*, 23:121 – 134, 02 2006. doi: 10.1002/cnm.887.

[124] S. Minaee, Y. Y. Boykov, F. Porikli, A. J. Plaza, N. Kehtarnavaz, and D. Terzopoulos. Image Segmentation Using Deep Learning: A Survey. *IEEE Transactions on Pattern Analysis and Machine Intelligence*, pages 1–1, 2021.

[125] T. Monaghan, S. Rahman, C. Agudelo, A. Wein, J. Lazar, K. Everaert, and R. Dmochowski. Foundational Statistical Principles in Medical Research: Sensitivity, Specificity, Positive Predictive Value, and Negative Predictive Value. *Medicina*, 57:503, 05 2021. doi: 10.3390/medicina57050503.

[126] H. Moradmand, S. Setayeshi, A. Karimian, M. Sirous, and M. Akbari. Comparing the Performance of Image Enhancement Methods to Detect Microcalcification Clusters in Digital Mammography. *Iranian journal of cancer prevention*, 5:61–8, 02 2012.

[127] P. Moreira, I. Peterlík, M. Herink, C. Duriez, S. Cotin, and S. Misra. Modelling Prostate Deformation: SOFA versus Experiments. *Mechanical Engineering Research*, 3:64–72, 2013.

[128] S. Muratov, C. Canelo-Aybar, J.-E. Tarride, P. Alonso, N. Dimitrova, B. Borisch, X. Castells, S. Duffy, P. Fitzpatrick, M. Follmann, L. Giordano, S. Hofvind, A. Lebeau, C. Quinn, A. Torresin, C. Vialli, S. Siesling, A. Ponti, P. Giorgi Rossi, and M. Broeders. Monitoring and evaluation of breast cancer screening programmes: Selecting candidate performance indicators. *BMC Cancer*, 20, 08 2020. doi: 10.1186/s12885-020-07289-z.

[129] J. Myong, B. Kang, S. Yoon, S. Kim, and Y. An. The Clinical Utility of a Adding Lateral Approach to Conventional Vertical Approach for Prone Stereotactic Vacuum-Assisted Breast Biopsy. *Korean journal of radiology : official journal of the Korean Radiological Society*, 14:568–75, 07 2013. doi: 10.3348/kjr.2013.14.4.568.

[130] T.-T. Nguyen, T.-H. Nguyen, and B.-V. Ngo. A GLCM Algorithm for Optimal Features of Mammographic Images for Detection of Breast Cancer. In *2021 International Conference on System Science and Engineering (ICSSE)*, pages 295–299, 2021. doi: 10.1109/ICSSE52999.2021.9538426.

[131] M. Nolden, S. Zelzer, A. Seitel, D. Wald, M. Müller, A. M. Franz, D. Maleike, M. Fangerau, M. Baumhauer, L. Maier-Hein, K. H. Maier-Hein, H.-P. Meinzer, and I. Wolf. The Medical Imaging Interaction Toolkit: challenges and advances: 10 years of open-source development. *International journal of computer assisted radiology and surgery*, 8(4):607—620, July 2013. ISSN 1861-6410.

[132] N. I. of Biomedical Imaging and Bioengineering. Magnetic Resonance Imaging. URL `https://www.nibib.nih.gov/science-education/science-topics/magnetic-resonance-imaging-mri`. Accessed: December 9, 2021.

[133] O.-E. Ørebæk and M. Geitle. Exploring the Hyperparameters of XGBoost Through 3D Visualizations. In *AAAI Spring Symposium Combining Machine Learning with Knowledge Engineering*, 2021. URL `https://api.semanticscholar.org/CorpusID:233354953`.

[134] T. Ortiz-Perez and W. Jr. Mammography Techniques, Positioning, and Optimizing Image Quality. *Breast Cancer Screening and Diagnosis*, pages 37–63, 08 2015. doi: 10.1007/978-1-4939-1267-4_3.

[135] D. Pandey, X. Yin, H. Wang, M.-Y. Su, J.-H. Chen, J. Wu, and Y. Zhang. Automatic and fast segmentation of breast region-of-interest (ROI) and density in MRIs. *Heliyon*, 4(12):e01042, 2018. ISSN 2405-8440. doi: https://doi.org/10.1016/j.heliyon.2018.e01042. URL `https://www.sciencedirect.com/science/article/pii/S2405844018327178`.

[136] S. PANDYA and R. G. MOORE. Breast Development and Anatomy. *Clinical Obstetrics and Gynecology*, 54(1), 2011. ISSN 0009-9201.

[137] J. E. Pasciak. The Mathematical Theory of Finite Element Methods (Susanne C. Brenner and L. Ridgway Scott). *SIAM Rev.*, 37:472–473, 1995.

[138] P. Patete, M. I. Iacono, M. F. Spadea, G. Trecate, D. Vergnaghi, L. T. Mainardi, and G. Baroni. A multi-tissue mass-spring model for computer assisted breast surgery. *Medical Engineering and Physics*, 35(1):47–53, 2013. ISSN 1350-4533. doi: https://doi.org/10.1016/j.medengphy.2012.03.008. URL `https://www.sciencedirect.com/science/article/pii/S1350453312000562`.

[139] A. C. Pelicano, M. C. T. Gonçalves, D. M. Godinho, T. Castela, M. L. Orvalho, N. A. M. Araújo, E. Porter, and R. C. Conceição. Development of 3D MRI-Based Anatomically Realistic Models of Breast Tissues and Tumours for Microwave Imaging Diagnosis. *Sensors*, 21(24), 2021. ISSN 1424-8220. doi: 10.3390/s21248265. URL `https://www.mdpi.com/1424-8220/21/24/8265`.

[140] C. Peng, W.-A. Lin, H. Liao, R. Chellappa, and S. K. Zhou. Chapter 6 - deep slice interpolation via marginal super-resolution, fusion, and refinement. In A. S. El-Baz and J. S. Suri, editors, *State of*

the Art in Neural Networks and Their Applications, pages 133–145. Academic Press, 2023. ISBN 978-0-12-819872-8. doi: https://doi.org/10.1016/B978-0-12-819872-8.00013-6. URL https://www.sciencedirect.com/science/article/pii/B9780128198728000136.

[141] S. Pereira, J. Hipwell, V. Mccormack, C. Tanner, S. Moss, L. Wilkinson, L. Khoo, C. Pagliari, P. Skippage, C. Kliger, D. Hawkes, and I. Silva. Automated registration of diagnostic to prediagnostic X-ray mammograms: Evaluation and comparison to radiologists' accuracy. *Medical physics*, 37:4530–9, 09 2010. doi: 10.1118/1.3457470.

[142] N. Perry, M. Broeders, C. de Wolf, S. Törnberg, R. Holland, and L. von Karsa. European guidelines for quality assurance in breast cancer screening and diagnosis. fourth edition—summary document. *Annals of Oncology*, 19(4):614–622, 2008. ISSN 0923-7534. doi: https://doi.org/10.1093/annonc/mdm481. URL https://www.sciencedirect.com/science/article/pii/S0923753419414348.

[143] R. Phellan, B. Hachem, J. Clin, J.-M. Mac-Thiong, and L. Duong. Real-time biomechanics using the finite element method and machine learning: Review and perspective. *Medical Physics*, 48 (1):7–18, 2021. doi: https://doi.org/10.1002/mp.14602. URL https://aapm.onlinelibrary.wiley.com/doi/abs/10.1002/mp.14602.

[144] A. Poellinger, U. Bick, T. Freund, S. Diekmann, B. Hamm, and F. Diekmann. Evaluation of 11-Gauge and 9-Gauge Vacuum-Assisted Breast Biopsy Systems in a Breast Parenchymal model. *Academic Radiology*, 14(6):677–684, 2007. ISSN 1076-6332. doi: https://doi.org/10.1016/j.acra.2007.02.013. URL https://www.sciencedirect.com/science/article/pii/S1076633207001304.

[145] H. Preibsch, A. Baur, B. M. Wietek, B. Krämer, A. Staebler, C. D. Claussen, and K. C. Siegmann-Luz. Vacuum-assisted breast biopsy with 7-gauge, 8-gauge, 9-gauge, 10-gauge, and 11-gauge needles: how many specimens are necessary? *Acta Radiologica*, 56(9):1078–1084, 2015. doi: 10.1177/0284185114549224. URL https://doi.org/10.1177/0284185114549224. PMID: 25232187.

[146] V. Rajagopal, A. Lee, J.-H. Chung, R. Warren, R. P. Highnam, P. M. F. Nielsen, and M. P. Nash. Towards Tracking Breast Cancer Across Medical Images Using Subject-Specific Biomechanical Models. In N. Ayache, S. Ourselin, and A. Maeder, editors, *Medical Image Computing and Computer-Assisted Intervention – MICCAI 2007*, pages 651–658, Berlin, Heidelberg, 2007. Springer Berlin Heidelberg. ISBN 978-3-540-75757-3.

[147] N. G. Ramião, P. S. Martins, R. Rynkevic, A. A. Fernandes, M. Barroso, and D. C. Santos. Biomechanical properties of breast tissue, a state-of-the-art review. *Biomechanics and modeling*

in mechanobiology, 15(5):1307—1323, October 2016. ISSN 1617-7959. doi: 10.1007/s10237-016-0763-8. URL https://doi.org/10.1007/s10237-016-0763-8.

[148] E. Reinhard. *Sampling, Reconstruction, Aliasing, and Anti-aliasing*, pages 1414–1422. Springer Berlin Heidelberg, 09 2023. ISBN 978-3-030-89861-8. doi: 10.1007/978-3-030-89862-5_416.

[149] C. Rochat, G. Baird, and A. Lourenco. Digital Mammography Stereotactic Biopsy versus Digital Breast Tomosynthesis–guided Biopsy: Differences in Biopsy Targets, Pathologic Results, and Discordance Rates. *Radiology*, 294:191525, 01 2020. doi: 10.1148/radiol.2019191525.

[150] U. Roessner, A. Nahid, B. Chapman, A. Hunter, and M. Bellgard. 1.33 - Metabolomics – The Combination of Analytical Biochemistry, Biology, and Informatics. In M. Moo-Young, editor, *Comprehensive Biotechnology (Second Edition)*, pages 447–459. Academic Press, Burlington, second edition edition, 2011. ISBN 978-0-08-088504-9. doi: https://doi.org/10.1016/B978-0-08-088504-9.00052-0. URL https://www.sciencedirect.com/science/article/pii/B9780080885049000520.

[151] K. Rohr. Elastic registration of multimodal medical images: A survey. *KI*, 14:11–17, 01 2000.

[152] O. Ronneberger, P. Fischer, and T. Brox. U-Net: Convolutional Networks for Biomedical Image Segmentation. In N. Navab, J. Hornegger, W. M. Wells, and A. F. Frangi, editors, *Medical Image Computing and Computer-Assisted Intervention – MICCAI 2015*, pages 234–241, Cham, 2015. Springer International Publishing. ISBN 978-3-319-24574-4.

[153] L. Roose, W. De Maerteleire, W. Mollemans, and P. Suetens. Validation of different soft tissue simulation methods for breast augmentation. *International Congress Series*, 1281:485–490, 2005. ISSN 0531-5131. doi: https://doi.org/10.1016/j.ics.2005.03.126. URL https://www.sciencedirect.com/science/article/pii/S053151310500333X. CARS 2005: Computer Assisted Radiology and Surgery.

[154] L. Roose, W. De Maerteleire, W. Mollemans, F. Maes, and P. Suetens. Simulation of Soft-Tissue Deformations for Breast Augmentation Planning. In M. Harders and G. Székely, editors, *Biomedical Simulation*, pages 197–205, Berlin, Heidelberg, 2006. Springer Berlin Heidelberg. ISBN 978-3-540-36010-0.

[155] F. Rosenblatt. The perceptron: A probabilistic model for information storage and organization in the brain [j]. *Psychol. Review*, 65:386 – 408, 12 1958. doi: 10.1037/h0042519.

[156] N. Ruiter. *Registration of X-ray Mammograms and MR-Volumes of the Female Breast based on Simulated Mammographic Deformation*. 01 2004. ISBN 978-3-89820-656-3.

[157] N. Ruiter, T. Müller, R. Stotzka, H. Gemmeke, J. Reichenbach, and W. Kaiser. Automatic image matching for breast cancer diagnostics by a 3D deformation model of the mamma. 47(s1b):644–647, 2002. doi: doi:10.1515/bmte.2002.47.s1b.644. URL https://doi.org/10.1515/bmte.2002.47.s1b.644.

[158] N. Ruiter, R. Stotzka, T.-O. Muller, H. Gemmeke, J. Reichenbach, and W. Kaiser. Model-based registration of X-ray mammograms and MR images of the female breast. *IEEE Transactions on Nuclear Science*, 53(1):204–211, 2006. doi: 10.1109/TNS.2005.862983.

[159] M. J. Rupérez, F. Martínez-Martínez, M. Martínez-Sober, M. A. Lago, D. Lorente, P. R. Bakic, A. J. Serrano-López, S. Martínez-Sanchis, C. Monserrat, and J. D. Martín-Guerrero. Modeling the Mechanical Behavior of the Breast Tissues Under Compression in Real Time. In J. M. R. Tavares and R. Natal Jorge, editors, *VipIMAGE 2017*, pages 583–592, Cham, 2018. Springer International Publishing. ISBN 978-3-319-68195-5.

[160] S. Said, P. Clauser, N. V. Ruiter, P. A. T. Baltzer, and T. Hopp. Image registration between MRI and spot mammograms for X-ray guided stereotactic breast biopsy: preliminary results. In C. A. Linte and J. H. Siewerdsen, editors, *Medical Imaging 2021: Image-Guided Procedures, Robotic Interventions, and Modeling*, volume 11598, page 115981C. International Society for Optics and Photonics, SPIE, 2021. doi: 10.1117/12.2581820. URL https://doi.org/10.1117/12.2581820.

[161] S. Said, P. Clauser, N. Ruiter, P. A. T. Baltzer, and T. Hopp. Image based registration between full X-ray and spot mammograms for X-ray guided stereotactic breast biopsy. In C. A. Linte and J. H. Siewerdsen, editors, *Medical Imaging 2022: Image-Guided Procedures, Robotic Interventions, and Modeling*, volume 12034, pages 614 – 621. International Society for Optics and Photonics, SPIE, 2022. doi: 10.1117/12.2611509. URL https://doi.org/10.1117/12.2611509.

[162] S. Said, M. Meyling, R. Huguenot, M. Horning, P. Clauser, N. V. Ruiter, P. A. T. Baltzer, and T. Hopp. MRI breast segmentation using unsupervised neural networks for biomechanical models. In H. Bosmans, N. Marshall, and C. V. Ongeval, editors, *16th International Workshop on Breast Imaging (IWBI2022)*, volume 12286, page 122860C. International Society for Optics and Photonics, SPIE, 2022. doi: 10.1117/12.2624245. URL https://doi.org/10.1117/12.2624245.

[163] S. Said, P. Clauser, N. V. Ruiter, P. A. T. Baltzer, and T. Hopp. Image-based registration between full X-ray and spot mammograms: analysis of registration accuracy in subgroups. In C. A. Linte and J. H. Siewerdsen, editors, *Medical Imaging 2023: Image-Guided Procedures, Robotic Interventions, and Modeling*, volume 12466, page 124661V. International Society for Optics and Photonics, SPIE, 2023. doi: 10.1117/12.2653978. URL https://doi.org/10.1117/12.2653978.

[164] S. Said, Z. Yang, P. Clauser, N. Ruiter, P. Baltzer, and T. Hopp. Estimation of the biomechanical mammographic deformation of the breast using machine learning models 11orcid (s): https://orcid.org/0000-0002-7441-6135 (s. said); https://orcid.org/0000-0002-4411-4056 (n.v. ruiter); https://orcid.org/0000-0001-7324-1735 (t. hopp). *Clinical Biomechanics*, page 106117, 2023. ISSN 0268-0033. doi: https://doi.org/10.1016/j.clinbiomech.2023.106117. URL `https://www.sciencedirect.com/science/article/pii/S0268003323002486`.

[165] A. Samani, J. Bishop, M. Yaffe, and D. Plewes. Biomechanical 3-D Finite Element Modeling of the Human Breast using MRI Data. *IEEE Transactions on Medical Imaging*, 20(4):271–279, 2001. doi: 10.1109/42.921476.

[166] U. Sara, M. Akter, and M. S. Uddin. Image Quality Assessment through FSIM, SSIM, MSE and PSNR—A Comparative Study. *Journal of Computer and Communications*, 2019.

[167] D. Sengupta, P. Gupta, and A. Biswas. A survey on mutual information based medical image registration algorithms. *Neurocomputing*, 486:174–188, 2022. ISSN 0925-2312. doi: https://doi.org/10.1016/j.neucom.2021.11.023. URL `https://www.sciencedirect.com/science/article/pii/S0925231221016970`.

[168] D. R. I. M. Setiadi. PSNR vs SSIM: Imperceptibility Quality Assessment for Image Steganography. 80(6):8423–8444, mar 2021. ISSN 1380-7501. doi: 10.1007/s11042-020-10035-z. URL `https://doi.org/10.1007/s11042-020-10035-z`.

[169] H. Si and K. Gärtner. Meshing Piecewise Linear Complexes by Constrained Delaunay Tetrahedralizations. In B. W. Hanks, editor, *Proceedings of the 14th International Meshing Roundtable*, pages 147–163, Berlin, Heidelberg, 2005. Springer Berlin Heidelberg. ISBN 978-3-540-29090-2.

[170] R. L. Siegel, K. D. Miller, H. E. Fuchs, and A. Jemal. Cancer statistics, 2022. *CA: A Cancer Journal for Clinicians*, 72(1):7–33, 2022. doi: https://doi.org/10.3322/caac.21708. URL `https://acsjournals.onlinelibrary.wiley.com/doi/abs/10.3322/caac.21708`.

[171] D. Silvia, L. Donata, L. A. Jesus, N. Luciana, S. P. Z. Esther, and U. T. Asli. Review and analysis of external quality assessment of breast cancer services in europe: Supporting information for the development of a european quality assurance scheme for breast cancer services. 2015.

[172] R. Sindi, C. Sá Dos Reis, C. Bennett, G. Stevenson, and Z. Sun. Quantitative Measurements of Breast Density Using Magnetic Resonance Imaging: A Systematic Review and Meta-Analysis. *Journal of Clinical Medicine*, 8(5), 2019. ISSN 2077-0383. doi: 10.3390/jcm8050745. URL `https://www.mdpi.com/2077-0383/8/5/745`.

[173] S. Sinsomboonthong. Performance Comparison of New Adjusted Min-Max with Decimal Scaling and Statistical Column Normalization Methods for Artificial Neural Network Classification.

International Journal of Mathematics and Mathematical Sciences, 2022:1–9, 04 2022. doi: 10.1155/2022/3584406.

[174] H. C. Smith, A. L. Chetlen, S. Schetter, J. A. Mack, M. Watts, and J. Zhu. Ppv(3) of suspicious breast MRI findings. *Academic radiology*, 21 12:1553–62, 2014.

[175] M. Smith. *ABAQUS/Standard User's Manual, Version 6.9*. Dassault Systèmes Simulia Corp, United States, 2009.

[176] P. C. Smole, C. Kaiser, J. Krammer, N. V. Ruiter, and T. Hopp. A comparison of biomechanical models for mri to digital breast tomosynthesis 3d registration. In P. M. F. Nielsen, A. Wittek, K. Miller, B. Doyle, G. R. Joldes, and M. P. Nash, editors, *Computational Biomechanics for Medicine*, pages 107–117, Cham, 2019. Springer International Publishing. ISBN 978-3-319-75589-2.

[177] J. A. Solves-Llorens, M. J. Rupérez, C. Monserrat, E. Feliu, M. García, and M. Lloret. A complete software application for automatic registration of X-ray mammography and magnetic resonance images. *Medical Physics*, 41(8Part1):081903, 2014. doi: https://doi.org/10.1118/1.4885957. URL `https://aapm.onlinelibrary.wiley.com/doi/abs/10.1118/1.4885957`.

[178] S. Song, Y. Zheng, and Y. He. A review of Methods for Bias Correction in Medical Images. *Biomedical Engineering Review*, 1(1), 2017. ISSN 2375-9151. doi: 10.18103/bme.v3i1.1550. URL `https://esmed.org/MRA/bme/article/view/1550`.

[179] C. Spick and P. A. T. Baltzer. Diagnostic Utility of Second-Look US for Breast Lesions Identified at MR Imaging: Systematic Review and Meta-Analysis. *Radiology*, 273(2):401–409, 2014. doi: 10.1148/radiol.14140474. URL `https://doi.org/10.1148/radiol.14140474`. PMID: 25119022.

[180] C. Studholme, D. Hill, and D. Hawkes. An overlap invariant entropy measure of 3D Medical Image Alignment. *Pattern Recognition*, 32(1):71–86, 1999. ISSN 0031-3203. doi: https://doi.org/10.1016/S0031-3203(98)00091-0. URL `https://www.sciencedirect.com/science/article/pii/S0031320398000910`.

[181] H. Sung, J. Ferlay, R. L. Siegel, M. Laversanne, I. Soerjomataram, A. Jemal, and F. Bray. Global cancer statistics 2020: Globocan estimates of incidence and mortality worldwide for 36 cancers in 185 countries. *CA: A Cancer Journal for Clinicians*, 71(3):209–249, 2021. doi: https://doi.org/10.3322/caac.21660. URL `https://acsjournals.onlinelibrary.wiley.com/doi/abs/10.3322/caac.21660`.

[182] R.-J. I. Sweeney, S. J. Lewis, P. Hogg, and M. F. McEntee. A review of mammographic positioning image quality criteria for the craniocaudal projection. *The British Journal of Radiology*, 91

(1082):20170611, 2018. doi: 10.1259/bjr.20170611. URL https://doi.org/10.1259/bjr.20170611. PMID: 29125335.

[183] D. Systèmes. Hyperelasticity. URL http://130.149.89.49:2080/v6.13/books/gsk/default.htm?startat=ch10s06.html. Accessed: January 15, 2022.

[184] D. Systèmes. Energy balance I, 2009. URL https://classes.engineering.wustl.edu/2009/spring/mase5513/abaqus/docs/v6.6/books/gsx/default.htm?startat=ch03s07.html. Accessed: January 15, 2022.

[185] D. Systèmes. Mass scaling, 2009. URL https://classes.engineering.wustl.edu/2009/spring/mase5513/abaqus/docs/v6.6/books/usb/default.htm?startat=pt04ch11s07aus63.html. Accessed: August 29, 2020.

[186] D. Systèmes. Energy balance II, 2017. URL https://abaqus-docs.mit.edu/2017/English/SIMACAEGSARefMap/simagsa-c-qsienergybal.htm. Accessed: January 15, 2022.

[187] D. Systèmes. Hyperelastic behavior of rubberlike materials, 2017. URL https://abaqus-docs.mit.edu/2017/English/SIMACAEMATRefMap/simamat-c-hyperelastic.htm. Accessed: January 28, 2022.

[188] D. Systèmes. Verify your mesh, 2017. URL https://abaqus-docs.mit.edu/2017/English/SIMACAECAERefMap/simacae-c-mgnconcpartitionverify.htm. Accessed: May 20, 2021.

[189] V. Sze, Y.-H. Chen, T.-J. Yang, and J. S. Emer. Efficient Processing of Deep Neural Networks: A Tutorial and Survey. *Proceedings of the IEEE*, 105(12):2295–2329, 2017. doi: 10.1109/JPROC.2017.2761740.

[190] R. T., M. A., T. C., and et al. Mammography views, May 2013. URL https://doi.org/10.53347/rID-15023. Accessed: April 19, 2023.

[191] C. Tanner, J. A. Schnabel, D. L. G. Hill, D. J. Hawkes, M. O. Leach, and D. R. Hose. Factors influencing the accuracy of biomechanical breast models. *Medical Physics*, 33(6Part1):1758–1769, 2006. doi: https://doi.org/10.1118/1.2198315. URL https://aapm.onlinelibrary.wiley.com/doi/abs/10.1118/1.2198315.

[192] C. Tanner, M. White, S. Guarino, M. A. Hall-Craggs, M. Douek, and D. J. Hawkes. Large breast compressions: Observations and evaluation of simulations. *Medical Physics*, 38(2):682–690, 2011. doi: https://doi.org/10.1118/1.3525837. URL https://aapm.onlinelibrary.wiley.com/doi/abs/10.1118/1.3525837.

[193] M. Thoma. A Survey of Semantic Segmentation. *arXiv e-prints*, art. arXiv:1602.06541, Feb. 2016.

[194] I. Thomassin-Naggara, I. Trop, L. Lalonde, J. David, L. Péloquin, and J. Chopier. Tips and techniques in breast MRI. *Diagnostic and Interventional Imaging*, 93(11):828–839, 2012. ISSN 2211-5684. doi: https://doi.org/10.1016/j.diii.2012.06.004. URL https://www.sciencedirect.com/science/article/pii/S2211568412002367.

[195] L. R. G. Treloar. The elasticity and related properties of rubbers. *Reports on Progress in Physics*, 36(7):755, jul 1973. doi: 10.1088/0034-4885/36/7/001. URL https://dx.doi.org/10.1088/0034-4885/36/7/001.

[196] S. Tripathy and T. Swarnkar. Unified Preprocessing and Enhancement Technique for Mammogram Images. *Procedia Computer Science*, 167:285–292, 2020. ISSN 1877-0509. doi: https://doi.org/10.1016/j.procs.2020.03.223. URL https://www.sciencedirect.com/science/article/pii/S1877050920306888. International Conference on Computational Intelligence and Data Science.

[197] P. van den Elsen, E.-J. Pol, and M. Viergever. Medical image matching-a review with classification. *IEEE Engineering in Medicine and Biology Magazine*, 12(1):26–39, 1993. doi: 10.1109/51.195938.

[198] B. Van der Velden, M. Janse, M. Ragusi, C. Loo, and K. Gilhuijs. Volumetric breast density estimation on MRI using explainable deep learning regression. *Scientific Reports*, 10, 10 2020. doi: 10.1038/s41598-020-75167-6.

[199] V. Verma and R. Aggarwal. A comparative analysis of similarity measures akin to the Jaccard index in collaborative recommendations: empirical and theoretical perspective. *Social Network Analysis and Mining*, 10:1–16, 2020.

[200] D. Veronda and R. Westmann. Mechanical characterization of skin—Finite deformations. *Journal of Biomechanics*, 3(1):111–124, 1970. ISSN 0021-9290. doi: https://doi.org/10.1016/0021-9290(70)90055-2. URL https://www.sciencedirect.com/science/article/pii/0021929070900552.

[201] G. G. Waade, R. Highnam, I. H. R. Hauge, M. F. McEntee, S. Hofvind, E. Denton, J. Kelly, J. J. Sarwar, and P. Hogg. Impact of errors in recorded compressed breast thickness measurements on volumetric density classification using volpara v1.5.0 software. *Medical Physics*, 43(6Part1):2870–2876, 2016. doi: 10.1118/1.4948503. URL https://aapm.onlinelibrary.wiley.com/doi/abs/10.1118/1.4948503.

[202] L. Wang, B. Platel, T. Ivanovska, M. Wenzel, and H. Hahn. Fully automatic breast segmentation in 3d breast mri. *Proceedings - International Symposium on Biomedical Imaging*, pages 1024–1027, 05 2012. doi: 10.1109/ISBI.2012.6235732.

[203] Z. Wang and A. Bovik. A universal image quality index. *IEEE Signal Processing Letters*, 9(3): 81–84, 2002. doi: 10.1109/97.995823.

[204] Z. Wang, E. Simoncelli, and A. Bovik. Multiscale structural similarity for image quality assessment. In *The Thrity-Seventh Asilomar Conference on Signals, Systems and Computers, 2003*, volume 2, pages 1398–1402 Vol.2, 2003. doi: 10.1109/ACSSC.2003.1292216.

[205] Z. Wang, A. Bovik, H. Sheikh, and E. Simoncelli. Image quality assessment: from error visibility to structural similarity. *IEEE Transactions on Image Processing*, 13(4):600–612, 2004. doi: 10.1109/TIP.2003.819861.

[206] P. Wellman, R. Howe, and K. Kern. Breast Tissue Stiffness in Compression is Correlated to Histological Diagnosis. 1999.

[207] J. Wu. Introduction to convolutional neural networks. 2017.

[208] S. Wu, S. Weinstein, B. M. Keller, E. F. Conant, and D. Kontos. Fully-Automated Fibroglandular Tissue Segmentation in Breast MRI. In A. D. A. Maidment, P. R. Bakic, and S. Gavenonis, editors, *Breast Imaging*, pages 244–251, Berlin, Heidelberg, 2012. Springer Berlin Heidelberg. ISBN 978-3-642-31271-7.

[209] K. Wuensch and J. Evans. Straightforward statistics for the behavioral sciences. *Journal of the American Statistical Association*, 91:1750, 12 1996. doi: 10.2307/2291607.

[210] H. Yin, L. Sun, G. Wang, T. Yamada, J. Wang, and M. Vannier. Imageparser: a tool for finite element generation from three-dimensional medical images. *Biomedical engineering online*, 3:31, October 2004. ISSN 1475-925X. doi: 10.1186/1475-925x-3-31. URL `https://europepmc.org/articles/PMC524181`.

[211] H. Yin, L. Sun, G. Wang, T. Yamada, J. Wang, and M. Vannier. Imageparser: A tool for finite element generation from three-dimensional medical images. *Biomedical engineering online*, 3: 31, 11 2004. doi: 10.1186/1475-925X-3-31.

[212] X.-X. Yin, Y. Jian, Y. Zhang, Y. Zhang, J. Wu, H. Lu, and M.-Y. Su. Automatic breast tissue segmentation in mris with morphology snake and deep denoiser training via extended stein's unbiased risk estimator. *Health Information Science and Systems*, 9, 04 2021. doi: 10.1007/s13755-021-00143-x.

[213] Y. Zhang, J.-H. Chen, K.-T. Chang, V. Y. Park, M. J. Kim, S. Chan, P. Chang, D. Chow, A. Luk, T. Kwong, and M.-Y. Su. Automatic Breast and Fibroglandular Tissue Segmentation in Breast MRI Using Deep Learning by a Fully-Convolutional Residual Neural Network U-Net. *Academic Radiology*, 26(11):1526–1535, 2019. ISSN 1076-6332.

[214] P. Zhou, W. Shi, J. Tian, Z. Qi, B. Li, H. Hao, and B. Xu. Attention-Based Bidirectional Long Short-Term Memory Networks for Relation Classification. In *Proceedings of the 54th Annual Meeting of the Association for Computational Linguistics (Volume 2: Short Papers)*, pages 207–212, Berlin, Germany, Aug. 2016. Association for Computational Linguistics. doi: 10.18653/v1/P16-2034. URL `https://aclanthology.org/P16-2034`.

[215] O. Zienkiewicz, R. Taylor, and J. Zhu. Chapter 10 - Incompressible Problems, Mixed Methods, and Other Procedures of Solution. In O. Zienkiewicz, R. Taylor, and J. Zhu, editors, *The Finite Element Method: its Basis and Fundamentals (Seventh Edition)*, pages 315–359. Butterworth-Heinemann, Oxford, seventh edition edition, 2013. ISBN 978-1-85617-633-0. doi: https://doi.org/10.1016/B978-1-85617-633-0.00010-1. URL `https://www.sciencedirect.com/science/article/pii/B9781856176330000101`.

[216] B. Zitová and J. Flusser. Image registration methods: a survey. *Image and Vision Computing*, 21(11):977–1000, 2003. ISSN 0262-8856. doi: https://doi.org/10.1016/S0262-8856(03)00137-9. URL `https://www.sciencedirect.com/science/article/pii/S0262885603001379`.

[217] H. Zolfagharnasab, S. Bessa, S. Oliveira, P. Faria, J. Teixeira, J. Cardoso, and H. Oliveira. A Regression Model for Predicting Shape Deformation after Breast Conserving Surgery. *Sensors*, 18:167, 01 2018. doi: 10.3390/s18010167.

[218] K. Zou, S. Warfield, A. Bharatha, C. Tempany, M. Kaus, S. Haker, W. Wells, F. Jolesz, and R. Kikinis. Statistical validation of image segmentation quality based on a spatial overlap index. *Academic radiology*, 11 2:178–89, 2004.

[219] A. R. Zubair and O. Alo. Grey Level Co-occurrence Matrix (GLCM) Based Second Order Statistics for Image Texture Analysis. page 93, 10 2019.